D0907204

Designing Teaching Strategies

An Applied Behavior Analysis Systems Approach

This is a volume in the Academic Press
EDUCATIONAL PSYCHOLOGY SERIES

Critical comprehensive reviews of research knowledge, theories, priniciples, and practices

Under the editorship of Gray D. Phye

Designing Teaching Strategies

An Applied Behavior Analysis Systems Approach

R. Douglas Greer

Psychology and Education, Columbia University
Teachers College, New York

ACADEMIC PRESS

An imprint of Elsevier Science

Amsterdam   Boston   London   New York   Oxford   Paris
San Diego   San Francisco   Singapore   Sydney   Tokyo

Images copyright © Getty Images 2002.

The sponsoring editor for this book was Nikki Levy, the editorial coordinator was Barbara Makinster, the book project manager was Molly Wofford. The cover was designed by Suzanne Rogers. Composition was done by Kolam Typesetters, India, and the book was printed and bound by The Maple-Vail Book Manufacturing Group, York, Pennsylvania.

This book is printed on acid-free paper.

Academic Press
An Elsevier Science Imprint
525 B Street, Suite 1900, San Diego, California 92101-4495, USA
http://www.academicpress.com

Academic Press
84 Theobalds Road, London WC1X 8RR, UK
http://www.academicpress.com

Library of Congress Catalog Card Number: 2002101650

International Standard Book Number: 0–12–300850–6

PRINTED IN THE UNITED STATES OF AMERICA
02   03   04   05   06   07   MM   9   8   7   6   5   4   3   2   1

Dedication

To my wonderful progeny, in the order in which they entered my life
—Angela, John, and not the least, Lissie.

Contents

5. TEACHER REPERTOIRES FOR STUDENTS FROM PRELISTENER TO EARLY READER STATUS

6. TEACHING PRACTICES FOR STUDENTS WITH ADVANCED REPERTOIRES OF VERBAL BEHAVIOR (READER TO EDITOR OF OWN WRITTEN WORK)

Functional Repertoires: Curricula from the Perspective of Behavior Selection and Verbal Behavior

7. BEHAVIORAL SELECTION AND THE CONTENT OF CURRICULUM

8. WRITING AND DESIGNING CURRICULA

Organizational Behavior Analysis: A Support System for Expert Pedagogy and Curricular Design

9. TEACHING AND MENTORING TEACHERS

10. THE SCHOOL PSYCHOLOGIST AND OTHER SUPPORTIVE PERSONNEL: A CONTEMPORARY BEHAVIORAL PERSPECTIVE

Preface

Teaching is the key. When teaching is treated as an art, good teaching is an accident. When teaching is treated as a science, good teaching can be replicated across many professionals in a reliable fashion. The research from the existing corpus of applied behavior analysis, and *20 years of applications of teaching as a strategic science* in our demonstration schools, clearly shows that the advancement of teachers in the components of what we term *teaching as a strategic science* determines the rate of student progress.

Over the past decade, we have researched and identified procedures that provide a system for teaching teachers, parents, supervisors, and graduate students in psychology and education the repertoires of teaching and the supervision of teachers as a science. We have developed schools in which students make remarkable progress because all their teachers and supervisors apply the science of pedagogy to all their students. These schools are centers in four different countries. They are schools that work for all children, especially for those who are left behind in most schools. A comprehensive description of this systems science of schooling is available for the first time in this book.

Designing Teaching Strategies: A Behavioral Systems Approach introduces new and advanced applications of applied behavior analysis that incorporates over 65 years of research in behavior analysis with the extensive set of new practices associated with a behavioral systems science of schooling. In a manner of speaking, it is a systems science for providing superior instruction—instruction that has worked for more than a decade across several demonstration schools for children who are difficult to teach. Our new pedagogy and schooling practices provide the critical measurement of learn units that predict student outcomes along with supervisory and instructional effectiveness and the analytic, instructional, supervisory, and organizational practices for developing and maintaining instruction that can teach all students. We also introduce cost–benefit measures of instruction.

The reader will need to be versed in the basic procedures of applied behavior analysis. The book is an advanced text that builds on the coverage of introductory and intermediate texts. The content addresses the complexity of building educational systems that work for all learners with the necessary levels of scientific complexity required to provide superior and accountable education. Both graduate students and scholars of teaching and behavior analysis, as well as those in other areas devoted to behavior change, will find a thorough treatment of the following procedures and benefits:

- A comprehensive and system-wide *science of teaching* — a postmodern–postmodern unabashedly scientific approach;
- Tested procedures that result in four to seven times more learning for all students;
- Tested procedures for *supervisors to use with teachers* that result in significant student learning;
- Tested procedures for providing the *highest accountability*;
- A systems approach for schooling problems that provide *solutions rather than blame*;
- Parent-approved and *parent-requested educational practices*;
- Means for psychologists to work with teachers and students to *solve behavior and learning problems*;
- A comprehensive *systems science of schooling*;
- An advanced and sophisticated *science of pedagogy and curriculum design*;
- Students who are not being served with traditional education *meeting or exceeding the performance of their more fortunate peers*;
- Supervisors *mentoring teachers and therapists to provide state-of-the-science instruction*;
- Parent education procedures creating a setting for *parents, educators, and therapists to work together* in the best interests of the student;
- Teachers and supervisors *measuring* as they teach produce *significantly better outcomes for students*;
- Systemic solutions to instructional and behavioral problems involving teachers, parents, and supervisors provide means *to pursue problems to their solution*;
- All tactics of behavior analysis are incorporated to analyze and determine *best-fit tactics for students and clients*; and
- A functional behavior selection treatment of curriculum to develop true functional repertoires of all aspects of curricula.

A science of teaching, as opposed to an art of teaching, can provide an educational system that treats the students and the parents as the clients.

Designing . . . describes how teachers can use *measurably effective learner-driven classroom instruction* that results in four to seven times more learning over control conditions. Teachers *and teachers of teachers* learn research-based protocols on how to isolate instructional problems and apply over 200 research-based tactics to their solution, regardless of the learning problem. Psychologists and educators will find procedures that allow them to get to the *real roots of aberrant behavior in the classroom using a systems approach* that draws on all of the players in the student's world—teachers, parents, supervisors, administrators, and professors in universities.

At a time when the demand for behavior analytic teachers and therapists exceeds the supply, *Designing* . . . provides a blueprint to train and *ensure quality applications of behavior analysis in education and child therapy settings*. Equally important, *Designing* . . . describes how to develop curricula based on behavior selection that result in functional learning outcomes rather than the common practice of teaching the structure of knowledge (i.e., "inert knowledge"). These curricula design protocols provide the *means to teach national and state standards such that students learn meaningful repertoires for problem solving, effective communication, academic literacy, and creating an enlarged community of interests*. The curriculum design and pedagogical science presented herein allows the educator to teach students to *function* effectively in the 21st century.

A reporter for *The Irish Times* recently described his impression on first visiting a school in Ireland that incorporates all of the features described in this text.

> Perched on the sloping hills behind University College Cork is Glasheen National School. On the outside it looks like an ordinary school but on the inside [in the CABAS Classrooms that are part of the Cork CABAS School] something extraordinary is taking place . . . The charts on the wall show how a group of . . . children [with severe academic and social delays] are making remarkable progress as part of a pilot education program . . . It is a tailor-made system of learning which responds to children's strengths and weaknesses . . . With applied behavior analysis the learning process is speeded up and in many cases . . . children [with severe academic and social delays] are able to catch up with their peers (Carl O'Brian, *The Irish Times*, July 21, 2002, p. 7).

The objective of our text is to provide other professionals with the wherewithal to provide the same "remarkable progress" for their students.

Acknowledgments

I acknowledge the assistance of Dolleen-Day Keohane, Pamela Osnes, and Marla Brassard for their careful and gentle proofing of this manuscript. Their insightful comments and encouragement was instrumental in its completion. I also thank Gary Phye for championing the book. In addition, I thank the several generations of my graduate students who labored through earlier drafts—their responses drove revisions and additions. Finally, I thank the hundreds of students in our CABAS Schools in the USA, Ireland, England, and Italy whose daily data taught us the systems approach that has served us all so well. Yes, the student does know best.

Advanced Applications of Applied Behavior Analysis to Teaching

.

Teaching as Applied Behavior Analysis: A Professional Difference

TERMS AND CONSTRUCTS TO MASTER

- Natural selection extended to behavior analysis
- Teaching as applied behavior analysis—The professional difference
- Prescientific and postscientific perspectives on student learning problems
- Pedagogy
- Teaching as environmental design

- Functional relationship between teaching and student learning
- Nine characteristics of teaching as applied behavior analysis
- Teaching expertise and financial rewards
- Science of the behavior of the individual

TOPICS

A Definition of Teaching and Pedagogy

Teaching as a Scientifically Based Profession

Characteristic Practices of Teaching as Applied Behavior Analysis

---

A DEFINITION OF TEACHING AND PEDAGOGY

Pedagogy is the traditional term for the study of the methods of teaching. While the word *pedagogy* is seldom used in descriptions of the methods of teaching in normative educational courses anymore, it is an accurate term and the activities of pedagogy are critical to expert teaching. Pedagogy is that component of teaching that *comprises interventions used by a teacher* to bring about student learning—in short, interventions that *occasion* learning. Our definition of pedagogy will incorporate what we know about the behavior of the individual with particular relevance to how the individual learns. In our enlarged definition, pedagogy refers to the instructional operations performed by a teacher or by an automated teaching device that result in a student learning a behavior, a response class, and a repertoire. The learning must have occurred as *a function* (that is as a "cause") *of, or a correlate of*, the instructional operations performed by the teacher. The teaching operations were either sufficient or necessary to the learning. Without them the student would not learn.

We use the term "normative" as a generic term for the prevalent view in education that "teaching is an art." We do not wish to characterize that view in a pejorative manner. Rather, we use it to help the reader differentiate prescientific approaches to pedagogy with a thoroughgoing scientific approach found in teaching as applied behavior analysis. In the practice of normative education eclecticism is considered desirable. Because normative education is so eclectic in nature it is difficult if not impossible to characterize the intellectual beliefs or epistemology associated with it. However, it is probably accurate to say that normative approaches to education are not tied to a view of teaching as a thoroughgoing strategic science. Hence we use the terms prescientific pedagogy and normative education interchangeably.

In our view, the presentation of material, as in a lecture or reading assignment, is not an instance of pedagogy, any more than having students watch a videotape is necessarily an activity that results in learning. The latter activity may set the stage for pedagogical activities by the teacher with the student that may, indeed, lead the student to learn from such presentations. However, pedagogy begins when the student responds to teacher presentations and *continues* when the teacher responds, in turn, to the student's response in ways that produce the desired outcome. Moreover, what the student learns under true pedagogical operations is what the teacher sets out to teach. When the student learns the correct response or chain of responses (e.g., a problem-solving task) *as a result* of the teacher's responding to the student's behavior, we say pedagogy has occurred (Greer, 1996).

Acts of pedagogy result in students learning *that which they could not do* before as a *function of* or as *a correlation with* the activities of pedagogy. Pedagogy comes fully into play *only* when the student is responding. It includes the teacher activities that occasion the student's response and teacher responses to the student's effort. If the student continues to learn simply by encountering the materials for which no special program of instruction was necessary, the student is learning but little teaching occurs. In the latter case the student may continue to learn as a result of prior learning and which in turn can be a result of prior pedagogy, not just chance.

Of course, maintaining conditions to motivate learning are also part of teaching. Teaching, as an act of pedagogy, takes place when the student encounters difficulty or when the teacher provides procedures or uses an automated device such that the student can perform that which he could not do before the intervention. When the student is not motivated, acts of pedagogy create motivation.

Our definition of pedagogy incorporates the design of *how* the student will encounter situations and stimuli to which he will respond and the differential consequences to the student's particular response in such a way that the student responds effectively (i.e., correctly) or more closely approximates effective responding to the situation. What we have described is superb individualized instruction. It is the kind of instruction that one seeks when one pays for a private tutor. *Of course, private tutoring is not necessarily individualized*, but the conditions of one-to-one tutoring are more likely to occasion individualized instruction. The application of the sciences of pedagogy and schooling can provide for frequent occurrences of individualization regardless of the ratio of students to teachers, provided that the students have the necessary prerequisites and teachers are strategic scientists of pedagogy.

Just as expert teaching requires optimum teacher interventions, it also requires *not intervening*. One of the goals of teaching is to teach students to be their own teacher. The sequence of experiences that students receive and the pedagogical operations associated with those experiences determines the students' attainment of self-instructional repertoires. The pedagogical acts and the curriculum that leads the student to self-teaching and self-discipline and the repertoires that allow the student to learn independently of the teacher are critical components of the expertise that we shall present.

Dynamic Nature of Teaching

Teaching is a dynamic interaction among four components: (a) the student, (b) the teacher, (c) the curriculum (or what is being taught), and (d) the learned repertoire (how to use it and when to use it). The study and development of teaching as behavior analysis provide in-depth treatment of the dynamic interaction of teaching. Applied behavior analysis is a strategic science. By strategic, we mean that specific findings and methods of the science are used differentially based on the moment-to-moment progress of the student. Thus, it is dynamic by nature. In a broad sense, the curriculum is the environment in which we want the student to be a part. That is not to say that the students "construct" their environments, but the part of the environment that affects or "controls" the students' behaviors is the unique world of environmental controls that exist for each student individually. For example, if the student does not speak the language that is being used, the controls for her behavior are not the same as those for an individual who speaks and responds to the language. When the student learns the language, she becomes part of that environment.

Prior to learning the language, she was oblivious to the function of the communication. The contingencies of experience and instruction bring her and the environment into contact such that her repertoire expands and her contribution, in turn, changes the environment. Thus, effective instruction or pedagogy is never static. Specific findings of the science must be applied to the student as the performance of the student dictates. Future chapters will provide scientific descriptors of this dynamic property of pedagogy. In the terminology of *the science of the behavior of the individual*, we refer to this as the vocabulary of the science. The use of these descriptors by someone identifying components of the science in moment-to-moment action is referred to as scientific *tacts* (i.e., the teacher makes verbal con*tacts* with the teaching activities using the scientific terminology). Indeed, it is this dynamic property of our science and its application that prompts our use throughout this text of the terms *strategic science and strategic applied behavior analysis*. The postbehavior analysis perspective is simply different from the prescientific one. Table I presents two different ways of characterizing the performance of students. The postscientific or behavior analytic view provides solutions to instruction, rather than categorization.

There are several sciences that contribute to *teaching as applied behavior analysis* that we also call a *strategic science of instruction*. First, there is the laboratory basis of the science, often termed *the experimental analysis of behavior*. Next, there is *applied behavior analysis* that encompasses applications of behavior analysis not only to education but also to medicine, business, manufacturing, therapy, parenting, and a host of other applied professions. In addition, the strategic science of teaching draws on a particular philosophy of science called *behavior selection*. The methodology of behavior analysis allows applications

TABLE I
Characterization of Learning Problems from Pre- and Postscientific Perspectives

Prescientific teachers: Normative education	Postbehavior analysis teachers: teaching as applied behavior analysis
The student is unmotivated.	The reinforcement or establishing operations are inadequate for the student.
The student has a learning disability.	Perquisite repertoires are not mastered or fluent and must be taught.
The child is incorrigible.	The instruction is inadequate in terms of the learn unit presentations; the controlling variables for behavior need to be shifted.
The child requires a multisensory approach.	The child is learning the wrong operant.
The problem is in the home.	Instruction in the school is the responsibility of individuals with pedagogical expertise and the school professionals are responsible for fixing the problem and assisting parents.

of *any research findings in educational settings* whenever they are relevant (e.g., findings in cognitive psychology, developmental psychology, psychophysics). In summary, the strategic science of pedagogy and its extension to the management of schooling is a science that builds on and contributes, in turn, to each of these sciences in a consistent and coherent manner driven by the individual needs of the student.

TEACHING AS A SCIENTIFICALLY BASED PROFESSION

Teaching is a common form of activity, both for our species and for that of many others. Because it is so common we often fail to appreciate its importance and its complexity. Yet the expertise of a teacher who can function as a strategic scientist of instruction is vastly different from the level of expertise demonstrated during untutored teaching interactions that occur between parents and children. Learning occurs much more frequently than instances of teaching, which is why inexpert teaching does not result in poor learning outcomes for children who are raised in privileged settings and who have no disabilities. Organisms (yes, that means we *Homo sapiens*, too) learn constantly from their environment regardless of the presence or absence of teaching by fellow organisms. Organisms learn because the consequences of behavior select adaptive repertoires of behavior for individuals (Donahoe, Burgos, & Palmer, 1993). Behaviors that work for the individual in a given set of circumstances become part of that organism's repertoire.

Learning has been researched extensively, while *teaching as it is typically conceptualized* has received relatively little attention. However, if teaching is defined as the identification and *arrangement of optimum learning environments for each individual*, the definition used in this book, a great deal is known about the activities of effective teaching as instructional operations and principles for individualized instruction.

The act of teaching must have represented a critical step in the evolution of our species. One can imagine a situation where it was critical to the survival of a group for one individual to teach others to perform some act, such as flushing game or planting seeds. Teaching must have been integrally related to the development of communicative behavior and, hence, language. While one may teach by example without speaking a word, the communicative act is inextricably tied to teaching, even when the communication is by example. Later we will highlight the relationship between teaching and communicative behaviors even more closely.

Great teachers are admired, but usually only posthumously. We are rarely given evidence of the learning outcomes of their instructional acts; thus we do not really know whether they were, indeed, effective teachers or simply inspirational producers of knowledge or even wisdom. Some surveys of public attitude suggest that the general public holds less respect for teaching as a profession than it once did. Certainly new teachers are not paid as much as recent graduates in business, engineering, or medicine. While monetary remuneration is not the most conspicuous variable that influences the choice of teaching as an occupation, the amount of money that society distributes to professions is some indication of the how useful a profession is for society. Society places a premium on professions that produce useful or financially remunerative outcomes.

Those of us who are educators are often critical of governments or school boards that seems to place such a small premium on schools and the learning of youngsters. However, on closer view, perhaps the government of "we the people" does care about children and learning, but perhaps "we" see less of a relationship between formal schooling and learning outcomes. Perhaps the less-than-generous commitment to schools reflects the belief that teachers have no special expertise at producing learning outcomes. Society is not convinced that there is a body of practices that the profession of teaching holds, beyond what any good parent does as a matter of course. Most of the public and many professional debates about education are devoted to curriculum, standards, policy, or what should or should not be taught. For example, many attempts at reforming education emphasize the need to raise standards—graduates should be more expert than they are. Others believe that the road to improving education is the development of new schooling arrangements such the decentralization or centralization of administration. Some believe, for example, that the solution to better reading is the teaching of phonetics; in math, the teaching of concepts rather than rote calculation; or in the means to literacy, the teaching of problem solving. Few talk about pedagogy or the act of teaching per se.

Perhaps the public's disenchantment with teaching is understandable given the poor performance of our high school graduates and the general quality of educational outcomes reported by the news media. Teachers who can document that they use measurably effective teaching procedures (e.g., have measurable outcomes of the effects of their teaching) and their parents know how beneficial the such teaching is for their students (*Irish Times*, pp. 3 and 12, January 15, 2000). These teachers feel privileged to spend their working hours engaged in the enterprise of teaching. Teachers, who do not see measurable improvements in their students as a result of what they do, are harassed by classrooms that are out of control and are likely to count the days until the end of the year or until retirement. Effective teachers use teaching operations that engender measurably superior educational results for each student individually. Ineffective teachers do not see measurable changes in their students. The empirically tested operations of pedagogy and their strategic application to specific learning difficulties are an applied science. Those who practice it expertly are strategic scientists of instruction who approach teaching as applied behavior analysis. This body of expertise results in outcomes that society will value and come to see as true professional expertise as the results of its application becomes more widely known. Currently, few teachers have such expertise.

While *normative* education may not *yet* realize the necessity for special expertise in pedagogy, teachers who are responsible for producing life-sustaining outcomes for students with disabilities or learning and behavior problems are constantly made aware of the need for expertise. Special education is in some measure the result of an implicit belief that special educators have a pedagogical expertise that regular or normative educators do not have (Greer & Dorow, 1976). An alternate view is that special education settings are locations to place classified students in order that normative educators can teach students who do not interfere with instruction in the regular classroom. This last view is untenable in a society that offers free public education.

Some might even argue that the majority of the public schools in our major cities are increasingly composed of students who have been relegated to special education. However, both with the call for inclusion and with the necessity for major improvements in schools increasingly mandated to teach all children, all teachers will need everything that this text and further research can provide.

This book is for teachers who teach *all* children, not just the children who come to school with much of the teaching already done or the children who begin school without disabilities. Our teachers are those who will be expert with those students *who need to be taught in order to learn*—students who are difficult to teach! Such teachers require a science of pedagogy and a science of schooling. If students do not need teachers who have the expertise we describe *in order to* learn perhaps no harm will be done when teaching is viewed as an art rather than a science.

How does one become an expert teacher, one who is priceless to their students and the community? How does one produce the learning outcomes for students that society will come to appreciate economically? The answer is found in the good news that not only is there *a science of pedagogy* but there is also *a systems science of schooling* that can produce from four to seven times more learning than the learning found in normative practice. The procedures and systems for developing and maintaining optimum teaching expertise are as yet known by only a few. When you master the contents of this book, there will be one more!

There are probably many reasons for why this new pedagogical expertise is not available on a broader basis. One possible explanation is that new innovations require long incubation periods before they are adapted widely—often decades. Another reason is that while much of the science has been known for some time, it has not been coalesced yet into a cogent account. Another possible reason is that systemic applications have occurred very recently and only in small systems of schooling. Still another and perhaps even larger impediment to widespread implementation is that the repertoires of teachers who are applied scientists of pedagogy are vastly more complicated than the repertoires used by teachers who use normative practices. Traditional teacher training prepared teachers to work with children who came to school with many of the critical components needed to learn already intact. Simply presenting material can work for some of these children. The body of knowledge on which our science is based is designed for *individualized instruction*, while the historical approach to public schooling has been based on group instruction *and the selection and separation of students who fail in normative classes*. Our scientific pedagogy is a science of individualized instruction. Individualized instruction provides optimum outcomes for students from different cultures, students with disabilities, and students with deficits in language experiences (Hart & Risley, 1996).

The demand for scientifically based teaching, however, already exceeds the supply. Catherine Maurice's account of her attempts to obtain appropriate services for her two children with autism, and the description of the striking success of her children when they were taught using applied behavior analysis, has provided thousands of parents with the knowledge that there is instruction out there that can truly save their child from the traditional prognosis associated with this disorder (Maurice, 1993). The parents at the few behavior analytic schools in the nation also know the value of expertise in teaching as applied behavior analysis. What is needed is more training in applied behavior analysis, training that ensures quality, and training that applies behavior analysis comprehensively to all aspects of teaching, schooling, and parenting. There is also a great deal we know beyond the basic applied behavior analysis—knowledge that is introduced in text form for the first time in this book. The purpose of this text is to make that knowledge available more widely in order to significantly improve the educational outcomes of *all children*.

The basic texts in applied behavior analysis are generic texts dealing with applications to all areas, including education, psychology, sociology, organizational behavior analysis, or behavioral medicine. There are also some very good texts that *introduce* teaching as applied behavior analysis. *There have been no texts to date that provide the advanced information on teaching or on organizing behavior analysis systems as applied behavior analysis*. Specialized research in behavior analysis, when combined with experience in applying these findings to schools, provides a level of pedagogical sophistication that was unavailable heretofore.

This text is designed to provide part of the technical, complex, and sophisticated expertise needed to provide the next stage of expertise to teachers who are to become strategic scientists and applied behavior analysts of instruction. The current educational system has a strong need for this expertise for teachers who work in schools that are failing as well as schools for students with disabilities. Still other schools need the expertise we describe in order to integrate special education students successfully into the regular education. In this text, teachers and administrators are provided with advanced applications of applied behavior analysis for teaching students who are experiencing difficulties learning—students that normative educators do not have the teaching operations to teach. This book is about providing those tools. The tools are strategic and tactical *repertoires* for teaching, analyzing instruction, and solving teaching problems. They are the *repertoires of teachers who are strategic scientists of pedagogy*. They include specialized and expert teaching actions and the advanced repertoires of an applied scientist. Teachers who master the repertoire will be expert in teaching those who other teachers cannot or will not teach.

What are the advantages of teaching as a strategic science of instruction? Most notably, the "method" of applied behavior analysis is not a method but rather a set of repertoires. What are these strategic and tactical operations that comprise the repertoire?

- First the operations, concepts, and practices of our science have been *tested extensively with diverse types of students*—children, adolescents, and individuals of all ages who needed or wanted remedial or accelerated instruction.
- Second, the procedures and findings of the sciences on which the teaching repertoire is based have *expanded* over the past few decades.
- Third, the *expansion of the practices* in classrooms and schools has come about as a *result of parent advocacy* rather than top-down policy decisions.
- Finally, the continuous measurements that are implicit in the approach provide true accountability. Perhaps it is the latter component, continual measurements of the outcomes for each individual student, that distinguishes what the sciences of pedagogy and schooling offer over all other approaches.

The key ingredient in the implementation of individualized instruction is the expertise of the teacher. While we shall cover the ingredients of *a system that supports optimum pedagogy*, the core of the text concerns the description of the components of a science of individualized instruction. These components provide the means for individualized instruction within classrooms that include large numbers of students with wide variations in their existing academic repertoires and behaviors. The legally mandated function of special educators in the United States is to provide *individualized educational programs*. The utility of a strategic science of instruction to the education of educators who can provide individualized instruction is obvious. Special educators are asked to provide their expertise in a variety of settings from self-contained classrooms (e.g., a classroom composed of a small number of students all of whom are designated as "having disabilities") to normative classrooms where special educators function as "consulting teachers." However, if a consulting teacher is to be effective, he or she must be just that—a consultant to the teacher in the normative classroom. This means that she is not simply a shadow or individual tutor for the student. Rather, she must see that the teacher in the normative classroom gains critical components of the teaching repertoires found in this text and in introductory texts in applied behavior analysis. The consulting teacher must teach, monitor, and motivate the normative teacher to function as a behavior analytic teacher. The operations of superior teaching must be *included* also. The first requirement for doing this is that our consulting teacher has exceptional expertise in teaching children and teachers.

Teachers who work with children in schools located in areas of poverty cannot succeed in these settings without expertise in teaching as a strategic science of individualized instruction. While this expertise alone will not fix all of the problems in failing schools, expertise will increase the prognosis of success for children of lower income families (Hart & Risley, 1996). Thus, the material in this text is needed by the legions of teachers who work in failing schools located in poverty-stricken communities. Also, in our experience in teaching hundreds of teachers who work in schools in financially prosperous communities, the operations described in this text have made significant differences in their students' prognosis.

A relatively new and growing trend is prereferral for children having difficulty in schools that use normative educational practices. That is, before the child is assigned to special education, one or more professionals work to change the behaviors of the student in such a manner that the special education referral is avoided. In this case our applied teacher-scientist could provide the *normative teacher* with tools of teaching as applied behavior analysis *before* the student is classified and separated from the mainstream. In their consulting role, both as a preventative solution and as a postcategorization effort to mainstream the student, special educators will need to provide normative educators with the necessary pedagogical expertise to rescue the student from teaching practices that do not work for that student. In many instances, the normative educator does not view his or her role as one that requires a change in *his or her* procedures. After all, was not the expectation that the student was sent to special education to be "cured." But as you have learned in you introductory courses in applied behavior analysis, behavior is a product of the environment and the students' existing repertoires. The normative educator will need to change her pedagogy if the effort is to be successful. Thus, a major role of our consulting teacher is to change the pedagogical practices in the regular education classroom. In such a role the consultant teacher can provide expertise to the regular educator for the potential benefit of *all* of the students in the regular education class.

The task of mainstreaming pedagogical expertise is difficult with normative teachers who are the special educator's peer. For many normative educators, the special educator is seen only as a tutor for the target student. Convincing regular educators to embrace scientifically sophisticated expertise places the special educator in the role of a nonadversarial coach for the regular educator. Hence special educators who are successful will need to draw on the expertise from the chapter on supervision as well as the chapters devoted to the science of individualized pedagogy.

Similarly, the chapter devoted to the role of support personnel (i.e., psychologists, social workers, speech therapists, and subject matter experts) in this text provides tactics and strategies for the consulting teacher to use to introduce expert pedagogy into the home as well as the normative classroom. In fact, without the expertise presented in this text the consulting teacher cannot provide the normative educator with results that are likely to convince her to adopt them. Without this expertise, the consulting teacher can only provide a tutoring role. Even the tutoring role will be unsatisfactory, if measurably superior analytical and teaching repertoires are not used. Thus, to be an effective consulting teacher the results of the *consultant teacher's interventions used must be so apparent that other teachers will seek his or her expertise.*

Prerequisite Repertoires for the Audience of This Text

This text presumes that the reader *has mastered the material presented in introductory texts and courses in applied behavior analysis.* The reader will need to be familiar with the principles of the science of the behavior of individual organisms, traditional procedures for measuring behavior in applied settings, and methods for determining functional relations in applied settings (e.g., experimental designs for individuals). You will need to know the subject matter typically mastered in a first semester course in applied behavior analysis. If you have read and can apply the findings of research articles in journals like the *Journal of Applied Behavior Analysis, Journal of Behavioral Education, Precision Teaching*, or the *Educational Treatment of Children*, you will be well prepared for the material that follows. If you do not have the necessary background you will need to master that material as you master the material presented herein. This is, possibly, the first advanced text in applications of behavior analysis to teaching. While this text has been tested and found to be effective for over 100 graduate students in their courses and in their applications in classrooms, the graduate students had all had a one semester course in applied behavior analysis before learning the material in this text.

We designed the text to meet the need for applications of the relevant sciences to teach children who require expert instruction. The material is graduate level in difficulty and complex, but it is eminently practical. The research evidence identifies the repertoires presented herein as those that are functionally or "causally" related to teachers' successes with teaching their students.

The precise use of the *vocabulary of the science of behavior analysis*, the sophisticated use of the *repertoires for applying the science*, and the *repertoires for analyzing instruction* are all critical to effective practice. We have attempted to avoid the use of meaningless jargon; rather, *the terms we present are necessary tools for a teacher who would be a strategic scientist and behavior analyst of instruction.* The contents of this volume constitute a new level of pedagogical expertise not heretofore available. The students who require teachers with the needed expertise are everywhere.

CHARACTERISTIC PRACTICES OF TEACHING AS
APPLIED BEHAVIOR ANALYSIS

What are the conspicuous characteristics of teachers who have adequate expertise in teaching as applied behavior analysis?

- All *instruction is individualized* whether the instruction is provided in a one-to-one setting or in groups;
- *Teachers continuously measure* teaching and student responses;
- *Graphs of the measures of student's performance are used for decisions* about which tactics are best for students at any given instructional decision point;
- Logically and empirically *tested curricula and curricular sequences are used;*
- The *principles* of *the basic science* of the behavior of the individual and *tactics from the applied research* are used to teach *educationally and socially significant repertoires;*
- Teaching is driven by the moment to moment responses of each individual student and existing research findings;
- The classroom is a positive environment—coercive procedures are avoided (e.g., reprimands are not used);
- Expertise in *the science is used to make moment to moment decisions based on the continuous collection of data and its visual summary in graphs;*
- *Teachers are strategic scientists of pedagogy* and applied behavior analysts;
- *The progress of students is always available* for view in the form of up to date graphs that summarize all of the students' responses to instruction.

These are the basic practices of the teacher who is a true professional. The practices of professionally competent teachers are the subject matter of this book. But they are not enough. There are many examples of extraordinary applications of behavior analysis to teaching in our history. The problem has been *how do we maintain and expand the behavioral expertise daily.* Thus, much of this text is devoted to describing *how* expert teaching as behavior analysis can be maintained day in and day out in an ecologically sound system. In addition, the text sets out a perspective on curriculum design and analysis growing out of behavior analysis and its epistemology. Finally, we review the obstacles that impede the implementation of comprehensive behavior analytic instruction and some solutions for overcoming those obstacles. Thus, the text provides:

(a) The *components of teaching as advanced applied behavior* analysis for all children who require that expertise;
(b) The *teacher instruction and mentoring procedures and applications of organizational behavior analysis that are needed* for designing, supporting, and maintaining high-quality behavior analytic instruction applied to all of the significant roles required to teach *all* children;
(c) the *curricular analysis and design procedures* that are required for building curricula composed of functional repertoires; and

(d) The *obstacles for implementing teaching as behavior analysis and some solutions*.

THE ORGANIZATION OF THE TEXT

The text is organized according to several sub categories of pedagogical expertise, curricular design expertise, and the organizational expertise needed to develop and maintain a high quality of teaching on a day-to-day basis.

Part I: The Components of Teaching as Advanced Applications of Applied Behavior Analysis

- Chapter 2 introduces the *basic measure of pedagogy* as the building block of all that follows.
- Chapter 3 provides an *overview of the repertoires of teachers* who are behavior analysts of instruction—their scientific vocabulary, their classroom practices, and their analytic expertise.
- Chapter 4 treats the *analytic repertoires of strategic scientists of instruction in depth*.
- Chapter 5 deals with the particular repertoires needed by *teachers* to develop students' *prelistener repertoires into speaker/listener repertoires*.
- Chapter 6 deals with the particular repertoires needed by teachers to develop their students' advanced *communicative repertoires—repertoires that incorporate reading, writing, calculating, problem solving, and learner independence*.

Part II: A Reconceptualization of the Analysis and Organization of Curricula from the Perspective of Behavior Selection and the Concepts of Verbal Behavior

- Chapter 7 introduces the perspective of our science on the curriculum as environmental control and thus *the reconstruction of curriculum* from a comprehensive perspective derived from behavior selection.
- Chapter 8 outlines ways of *delivering the curriculum* via classroom organization and management of comprehensive individualization, regardless of classroom size.

Part III: An Organizational and Professional Support System to Teach and Support Expert Pedagogy and Curricular Design

- Chapter 9 provides *operations that the teacher coach, supervisor, and consulting teacher* can use for inclusion of students and components of the science into normative classrooms or the development of Comprehensive Applications of Behavior Analysis to Schools and School Systems.

• Chapter 10 provides *tools for school psychologists and parent educators to* provide comprehensive and high-quality schooling for all members of an educational process in the best interests of the learner.

References

Donahoe, J. W., Burgos, J. E., & Palmer, D. C. (1993). Selectionist approach to reinforcement. *Journal of the Experimental Analysis of Behavior, 58*, 17–40.

Greer, R. D. (1996). The educational crisis. In M. A. Mattaini & B. Thyer (Eds.), *Finding solutions to social problems: Behavioral strategies for change* (pp. 113–146). Washington, DC: American Psychological Association.

Greer, R. D., & Dorow, L. G. (1976). *Specializing education behaviorally*. Dubuque, IA: Kendall Hunt.

Hart, B., & Risley, T. (1996). *Meaningful differences in the everyday life of America's children*. New York: Paul Brookes.

Maurice, C. M. (1993). *Let me hear your voice*. New York: Knopf.

CHAPTER 2

The Learn Unit: A Natural Fracture of Teaching

TERMS AND CONSTRUCTS TO MASTER

- Operants
- Learn unit
- Respondents
- Sequencing learn units
- Antecedent
- Successive approximation
- Response
- Learn unit prerequisites
- Consequence (postcedent)
- Generalized reinforcers
- Setting events
- Prosthetic reinforcers
- Setting stimuli
- Natural conditioned reinforcers
- Consequence
- Establishing operation
- Reader, writer, writer as own reader repertoires
- Operant chamber
- Intraresponse time
- Programming
- Vocal learn units
- Frame of programmed instruction
- Textual learn units
- Programmed Instruction
- Essay learn units
- Teaching machine
- Three-term contingency
- Reader repertoire
- Writer repertoire
- Scripted curriculum
- Number correct/incorrect
- A natural fracture
- Engaged academic time and opportunity to respond
- Individualized instruction
- Basic unit of pedagogy
- Criterion referenced objectives
- Cost–benefits analysis
- Nucleus operant
- Instructional history
- Phylogenetic contributions

- Role of the target S^d
- Correction operations
- Instructional control
- The student's three-term contingency
- Insertion of learn units

- Interlocking operants
- Conversational unit
- Speaker as own listener
- Guided notes for presentations

TOPICS

A Measure for Teaching

The Basic Unit of Pedagogy

The Research Base

Analyses of the Components of the Student's Three-Term Contingency in the Learn Unit

The Presence and Absence of Learn Units in Educational Practice

Other Literature on Operant Episodes

The Converging Literature

Student Progress and Changes in the Location and Frequency of Learn Units

The Learn Unit as an Analytic Tool

A MEASURE FOR TEACHING

Education requires a measure that contacts the *natural fractures of instruction*—one that includes both the behaviors of the teacher or teaching device and the behaviors of the student. The measure must provide immediate feedback to the teacher, be a valid prediction of long-term outcomes, and be an integral part of instruction. The identification and acceptance of such a measure will serve to establish strong practices and effective practitioners. We believe that the research literature has converged on just such a *natural fracture* of pedagogy.

A *natural fracture* is a unit of a compound that separates naturally from other components as a result of lawful conditions. It is an absolute unit (Johnston & Pennypacker, 1992). For example, in geology the break in geological strata on the face of a cliff is one instance of a natural fracture. DNA material, zero centigrade, and atoms are others. The identification of, and the quest for, natural fractures are more common in the natural sciences than in the social sciences. One of the characteristics of behavior analysis that aligns it with the natural sciences is a preference for the identification of natural fractures. Once identified, natural fractures of behavior (e.g., the operant) become direct measures of absolute units of behavior as opposed to artifacts or inferences about behavior (Johnston & Pennypacker, 1992; Sidman, 1960). Measures of con-structs such as IQ, "locus of control," personality, or engaged time are examples of measures that are not natural fractures. The operant and the respondent, on the other hand, are natural fractures of behavior.

The identification of measures that contact the natural dimensions or frac-tures of the world, including behavior, typically functions to produce rapid progress in the sciences (Mach, 1960; Skinner, 1938, 1953). B. F. Skinner (1938) identified the operant early on in the history of the science of the behavior of the individual. Instead of a natural fracture, Skinner might have measured the by-products of the operant such as the number of time intervals an organism was "engaged" in an operant chamber or the time that an organism was engaged in reaching the end of the maze (actually a prevalent measure at the time). It is difficult to imagine how our current science of behavior could have evolved based on "on task" measures of an organism. It was the identifi-cation of the operant as a natural fracture rather than an artifact of behavior/ environment interaction that led to rapid progress.

At present, much of education relies on measures of artifacts of behavior or inferences about behavior such as ratings of students or teachers' "percep-tions," engaged academic time, and the presence or absence of services (e.g., contact hours of teaching, the number of periods taught) (Brophy & Good, 1986; Hamburg, 1992; Stallings, 1985). Other common measures are *scores* on standardized tests, but these are not absolute units. They do not relate to the specific instruction received by the student, nor do they show the moment-to-moment behavior changes of the student. However, we shall show how some of this research on artifacts of behavior together with research in behavior analysis support convergence on a natural fracture of pedagogy—the *learn unit*.

THE BASIC UNIT OF PEDAGOGY

The learn unit consists of the least divisible component of instruction that incorporates *both* student and teacher interaction and it predicts new stimulus control for the student (Greer, 1994a). It is present when student learning occurs in teaching interactions, and when it is absent student learning does not occur. It is a countable unit of teacher and student interaction that has the potential to change the behavior of *each* party individually or jointly. That is, the goal for both parties is for the student to come under the control of the target S^d (stimulus discriminative) and the consequence of behavior. The three-term contingency or operant (antecedent, behavior, and consequence) is what the student is to learn. The teacher organizes the student's environment such that the student's operant emerges. The teacher also has a series of operant behaviors that must *interlock with the student's three-term contingency* in order for the operant to emerge. The learn unit also needs to occur within the motivational conditions in which the student's operant will be needed (i.e., the operant is to become part of the student's repertoire).[1] In order for this to happen, the teacher must respond in certain ways to the presentation of the student S^d and to the resulting behavior or its absence from the student. In effect, the teacher "learns" from the response of the student—that is, the teacher learns what to do next from the student's performance.

A basic scenario for the interlocking three-term contingencies of the teacher and student are as follows. The S^d for the teacher to present an S^d to the student is the student's attentive behavior. Presentation of an S^d by the teacher to the student is the teacher's response and the S^d for the student's response. The behavior or behavioral product of the student to the teacher presentation of the student's S^d serves as a consequence for the teacher's behavior and an S^d for the teacher's next response to the student (e.g., reinforcement of the student behavior or correction of an incorrect response), while the teacher's response is the consequence for the student.[2] The completion of this single learn unit functions as a consequence for the teacher's last response. The learn unit is a measure of the symbiotic relationship between the behavior of a teacher and a student.

There are many possible interactions involved in a single learn unit. For example, if the student is not attentive, another set of three-term contingencies is emitted by the teacher to bring the student's behavior under instructional control; thus the final presentation to the student would involve more teacher interactions than that described in the basic unit. Several examples are presented in Table 1. They all presume that the student has previously mastered the prior skills necessary to begin learning the new operants.

[1]We use the term *repertoire* to refer to a collection of behaviors, S^ds, consequences, and setting events that belong to a particular response class and, in turn, a collection or categorical grouping of response classes. For example, the term pure mand refers to a class of verbal operants that is under the control of deprivation and the presence of a listener and a without a verbal antecedent which is reinforced by the phenomenon specified by the topography of the behavior (Skinner, 1957). In addition, the response class of the mand is part of a still larger class of operants that includes speaker behavior. We use the term repertoire to designate the latter.

[2]When the student emits an incorrect response, the correction in the learn unit provides a critical component of the learn unit according to the evidence to date. We do not know why this is the case, but the correction may function as a prompt when done well. That is, when the student is attending to the S^d and repeats the correct response this sequence of events functions to evoke the correct response for future learn units. That is, the student is provided with a prompted response and when the target s^d is presented again, and the student emits the correct response, the response is reinforced. The same operation occurs in Programmed Instruction wherein the correction operation prompts the correct responses. See Stephens (1967) for an extensive review of the research on Programmed Instruction and the role of the correction. In most learn units the student is required to emit a corrected response after seeing or hearing the teacher's correction. If the reinforcement is adequate for correct responding, corrections are typically not reinforced with prosthetic reinforcements in order that the effects of differential reinforcement of correct responses are maintained. For a few types of responses (e.g., responses to "sit still," "look at me") the teacher correction is conspicuous planned ignoring.

TABLE I
Example of Learn Units—Interlocking Contingencies between the Teacher and Student

Event	Operant components
Examples 1	
Correct student response	
1. Attending student.	Teacher S^d
2. Teacher says "spell cat."	Teacher behavior
	Student S^d
3. Student responds "c-a-t."	Student behavior
	Teacher consequence
	Teacher S^d
4. Teacher responds "good job" and records the student's response.	Teacher behavior
	Student consequence
5. Completion of learn unit.	Teacher consequence
Incorrect student response	
1. Attending student.	Teacher S^d
2. Teacher says "spell cat."	Student S^d
	Teacher behavior
3. Student responds "k-a-t."	Student behavior
	Teacher consequence
	Teacher S^d
4. Teacher says "c-a-t spells cat." Student echoes teacher correction. Teacher records the student's response.	Student consequence
	Teacher behavior
5. Completion of the learn unit.	Teacher consequence
Example 2	
Correct student response	
1. Attending student.	Teacher S^d
2. Teacher says "recite the 3s multiplication table from 0 to 10."	Student S^d
	Teacher behavior
3. Student responds "0,3,6,9,12,15,18,21,24,27,30."	Student behavior
	Teacher consequence
	Teacher S^d
4. Teacher says "great" and records the student's response.	Teacher behavior
	Student consequence
5. Completion of learn unit.	Teacher consequence
Incorrect student response	
1. Attending student.	Teacher S^d
2. Teacher says "recite the 3s multiplication table from 0 to 10."	Student S^d
	Teacher behavior

(continues)

TABLE 1 (*continued*)

3. Student responds "0,3,6,8,11,14,17,20,24,27,30."	Student behavior
	Teacher consequence
	Teacher S^d
4. Teacher records the student's response and says "0,3,6,9,12,15,18,21,24,27,30."	Student consequence
	Student echoes teacher correction.
	Teacher behavior
5. Completion of the learn unit.	Teacher consequence

Example 3

Correct student response

1. Attending student.	Teacher S^d
2. Teacher says "write a 5 page composition comparing and contrasting the characters in two of Shakespeare's tragedies." The teacher provides the student with a list of 10 critical variables to include in the composition.	Teacher behavior Student S^d
3. Student includes the 10 critical variables correctly throughout the composition.	Student behavior Teacher consequence Teacher S^ds
4. Teacher reads the composition and identifies the variables with check marks. Teacher returns the corrected composition to the student for review. Teacher records the student's responses.	Teacher behavior Student consequences Student S^d
5. Student reviews the composition based on teacher consequences to each of the 10 critical variables. Completion of learn unit.	Student behavior Teacher consequence Student consequence

Incorrect student response

1. Attending student.	Teacher S^d
2. Teacher says "write a 5 page composition comparing and contrasting the characters in two of Shakespeare's tragedies." The teacher provides the student with a list of 10 critical variables to include in the composition.	Teacher behaviors Student S^ds
3. Student includes 5 of the 10 critical variables correctly throughout the composition.	Student behaviors Teacher consequences Teacher S^ds
4. Teacher reads the composition and identifies the correct 5 variables with check marks. Teacher indicates 5 missed variables on the front of the composition. Teacher returns the corrected composition to the student for review and revisions. Teacher records the student's initial responses.	Teacher behaviors Student consequences Student S^ds
5. Student reviews the teacher consequences and revises the composition to include all 10 variables based on teacher consequences to each of the 10 critical variables. Completion of 10 learn units.	Student behaviors Teacher consequences Student consequences

(continues)

TABLE 1 *(continued)*

Example 4

Correct student response	
1. Attending student.	Teacher S^d
2. Teacher says "read the next paragraph." Teacher times the student's responses.	Teacher behaviors Student S^d
3. Student reads the text.	Student behaviors
	Teacher consequence
	Teacher S^d
4. The teacher provides praise on a variable reinforcement schedule throughout the reading session. Teacher stops timing the session and says "nice reading." Teacher counts the number of words read and records the accuracy and rate of the student's response and has the student correct any errors while attending to the words missed.	Teacher behaviors Student consequence
5. Completion of the learn unit	Teacher consequence
Incorrect Student Response	
1. Attending student.	Teacher S^d
2. Teacher says "read the next paragraph." Teacher times the student's responses.	Teacher behavior Student S^d
3. Student reads the text.	Student behaviors
	Teacher consequence
	Teacher S^{d}s
4. The teacher provides the correct response for incorrect responses and student repeats. Teacher stops timing. Teacher counts the number of words read and records the accuracy and rate of the student's response.	Teacher behaviors Student consequences
5. Completion of the learn unit	Teacher consequence

THE RESEARCH BASE

Learn units must have all student components [establishing operations, discriminative stimuli (S^{d}s), response opportunities, and consequences that reinforce or correct]. The teacher must emit the necessary behaviors that set the occasion for the contingency for the student. The teacher's behaviors are themselves two or more three-term contingencies. According to the existing research a learn unit must include the following.

    (a) *The teacher must provide a consequence to the student's response* or lack of response in the form of either a reinforcement operation or a correction operation *contingent on the student's response* or lack thereof. The S^d presentation for the student must be unambiguous (Albers & Greer, 1991; Diamond, 1992; Ingham & Greer, 1992).

(b) *The students must observe (e.g., see, hear, touch, taste, or smell or a combination thereof) the S^d* to which they respond in correction or reinforcement operations that they receive from the teacher in either written or spoken forms (Hogin, 1996; Hogin & Greer, 1994).

(c) *The students must respond or have the opportunity to do so* (Fisher & Berliner, 1986; Greenwood, Hart, Walker, & Risley, 1994; Greenwood, Carta, Arreaga-Mayer, & Rager, 1991; Heward, 1994).

(d) Better student performance results from faster rates of intact learn unit presentations (Carnine & Fink, 1978; Ingham & Greer, 1992).

(e) *Greater numbers of learn unit presentations*, as opposed to presentations that are *not* learn units, *result in significantly higher rates of correct responses and higher numbers of instructional objectives* for students (Greer, McCorkle, & Williams, 1989; Heward, 1994; Ingham & Greer, 1992; Selinske, Greer, & Lodhi, 1991).

(f) *Replacing teacher/student interactions that are not learn units with interactions that are learn units increases student correct responses* from four to seven times (Albers & Greer, 1991; Diamond, 1992; Ingham & Greer, 1992; Selinske et al., 1991).

In the studies cited above and in those cited in Table 2, the various components of the learn unit were experimentally manipulated and identified as necessary components. Teacher actions that were learn units were compared to teacher actions that were not learn units. The results of all of these studies, taken together, affirmed that the presence of learn units resulted in educationally significant increases in correct responding (e.g., from four to seven times higher than baseline conditions) and the attainment of significantly more educational objectives.

TABLE 2
Summary of the Research Converging on the Learn Unit

Literature and issues	Representative references
(1) Twenty years of educational research with no significant identification of measures that predict schools with better student outcomes. (However, programmed instruction was an effective "method".)	(a) See Stephens (1967) for a comprehensive review of the research literature through 1966. Also, see the journal, *Review of Educational Research*, published by the American Educational Research Association.
(2) "Engaged academic" time correlates reliably with academic achievement with socioeconomic status controlled statistically; no teaching procedures were identified in this literature concerning how teachers could increase "engaged academic time."	(b) Berliner (1984); Brophy & Good (1986); Fiscer & Berliner (1985); Stallings (1980, 1985)

(continues)

TABLE 2 *(continued)*

(3) Applied behavior analysis identifies teacher consequences that result in:

- Student "on task" and student compliance
  - Hall, Lund, & Jackson (1968); Madsen, Becker, & Thomas, (1968)

- Increases in correct academic responding
  - Kirby & Shields (1997); Lindsley (1991)

- Student consequences to teachers or tutors that change teacher behavior
  - Sherman & Cormier (1974); Polirstok & Greer (1977); Greer, & Polirstok (1982)

(4) Theoretical and research literature in behavior selection (radical behaviorism) and interlocking operants:

- Skinner (1957) and verbal episodes, verbal behavior as social behavior
  - Verbal episodes between individuals and self-talk

- Behavior/behavior relationships
  - Bijou (1970); Goldiamond (1974b); Staats (1981); Guerin (1994)

- Research on interlocking verbal operants
  - Donley & Greer (1993); Lodhi & Greer (1989)

(5) Applied behavior analysis identifies a functional relationship between academic responding and student achievement:

- Academic responding relates to "engaged time"
  - Greenwood, Carta, Arreaga-Mayer, & Rager (1991); Greenwood, Hart, Walker, & Risley (1994); Heward (1994)

(6) Research on learn units as interlocking operants between teacher/tutor and student/tutee as best predictor of achievement.

- Albers & Greer (1991); Babbit (1986); Diamond (1992); Dorow, McCorkle, Williams, & Greer (1989); Greenwood, Delquardri, & Hall (1989); Greer (1994a); Greer, McCorkle, & Williams (1989); Hogin (1996); Heward (1994); Ingham & Greer (1992); Lamm & Greer (1991)

(7) Research on learn unit as independent variable or part of independent variable:

- Increased learn units decrease self-injurious and assaultive behavior and related to the matching law
  - Kelly & Greer (1996); Martinez & Greer (1997)

- Learn units used to test relationship between fluency (mastery performance at a fast speed) and maintenance
  - Chu (1998); Lindhart-Kelly & Greer (1997)

- Learn units used for contingency analyses for students resulted in significantly improved outcomes for students
  - Keohane (1997)

Several lines of inquiry identified essential components of effective peda-gogy, all of which are incorporated in the learn unit. Numerous studies found that the student must respond or have the opportunity to do so (Heward, 1994).[3] These *opportunities to respond* (Greenwood et al., 1994) are appropriately presented and "consequated" by the teacher in the intact learn unit.[4] We have found in the above-cited studies and in others (e.g., Dorow, McCorkle, Wil-liams, & Greer, 1989; Greer, 1994a, 1994b; Greer et al., 1989; Lamm & Greer, 1991; Singer & Greer, 1997) that learn units were stronger predictors of student learning than response opportunities alone without teacher–student inter-actions that did not include all components of the learn unit. Based on our reading of the research, *learn units are the strongest known physical dimensions of effective teaching.* Unlike measures of engaged time, active student responding, or teacher measures that do not incorporate student responding, learn units are direct measures of student and teacher behaviors and are the natural fractures of the teaching process rather than by-products of effective teaching.

ANALYSES OF THE COMPONENTS OF THE STUDENT'S THREE-TERM CONTINGENCY IN THE LEARN UNIT

The learn unit includes the components of the interlocking operants. They include the S^d, response, and consequence for both the student and the teacher and the manner in which they interact. Studies that investigated each of these are described below.

The Role of the S^d

Hogin (1996) found that when the correction component of the learn unit required the students to observe their responses to math problems (the students' S^d) viewing only their responses and the consequence alone, they did not master the math computation operations. When they *observed the S^d along with the response and consequences* they mastered the operations. Hogin's

---

[3]Heward (1994) provides a description of the research concerned with "engaged time," "on task," "engaged academic time," and "active student responding," relative to opportunity to respond. Fisher and Berliner (1985) edited a volume devoted to research in instructional time that makes the case for something like the learn unit. Interested readers will find both sources to be necessary components in the evolution from the measurement of by-products to the measurement of outcomes and behavioral processes via the learn unit.

[4]The term *consequate* is not an accepted verb in most dictionaries. However, it is used as a verb in the Precision Teaching literature and is useful in our explanation also, since a consequating operation may or may not have an effect on the preceding response. Use of the term obviates the need to differentiate between reinforcing or punishing operations and reinforcement and punishment effects (Catania, 1998). Vargas (1988) suggested a particularly apt term—*postcedent*.

study demonstrated the importance of the S^d for the student and related teacher behaviors in the interlocking three-term contingencies that constitute the learn unit. This study as well as one by Albers and Greer (1992) demonstrated that consequences need not occur at the moment the students' behaviors are emitted if the instruction involves written verbal contingencies between the teacher and the student. Once the teacher saw to it that the students *observed* their responses to the S^d and the consequence, learning occurred. In the Albers and Greer (1992) study, the results showed that learn units were more effective than teacher instructions that were not learn units regardless of whether they were consequated in written form up to several days later or they were consequated immediately in vocal form. Both written antecedents and written responses with delayed written consequences, or vocal responses to written antecedents with immediate vocal consequences, led to superior performance compared to vocal or written presentations that did not include the intact learn unit.

The Response Component

Greenwood et al. (1994) described the literature documenting the importance of opportunity to respond. Their program of research showed that increasing students' response opportunities and increasing academic antecedents (i.e., math problems, reading comprehension questions) alone resulted in educationally significant increases in students' correct responses compared to instruction not requiring students to respond. Their work showed that it was the students' emission of relevant academic responses rather than simply "academic engagement" that was responsible for better predictions of student achievement. Heward (1994) replicated the findings described in Greenwood et al.'s (1994) research program and related them to the "active student responding" literature in educational research. The necessity for student responding is well documented in the literature (Greenwood et al., 1991).

Reinforcement and Corrections

As to the role of the consequence for both the student and the teacher, the literature in applied behavior analysis is replete with studies documenting the role of the consequence and need not be reviewed here. However a few studies directly compared learn units, which include consequences, with the opportunity to respond without consequences. Diamond (1992) and Albers and Greer (1992) found that learn unit presentations resulted in better performance on math problems than did opportunities to respond that did not include the intact learn unit. Recall that other educational research documented the importance of "engaged" academic responding to effective schooling. Classrooms in which students listen to teachers present material with little student responding do not provide the necessary responses opportunities for students to learn.

The Three-Term Contingency

The consequence and the student response are not the only components that the student needs to experience. Rather, it is the S^d control over the response that is being established by the consequence. Thus all three components for the student and the teacher are part of the designation of learn unit. If any part is missing for either the student or the teacher (Ingham & Greer, 1992) the unit is not complete.

In summary, the *interlocking three-term contingencies* of the teacher and the student is a learn unit which is a *predictor* of learning either in written and delayed forms or in vocal and immediate forms. In the case of delayed consequences, it appears that when the students observe the S^d and the correct responses along with the teacher's consequence in written form directed to a specific student response (i.e., a check mark, a smiling face, a laudatory written statement), a learn unit has occurred. When the students are incorrect, they observe the S^d, their incorrect responses, a prompt for a correct responses (i.e., ''did you forget to ____?''), or an *x* mark, *and* they redo the problem *until* it is correct.

THE PRESENCE AND ABSENCE OF LEARN UNITS IN EDUCATIONAL PRACTICE

In well-designed classroom instruction, as in the operant laboratory, the responses of learners are (a) shaped by design via preplanned exercises or (b) captured, as when a parent or teacher capitalizes on an incidental situation to instruct or create operant units. It is a design, or positioning strategy, that results in the student responding effectively in settings in which the operant is needed in the world at large (Schwartz, 1994; Stokes & Baer, 1977). Some teacher strategies draw on changes in the motivational conditions associated with the student's learning, as when the teacher uses an interrupted chain or a brief motivational tactic to enhance an item or event used in a reinforcement operation (Hart & Risley, 1980; Michael, 1983; Schwartz, 1994). When one can manipulate establishing operations as part of the instructional process there is no need to wait for the optimal moment. Such instructional design has all of the advantages of incidental instruction, where the contextual or motivational situations are ideal, and all of the advantages of massed instruction where the number and rate of learning opportunities are sufficient to allow the student to become fluent (Greenwood et al., 1994).

Learn Units and Programmed Instruction

Programmed Instruction is one of the few educational innovations that was shown to be reliably effective in the early research on education (Stephens,

1967). Programmed Instruction is an example of a simulation of the learn unit for instruction done independently of the teacher. An automated and learner-controlled sequence of steps called frames leads the student to construct target responses (Holland & Skinner, 1961; Skinner, 1968). The student then observes her response as a reinforcement operation or correction. Some teaching machines had chutes to deliver candy or tokens to increase the probability that reinforcement operations (confirmation of correct responses) would *function* to reinforce (Skinner, 1968). When responses are initially incorrect, the student is required to correct the response before moving on. The frames include the antecedent or question, the student's response, and a consequence to the student's response or pedagogical three-term contingency.

Scripted Instruction, Guided Notes, and Learn Units

When teachers present instruction in scripted sequences based on logical analyses (Engelmann & Carnine, 1982) or in scripted instruction based on task analyses (Greer, 1986, 1990), operant units are scripted for the teacher. These "scripts" specify teacher behavior, student behavior, and its measurement in some cases, as well as the sequence of steps and objectives. They script for the teacher the components that are included in the frame in Programmed Instruction. Teaching scripts can specify learn units with students individually or with groups of students. Still other teachers or professors present guided notes for lectures. When presented as learn units they are more effective than traditional lecture (Bahadorian, 2000).

Incidence of Learn Units in Common Practice

Instruction is presented in less exacting formats in schools at present according to the research (Fisher & Berliner, 1985; Greenwood et al., 1991, 1994; Greer, 1994a, 1996b). Most teachers instruct *without regard for the operant*. That is, the presentation is typically improvised in lecture format (e.g., antecedent presentations with infrequent response opportunities) (Greenwood et al., 1991; Greer, 1994a). A teacher may spot a student having difficulty with a specific operation. The teacher may then ask the student a series of "leading" questions (i.e., antecedents), followed by student responses, which the teacher either corrects or reinforces (or does not consequate in most situations) (Albers & Greer, 1991; Ingham & Greer, 1992; Lamm & Greer, 1991; Selinske et al., 1991). When the teacher corrects or reinforces the student, operants are more likely to be formed by the student.

In some lectures or lessons, the teacher presents an extensive set of antecedents; at some point a student or students will be questioned. If the teacher consequates the student responses, an operant unit may be acquired. To date, the research shows that unless teachers are explicitly taught to do so, they do not typically provide units that are complete operants (Greenwood et al., 1994; Greer, 1994a; Madsen, Becker, & Thomas, 1968). That is, they do not carry out learn units. New discriminative stimulus control for the student is unlikely when learn units are not present (Diamond, 1992; Greenwood, Delquardri, & Hall, 1989; Greer, 1994a; Heward, 1994; Ingham & Greer, 1992; Kirby & Shields, 1972).

The presence or absence of learn units can be observed in any instructional setting. A trained observer, using procedures from the research literature, can measure intact learn units, by monitoring the presence or absence of the written, gestural, graphic, or vocal three-term contingencies of both student and teacher (Ingham & Greer, 1992; Greer, 1994). An observer can also determine whether or not the operants of each party interlock (Ingham & Greer, 1992). In order to interlock, the joint operants must occur in sequence and in a timely fashion according to the research that I cite in this chapter (Table 2). The number in time (rate) of *interlocking operants* between a teacher and students provides a measure of teaching that greatly increases the likelihood that the student will learn. While the learn unit alone cannot predict good outcomes, apparently it is necessary for good outcomes in those instances we have studied to date.

The research that we cited showed *that learn units, whether planned or incidental, are found in effective instruction* in various conformations, *but are not found in ineffective instruction*. Learn units, or their lack thereof, are observable in preschool classrooms, graduate classes, tutoring settings, group lectures, lecture– discussions, courses that use the Personalized System of Instruction, automated instruction, 1-min timings, one-to-one instruction, laboratory courses, writing exercises, and problem-solving projects in various behavioral topographies (e.g., spoken, written, or otherwise).

Choice of the Term

While we are not wedded to the *learn unit* nomenclature as the term for the explicit interaction between the teacher and student we describe, it does seem appropriate since the behavior of the student and the teacher changes based on the behavior of the other party. Both stand to "learn" from the interaction. This symbiotic "learning" relationship is characteristic of our inductive science; Skinner wrote that "the pigeon knows best" and Keller paraphrased this when he stated that "the student knows best "(Skinner, 1938; Keller, 1968). Thus, we think the *interactive nature of effective teaching* is communicated by our term. However, other terms may have more appeal. Regardless of the term that is eventually used, it is the identification of the unit that is important.

OTHER LITERATURE ON OPERANT EPISODES

An interest in the potential explanatory role of interlocking operants or operant episodes is not without a history. Staats used interlocking operants, as well as respondent/operant relationships, to interpret complex psychological constructs as behavioral interactions (Staats, 1981). Goldiamond incorporated verbal episodes that were interlocking operants in his analysis of systems (Goldiamond, 1974). Patterson identified reciprocal operant relationships between the behaviors of children and parents in his research on parenting (Patterson, 1982). Bijou suggested this perspective in his seminal paper on applying behavior analysis to the classroom (Bijou, 1970).

We began to see the convergence on the interlocking operants in the early analyses of the relationship between the behaviors of teachers and students. Hall, Lund, and Jackson (1968) showed the powerful effects of teacher attention on student behavior. They changed teacher consequences to students and the behavior of students changed. They treated teacher behavior as the independent variable and student behavior as the dependent variable. Their study was followed by many others that replicated the effect of teachers' consequences (e.g., Greer, Dorow, Miller, 1976; Greer, Dorow, Wachhaus, & White, 1973; Madsen et al., 1968).

Subsequent studies treated the student behaviors as the *independent variable* and the teachers' behaviors as the *dependent variable*. Sherman and Cormier (1974) found that when they changed one student's behavioral consequences to a teacher's behaviors, the behavior of the teacher also changed. Polirstok and Greer (1977) found that the interactions of four teachers with a student were changed as a *function* of changing the behavioral consequences of the student with the four teachers. Changes in, and the maintenance of, tutor interactions with tutees (e.g., use of contingent praise) were found to be a function of tutee behaviors following tutor reinforcement operations with tutees (Greer & Polirstok, 1982; Polirstok & Greer, 1986). Thus, changes in the behavior of students can change teacher behavior just as changes in the behavior of teachers can change students' behaviors. *Either party involved in the interaction has the potential to change that of the other party involved.* The relationships between the operants of each party are reciprocal—they are interlocking under the conditions described in this and other research.

Skinner suggested that interlocking verbal operants were part of the multiple controls of conversation which he termed *verbal episodes* between individuals (Skinner, 1957). Moreover, he suggested that interlocking verbal episodes could occur when one talks to ones self aloud or covertly. In both of these cases the individual acts as *both a speaker and a listener*. Each role involved a complete three-term contingency. Research with young children confirmed Skinner's theory of interlocking verbal operants in children's' self-talk (i.e., conversational units) (Lodhi & Greer, 1989). Still other research confirmed the presence of interlocking verbal operants between older children (Donley & Greer, 1993).

There is a body of research as well as theory on interlocking operants. The research validating the notion of interlocking operants included: parent/child interaction (Patterson, 1982); teacher to student interaction (Hall et al., 1968);

student to teacher interaction (Sherman & Cormier, 1974); student to student interaction (Greer & Polirstok, 1982); speaker to speaker interaction (Donley & Greer, 1993); and speaker as own listener interaction (Lodhi & Greer, 1989).

THE CONVERGING LITERATURE

The data from multiple sources that we cite involved analyses of what it is that effective teachers do. Heward (1994) provided a detailed review of how these disparate research findings converged and need not be reviewed in detail here.[3] The research showed that teachers who were effective presented student behavior opportunities and consequated student behaviors; moreover, effective teachers performed these behaviors in ways that resulted in their students emitting faster rates of and higher numbers of correct responses (and correspondingly lower rates of incorrect responses) than did teachers who were less effective (Albers & Greer, 1991; Greenwood et al., 1991; Greer, 1996a; Stallings, 1980, 1985).

STUDENT PROGRESS AND CHANGES IN THE LOCATION AND FREQUENCY OF LEARN UNITS

Learn units vary in their topography. Some involve vocal S^ds by the teacher, while others take textual form. Student responses within the learn unit may be vocal, written, typed, drawn, or otherwise (e.g., dance, music, physical education). When the student produces a product of behavior (e.g., essay, report, musical composition/performance, graph), the learn units become the components of the product *to which the teacher responds*. Each statement of praise or correction (e.g., ''rewrite this sentence in the active voice'') constitutes a learn unit when the consequence of the teacher is responded to or observed by the student (e.g., the student corrects an incorrect response based on teacher corrections or receives delayed delivery of reinforcement for a correct response). Some of the typical written or drawn reinforcement operations include check marks, smiling faces, and approval comments. Corrections call for the student to rework the response *while observing the relevant S^d (Catania, 1998; Hogin, 1996).*

    The frequency of and the point at which a learn unit is to maintain the student's progress is associated with the existing verbal repertoires of the student. The role of the existing verbal repertoires is illustrated by the contrast in the frequency of learn units needed to maintain the progress of students who only have speaker repertoires, compared with students who have advanced combinations of speaker, reader, and writer repertoires. Responses to presentations of each word will call for a learn unit when the student is initially learning textual responding. However, once the student has mastered the reading of discrete words, the responses of the student to *all* of the words in a long prose passage or list of words constitutes a learn unit. The learn unit for a student initially learning the words consists of each word or even each morpheme. However, for a student who has already mastered the individual words at a slow rate, and for whom the goal is mastery with a rate-of-responding criterion (e.g., fluency), a learn-unit consists of correctly responding to all of the words on the page in a predetermined unit of time.

As the student advances, in the process of writing a report or performing complex mathematical operations to solve a problem, the incidences of learn units change based on the students' increased expertise. When the learn unit is not needed, the subcomponent operants are already mastered —responding and antecedent control is under the control of moment-to-moment effects. Our view of this is consistent with Catania's statement that what an organism learns is the S^d, the behavior, and *the consequence* (Catania, 1998). When a student has not mastered the subcomponents, he or she requires learn unit interactions in order to progress.

Students require an instructional history involving thousands of learn units prior to accomplishing tasks independently that involve chains of numerous responses. The *progressive* mastery of instructional material (e.g., responses involving the reading of phonetic units aloud or the writing of words, sentences, paragraphs, or chapters) is important in the effective and *successive decline* in numbers of learn units required per learning objective and correspondingly a progressive increase in the numbers of responses that can be done independently before a learn unit is required by the student. That is, the learn unit at level one is the phonetic sound, at level two the word, level three the sentence, and level four the paragraph. Each advance leads to greater numbers of S^d–*response–consequence units* that the student emits successfully *before a learn unit is required from the teacher*. Those subcomponent responses of the student which do not require teacher learn units will consist of operants mastered previously and may include student-controlled learn units (i.e., the student functions as his or her own reader/writer or self-edits at the sophistication of the eventual target reader). The student with advanced repertoires will require fewer teacher-controlled learn units.

It is evident that learn units are required more frequently in the initial levels of instruction for a particular repertoire and become less frequent and less needed as the component operants comprising the repertoire are mastered. The operants that are mastered come under the control of the immediate contingencies for the student. At that point, learn units are needed only in those instances when the responses are not within the control of the natural contingencies. A kind of naturalistic thinning of the teacher's reinforcement and correction operations takes place as the student advances. However, if a more complex task occurs before the student has mastered the subcomponents of the repertoire, or before the student is fluent, as our recent study on the effects of fluency on maintenance suggests, the student will flounder

(Lindhart-Kelly & Greer, 1997). Johnson and Layng (1994) have proposed that more complex skills emerge as a function of certain levels of fluency with component skills in a process they describe as *adduction*. Thus, a particular rate of correct responding of the student may determine when and what type of learn unit is needed or not needed. We shall provide more information on these issues later.

We have known for some time that the shaping of chains of behavior is best accomplished by inserting a prosthetic reinforcer at progressively later points in the chain of behaviors, and the thinning of learn units with increases in learner sophistication is simply a specific instance of this general principle.

In addition, learn units are required at different points in the students' educational progress based on the verbal sophistication of the student (e.g., whether they have the repertoires of a speaker, listener, reader, writer, or self-editor). Learn units change in frequency based on what the student has mastered to date. It is probably just as important to omit the presentation of learn units when a student has mastered certain repertoires, as it is to insert a learn unit when it is needed by the student.

THE LEARN UNIT AS AN ANALYTIC TOOL

One of the serendipitous outcomes of the implementation of behavior analysis in a comprehensive fashion in schools that use CABAS (Comprehensive Application of Behavior Analysis to Schooling) has been the development of teachers who can function as strategic engineers or in some cases strategic scientists of instruction (Greer, 1996b; Keohane, 1997). In these schools, the teachers and other professionals monitor all of their students' responses to learn units. At the same time, the teachers receive instruction devoted to the progressively more complex application of behavior analysis to decisions about teaching (Greer, 1996b). Much of that instruction involves teaching teachers how to perform contingency analyses (Malott & Heward, 1995) and functional analyses of instruction using the learn unit. These teachers learn to make decisions about which tactics from the science need to be applied given a specific student's data pattern and history. When these decisions call for applications of existing research-based operations the teachers are functioning as engineers and in those cases when existing tactics do not work they introduce new procedures and test them through functional or experimental analyses. In the latter cases the teachers act as *strategic scientists*.

The literature in applied behavior analysis supplies a large number of research-based tactics to use with students as a first course of action or as needed. That is, some of the tactics are standard—they are done as the first course of action. For example, instruction calls for the use of learn units as a typical operation for most instruction. However, other tactics are used based on the performance and data history of the student (e.g., tactics associated with the S^ds, responses, consequences, setting events, and instructional history) of both the teacher and the student in the learn unit interaction. When the student is floundering, the first question teachers' need ask is, "are learn units present?" If they are not, simply presenting the intact learn unit often eliminates the problem. If all of the components that constitute the learn units are present, then the teacher can engage in theoretical contingency analyses based on the data history of the students or functional analyses, if needed. That is, if there are no components of the learn unit missing or in error, the source of the difficulty will lie in other areas surrounding the learn unit, such as motivational events (Michael, 1983), the student's instructional history, or even physiological conditions (e.g., hearing or visual impairments). The results of these analyses are used by the expert teacher to *select tactics* from the applied research repository that fit the particular instructional problem. A related use of functional analyses is found in the applied literature concerned with self-injurious and assaultive behaviors (Iwata, Dorsey, Slifer, Bauman, & Richman, 1982).

For example, when the learn unit is not at fault, determined through particular observations identified in research by Ingham and Greer (1992), a series of technical questions should be asked about the related components of the student's operant, the student's instructional history, or existing establishing operations. The probable answers to those questions suggest, in turn, tactics from the literature that might be effective. If the S^d is not in the child's repertoire, a *simultaneous prompting procedure* can be used (Johnson, Schuster, & Bell, 1996; Wolery, Holcombe, Billings, & Vassilaros, 1993). If the problem is a particular listener response, when other listener responses are present, *behavioral momentum* can be used. If the S^d control is faulty an intrinsic or *extrinsic stimulus prompt* may suffice (Cooper, Heron, & Heward, 1987). If the problem is motivational, an *additional establishing operation* can be useful (e.g., interrupted chain, incidental presentations) (Schwartz, 1994; Sundberg, 1993). Alternately, if the functional responses of the student (e.g., tacting or exemplar identification) are presumed to be related to some other response because the response *topography* is the same, when in fact the two responses are *functionally independent* (e.g., manding or exemplar construction) then the function must be incorporated into the learn unit. In the latter case the two responses are independent but the instruction has presumed that they are related (e.g., multiple choice versus construction or short essay, exemplar producing versus exemplar identification) (Donahoe & Palmer, 1994; Lamarre & Holland, 1985; Lindsley, 1991).

Is there any evidence to suggest that teachers who are taught to use the learn unit to make these analyses produce better outcomes for their students? Keohane's (1997) experiment compared the number of learn units required by three target and three generalization students before three teachers learned to use the learn unit to analyze the probable causes for student plateaus as described in the preceding paragraph. After the baseline when teachers were taught to use the analysis operations, the teachers made only a few errors in decisions about changing interventions with target and generalization students. The teachers' correct decisions about the choice of, and changes in, tactics resulted in significantly fewer learn units required by the target and generalization children to achieve educational objectives than were required of them in the baseline. The study used a multiple baseline across the three target and the three generalization students for 18 months. A follow-up probe, 1 year later, showed the teachers were still making few decision errors. Of course, this finding and the other theoretical extensions of the learn unit call for replications and new research.

Functions of Learn-Units

Based on the research to date, and our extrapolations from the data, the learn unit serves several functions.

(1) It is a measure that predicts learning for the types of instruction tested in the existing research.
(2) Because it predicts learning, the learn unit is a basic measure of effective teaching and can be used to discriminate between effective and ineffective teaching.
(3) Thus, the learn unit provides a database for what teachers need to learn in order to be effective—a scientifically based curriculum for teacher training.
(4) The learn unit, together with its context (i.e., establishing operations and the student's instructional history), provides verbally governed strategies that, in turn, provide operations for solving instructional problems through contingency analyses.

Because the learn unit is the basic measure and building block of effective instruction, the reader will need to master the construct and its applications described in this chapter. Future chapters will describe the application of learn units with a range of students.

References

Albers, A., & Greer, R. D. (1991). Is the three term contingency trial a predictor of effective instruction? *Journal of Behavioral Education*, *1*, 337–354.

Babbit, R. (1986). Computerized data management and the time-distribution of tasks performed by supervisors in a data based educational organization (Doctoral dissertation, Columbia University, 1986). *Dissertation Abstract International*, *47*, 3737a.

Bahadorian, A. J., (2000). *The effect of the learn-unit on student's performance in two university courses*. Unpublished Ph.D. dissertation, Columbia University.

Berliner, D. C. (1984). The half-full glass: A review of research on teaching. In P. L. Hosford (Ed.), *Using what we know about research* (pp. 51–77). Alexandria, VA: Association for Supervision and Instruction.

Bijou, S. (1970). What psychology has to offer education—Now! *Journal of Applied Behavior Analysis, 3*, 65–71.

Brophy, J., & Good, T. (1986). Teacher behavior and student achievement. In I. M. C. Wittrock (Ed.), *Handbook on research on teaching* (3rd ed., pp. 328–375). New York: Macmillan.

Carnine, D. W., & Fink, W. T. (1978). Increasing rates of presentation and the use of signals in elementary classroom teachers. *Journal of Applied Behavior Analysis, 11*, 35–46.

Catania, A. C. (1998). *Learning* (fourth ed.). Upper Saddle River, NJ: Prentice Hall.

Chu, H. C. (1998). *Functional relations between verbal behavior or social skills training, and aberrant behaviors of young autistic children.* Unpublished Ph.D. dissertation, Columbia University, New York.

Cooper, J., Heron, T., & Heward, W. (1987). *Applied behavior analysis.* Columbus, OH: Merrill.

Diamond, D. (1992). *Beyond time on task: Comparing opportunities to respond and learn units to determine an accurate means of measuring educational gains.* Unpublished paper, Teachers College Columbia University.

Donahoe, J. W., & Palmer, D. C. (1994). *Learning and complex behavior.* Boston, MA: Allyn and Bacon.

Donley, C. R., & Greer, R. D. (1993). Setting events controlling social verbal exchanges between students with developmental delays. *Journal of Behavioral Education, 3*, 387–401.

Dorow, L. G., McCorkle, N., Williams, G., & Greer, R. D. (1989). *Effects of setting performance criteria on the productivity and effectiveness of teachers.* Paper presented at the International Conference of the Association for Behavior Analysis, Nashville, TN.

Engelmann, S., & Carnine, D. (1982). *Theory of instruction: Principles and applications.* New York: Irvington.

Fisher, W. F., & Berliner, D. C. E. (Eds.). (1985). *Perspectives on instructional time.* New York: Longham.

Goldiamond, I. (1974). Towards a constructional approach to social problems. *Behaviorism, 2*, 1–84.

Greenwood, C. R., Carta, J., Arreaga-Mayer, C., & Rager, A. (1991). The behavior analyst consulting model: Identifying and validating naturally effective instructional models. *Journal of Behavior Education, 1*, 165–192.

Greenwood, C. R., Delquardri, J., & Hall, R. V. (1989). Longitudinal effects of class wide peer tutoring. *Journal of Educational Psychology, 81*, 371–383.

Greenwood, C. R., Hart, B., Walker, D. I., & Risley, T. (1994). The opportunity to respond and academic performance revisited: A behavioral theory of developmental retardation and its prevention. In I. R. Gardner et al. (Eds.), *Behavior analysis in education: Focus on measurably superior instruction* (pp. 213–224). Pacific Groves, CA: Brooks/Cole.

Greer, R. D., Dorow, L. G., Wachhaus, G., & White, E. (1973). Adult approval and students' music selection behavior. *Journal of Research in Music Education, 21*, 293–299.

Greer, R. D., Dorow, L. G., & Miller, M. A. (1976). Graduate student initiated responses as a function of teacher approval. In R. D. Greer and L. Dorow (Eds.), *Specializing education behaviorally.* Dubuque, IA: Kendall/Hunt.

Greer, R. D. (1986, 1990). *Teaching operations for verbal behavior.* Yonkers, NY: CABAS and the Fred S. Keller School.

Greer, R. D. (1994a). The measure of a teacher. In I. R. Gardner et al. (Eds.), *Behavior analysis in education: Focus on measurably superior instruction.* Pacific Groves, CA: Brooks/Cole.

Greer, R. D. (1994b). A systems analysis of the behaviors of schooling. *Journal of Behavioral Education, 4*, 255–264.

Greer, R. D.(1996a). *Acting to save our schools (1984–1994).* In W. Ishaq & J. Cautela (Eds.), *Contemporary issues in behavior therapy: Improving the human condition.* New York: Plenum.

Greer, R. D. (1996b). The educational crisis. In M. A. Mattaini and B. A. Thyer (Eds.), *Finding solutions to social problems: Behavioral strategies for change* (pp. 113–146). Washington, DC: American Psychological Association.

Greer, R. D., McCorkle, N. P., & Williams, G. (1989). A sustained analysis of the behaviors of schooling. *Behavioral Residential Treatment, 4,* 113–141.

Greer, R. D., & Polirstok, S. R. (1982). Collateral gains and short term maintenance in reading and on-task responses by inner city adolescents as a function of their use of social reinforcement while tutoring. *Journal of Applied Behavior Analysis, 15,* 123–139.

Guerin, B. (1994). *Analyzing social behavior.* Reno, NV: Context Press.

Hall, R. V., Lund, D., & Jackson, D. (1968). Effects of teacher attention on study behavior. *Journal of Applied Behavior Analysis, 1,* 1–12.

Hamburg, D. A. (1992). *Today's children: Creating a future for a generation in crisis.* New York: Times Books.

Hart, B., & Risley, T. R. (1980). In vivo language intervention: Unanticipated general effects. *Journal of Applied Behavior Analysis, 13,* 407–432.

Heward, W. L. (1994). Three low tech" strategies for increasing the frequency of active student response during group instruction. In I. R. Gardner et al. (Eds.), *Behavior analysis in education: Focus on measurably superior instruction* (pp. 283–320). Pacific Groves, CA: Brooks/Cole.

Hogin, S. (1996). *Essential contingencies in correction procedures for increased learning in the context of the learn unit.* Unpublished Ph.D. dissertation, Columbia University, NY.

Hogin, S., & Greer, R. D. (1994, March). *CABAS for students with early self-editing repertoires.* Paper presented at the International Conference of the Association for Behaviorology, Guanajauta, Mexico.

Holland, J. G., & Skinner, B. F. (1961). *The analysis of behavior.* New York: McGraw–Hill.

Ingham, P., & Greer, R. D. (1992). Changes in student and teacher responses in observed and generalized settings as a function of supervisor observations. *Journal of Applied Behavior Analysis, 25,* 153–164.

Iwata, B. A., Dorsey, M. F., Slifer, K. J., Bauman, K. E., & Richman, G. S. (1982). Toward a functional analysis of self-injury. *Analysis and Intervention in Developmental Disabilities, 2,* 3–20.

Johnson, K. R., & Layng, T. V. (1994). The Morningside Model of Generative Instruction. In I. R. Gardner et al. (Eds.), *Behavior analysis in education: Focus on measurably superior instruction* (pp. 283–320). Pacific Groves, CA: Brooks/Cole.

Johnson, P., Schuster, J., & Bell, J. (1996). Comparison of simultaneous prompting with and without error correction in teaching science vocabulary words to high school students with mild disabilities. *Journal of Behavioral Education, 6,* 437–458.

Johnston, J., & Pennypacker, H. (1992). *Strategies and tactics of human behavioral research* (2nd ed.). Hillsdale, NJ: Erlbaum.

Kelly, T. M., & Greer, R. D. (1996). *A functional relationship between learn units and decreases in self-injurious and assaultive behavior.* Manuscript submitted for publication.

Keller, F. S. (1968). Goodbye teacher... *Journal of Applied Behavior Analysis, 1,* 79–90.

Keohane, D. (1997). *A functional relationship between teachers use of scientific rule governed strategies and student learning.* Unpublished Ph.D. Dissertation, Columbia University, New York.

Kirby, F. D., & Shields, F. (1972). Modification of arithmetic response rate and attending behavior. *Journal of Applied Behavior Analysis, 5,* 79–84.

Lamarre, J., & Holland, J. G. (1985). The functional independence of mands and tacts. *Journal of the Experimental Analysis of Behavior, 43,* 5–19.

Lamm, N., & Greer, R. D. (1991). A systematic replication of CABAS in Italy. *Journal of Behavioral Education, 1,* 427–444.

Lindhart-Kelly, R., & Greer, R. D. (1997). *A functional relationship between mastery with a rate requirement and maintenance of learning.* Manuscript submitted for publication.

Lindsley, O. R. (1991). Precision Teaching's unique legacy from B. F. Skinner. *Journal of Behavioral Education, 1,* 253–266.

Lodhi, S., & Greer, R. D. (1989). The speaker as listener. *Journal of the Experimental Analysis of Behavior, 51,* 353–359.

Mach, E. (1960). *The science of mechanics* (T. J. McCormack, Trans.). Lasalle, IL: Open Court.

Madsen, C. H. J., Becker, W. C., & Thomas, D. R. (1968). Rules, praise, and ignoring: Elements of elementary classroom control. *Journal of Applied Behavior Analysis, 1*, 139–150.

Malott, R. W., & Heward, W. L. (1995). Saving the world by teaching behavior analysis: A behavioral systems approach. *The Behavior Analyst, 18*, 341–354.

Martinez, R., & Greer, R. D. (1997). *Reducing aberrant behaviors of autistic through efficient instruction: A case of matching in the single alternative environment.* Unpublished doctoral dissertation, Columbia University, New York.

Michael, J. (1983). Distinguishing between discriminative and motivational functions. *Journal of the Experimental Analysis of Behavior, 37*, 149–155.

Patterson, G. R. (1982). *Coercive process.* Eugene, OR: Castela.

Polirstok, S. R., & Greer, R. D. (1986). A replication of collateral effects and a component analysis of a successful tutoring package for inner-city adolescents. *Educational Treatment of Children, 9*, 101–121.

Polirstok, S. R., & Greer, R. D., (1977). Remediation of a mutually aversive interaction between a problem student and four teachers by training the student in reinforcement techniques. *Journal of Applied Behavior Analysis, 10*, 573–582.

Schwartz, B. (1994). *A comparison of establishing operations for teaching mands.* Unpublished doctoral dissertation, Columbia University, New York.

Selinske, J., Greer, R. D., & Lodhi, S. (1991). A functional analysis of the Comprehensive Application of Behavior Analysis to Schooling. *Journal of Applied Behavior Analysis, 13*, 645–654.

Sherman, J., & Cormier, M. (1974). An investigation of the effect of student behavior on teacher behavior. *Journal of Applied Behavior Analysis, 7*, 11–21.

Sidman, M. (1960). *Tactics of scientific research.* New York: Basic Books.

Singer, J., & Greer, R. D. (1997, May). *A functional analysis of the role of the correction operation in the learn-unit.* Paper presented at the annual conference of the International Association for Behavior Analysis, Chicago, IL.

Skinner, B. F. (1938). *The behavior of organisms.* New York: Appleton-Century-Crofts.

Skinner, B. F. (1953). *Science and human behavior.* New York: Macmillan.

Skinner, B. F. (1957). *Verbal behavior.* Cambridge, MA: B. F. Skinner Foundation.

Skinner, B. F. (1968). *The technology of teaching.* New York: Appleton-Century-Crofts.

Staats, A. W. (1981). Paradigmatic behaviorism, unified theory construction methods, and the zeitgeist of separatism. *American Psychologist*, 346–378.

Stallings, J. (1980). Allocated academic learning time revisited, or beyond time on task. *Educational Researcher, 9*, 11–16.

Stallings, J. A. (Ed.). (1985). *Instructional time and staff development.* New York: Longham.

Stephens, J. M. (1967). *The process of schooling.* New York: Holt, Rinehart, & Winston.

Stokes, T. F., & Baer, D. M. (1977). An implicit technology of generalization. *Journal of Applied Behavior Analysis, 10*, 349–367.

Sundberg, M. L. (1993). The application of establishing operations. *The Behavior Analyst, 16*, 211–214.

Vargas, E. A. (1988). Event governed and verbally governed behavior. *The Analysis of Verbal Behavior, 6*, 11–22.

Wolery, M., Holcombe, A., Billings, S. S., & Vassilaros, M. A. (1993). Effects of simultaneous prompting and instructive feedback. *Early Education and Development, 4*, 20–31.

The Repertoires of Teachers Who Are Behavior Analysts

TERMS AND CONSTRUCTS TO MASTER

- Repertoires of teaching
- Contingency-shaped teaching repertoires
- Verbal behavior about the science as a repertoire of the strategic science of teaching
- Verbally mediated behavior and rule governed behavior as a repertoire of the strategic science of teaching
- *In situ* instruction
- Presentation of accurate learn units
- Teacher performance rate/accuracy observation procedure (TPRA)
- Appropriate contingency-shaped behavior
- Inappropriate contingency-shaped behavior
- Levels of teacher skills
- Principles of behavior
- Strategies of behavior analysis and pedagogy
- Tactics of behavior analysis and pedagogy
- Transition time and inter-learn unit time (latency between learn unit presentations)
- Tact

- Mand
- Verbal mediation as a controlling variable for analytic repertoires
- Unambiguous antecedent presentations of the student's discriminative stimulus (faultless presentations of the target student S^d)
- Teacher antecedent
- Response
- Postcedent or consequence in the operant and the learn unit
- Mastery and "fluency" (rate of responding as a component of mastery)
- Criterion-referenced objectives
- Successive approximation
- Stimulus control
- Unambiguous exemplar presentations
- Response classes
- Teachers and behavior analysts as transducers of data
- Scientific descriptions or characterizations of student and teacher behaviors (scientific tacts)

TOPICS

The Repertoires of the Teacher as Strategic Scientist

THE REPERTOIRES OF THE TEACHER AS STRATEGIC SCIENTIST

The teacher who is a strategic scientist of instruction

- *applies* the basic strategies and tactics of the science of behavior to all aspects of teaching automatically in the classroom (contingency-shaped teaching repertoire);
- *describes* the setting events, the behaviors of students, and teacher behaviors that are occurring using the *terms of the vocabulary* of the science (verbal behavior about the science repertoire); and
- *analyzes* instructional interactions and derives solutions to instructional problems from the perspective of the science (verbally mediated repertoire).

Each of these three repertoires has many gradations or levels of expertise. Because the science is complex and always growing, teachers who are learning to be strategic scientists of instruction, as well as those who are recognized experts, must continuously acquire better classroom practice skills, mastery of new concepts from the vocabulary of the science, and greater analytic expertise.

The three repertoires of scientists/teachers are best described using terms from the science as follows.

1. *Accurate contingency-shaped behaviors of teaching* (to refer to teaching practices acquired in the classroom);
2. *Verbal behavior about the science* (to refer to the use of the vocabulary of the concepts or terms of the science and correspondence between the terms and operations of teaching and learning); and
3. *Verbally mediated repertoires* (to refer to expertise in analyzing and solving instructional problems).

Strategic scientists who are applied behavior analysts of instruction have advanced expertise in all three repertoires. Moreover, all three are anchored to the findings, methodological practices, and epistemology of the science of behavior. This chapter is devoted to an introduction to these teaching repertoires.

Subsequent chapters will describe the repertoires in more detail as they apply to the instruction of students with different repertoires. The chapter on supervision and teacher consultation prescribes operations that the professional who is responsible for teaching and motivating the teacher (e.g., supervisor or professor) can perform to assist and support the teacher's increased sophistication and continued use of these repertoires of teaching.

Supervisors, consultant teachers, and psychologists or other school support personnel as well as teachers must master the three repertoires of pedagogy. The repertoires constitute the core of scientific teaching practices and conceptions of the process. We use the term *repertoire* to refer to a collection of behaviors, discriminated stimuli (S^ds), consequences, and setting events that belong to a particular response class and, in turn, a collection or categorical groupings of response classes.

The Contingency-Shaped Repertoires of Teaching

Contingency-shaped behaviors are those behaviors that are *reinforced or punished directly by contingencies in the environment*. Repertoires that are *contingency shaped* are "selected out" or taught directly by the antecedents and consequences of responses directly. Learning to ride a bicycle by repeated experience of staying on the bicycle is an example of a contingency-shaped behavior. Learning to hammer a nail, slice meat, type, and drive a car by experiencing the *consequences* of various actions selects out or shapes more efficient repertoires over time. Operating a word processing program in a computer fluently without giving the operations "a second thought" or consulting a manual is still another example. In some areas of scientific psychology fluent contingency-shaped behavior is described using the term "automaticity." The learner learns to behave automatically and "without having to think about what to do." In our case, the repertoire refers to using the best practices of the science of teaching automatically and fluently.

When a teacher learns classroom-teaching practices directly as a result of the responses of students she is acquiring contingency-shaped repertoires. They are the automatic and "unpremeditated" responses that teachers make to students.

Unfortunately, teachers who use procedures by trial and error will likely learn teaching practices that are punitive or ineffective for their students. That is, the teachers will learn behaviors that are of short-term benefit for the teacher but of long-term damage to the student's learning, as the research shows (White, 1975). Rather than engage in positive reinforcement operations, they will spend 80% of their time punishing or negatively reinforcing behavior (Madsen, Becker, & Thomas, 1968). Teachers *who are not taught* to use contingency-shaped repertoires based on the best scientific practices by specially designed classroom experiences typically learn to:

- Present flawed antecedents or instructions;
- Neglect to allow the student the opportunity to respond;
- Neglect to reinforce or correct the target response;

- Do not remediate student learning problems immediately;
- Do not present learn units;
- Do not teach to mastery; and
- Presume that their lectures function to teach.

These teachers are likely to have classrooms that are characterized by long periods of transition time (e.g., curriculum and instruction sequences are not immediately available to the student), frequent occurrences of aberrant or noncompliant behavior, and poor instructional outcomes for the students. The time between learn units (i.e., inter-learn unit latency) will be prolonged. This is not to suggest that a few teachers do not develop effective behavioral tactics unwittingly (Greenwood, Carta, Arreaga-Mayer, & Rager, 1991); but, according to the research, such teachers are the exception rather than the rule.

This state of affairs is understandable. In order to *ensure* that the contingency-shaped teaching practices of teachers are consistent with the science of pedagogy, teachers are best taught practices that are scientifically sound through in-class instruction. We refer to teaching the teacher as he or she teaches in the classroom as *in situ* teaching. Novice teachers who do not have the benefit of expert instruction and expert examples in the classroom will acquire inaccurate or flawed contingency-shaped repertoires. The set of procedures for teaching teachers in their classroom setting that were developed in the CABAS model and the related research have been found useful to teach accurate and flawless contingency-shaped teaching operations. They were developed from and built on studies from the applied scientific literature.

We should not expect teachers who are new in the field (i.e., first-year teachers or student teachers) or any teachers who *have not been trained behaviorally* to perform in the classroom consistent with the best practices of the science. In order to learn the appropriate classroom practices they should begin instructing a classroom *initially under the complete supervision and support of a qualified supervisor or teacher consultant (one who is a master strategic scientist and applied behavior analyst of instruction)*. They are initially assisted in basic assessments, setting objectives and priorities, developing or locating scripted or programmed curricula, and presenting learn units (antecedent and consequence) by a teacher coach (supervisor or teacher consultant with the necessary expertise). They are taught:

- When and how to ignore student behavior;
- To reinforce positively, contingently, and frequently for social and academic responding;
- To present intact learn units at increasingly faster rates;
- To supervise teaching assistants and peer tutors;
- To provide individualized instruction regardless of the size of the class; and
- To record and display all student responses in a reliable fashion.

They are taught *until* the basic scientifically *validated responses of the teacher are automatic.*

The coach teaches the teacher by demonstrating or modeling correct learn unit presentations, followed by having the teacher attempt learn unit presentations, followed in turn by the supervisor giving reinforcement or correction operations for the teacher's behavior. While lessons or programs (blocks of learn units) are presented by the supervisor, the teacher collects data on student responses until the supervisor and teacher agree on correctness and incorrectness of student behavior. The teacher then teaches while the supervisor observes teacher and student learn units or their absence thereof. Observations include supervisor feedback detailing reinforcement and corrections operations by the teacher. Coaching continues until a teacher emits accurate presentations and evokes successful student responses to learn units.

During the same period of time, teachers are taught to maintain the instructional engagement of all of their students. Teacher assistants are also taught to ensure independent engagement of all members of the class while the teacher is working individually with students. These teacher-training procedures are described in detail in the chapters that follow.

Teachers who must learn these practices without benefit of expert coaching will need to follow the instructions in this and the prerequisite behavioral texts. In doing so, their behavior will initially be *rule governed or verbally mediated.* When the practices that are initially dictated by verbal instructions come under the control of their effects on student behavior they will qualify as contingency shaped. And if they acquire the characteristics of automaticity and accuracy they will be fluent.

In this manner the teacher is taught contingency-shaped teaching behaviors that are consistent with the sciences of behavior and pedagogy. Rather than responding to disruptive behavior by disapproval that functions as reinforcement of the "bad" behavior of students (a typical teacher response), the teacher learns to reinforce appropriate behavior, thereby coming under the control of contingencies in the classroom that are consistent with good teaching practices. In this particular case, the misbehavior of students prompts a well-trained teacher to reinforce the behavior of other appropriately behaving students, while ignoring the misbehavior. The teacher automatically brings the classroom under instructional control, presents errorless or flawless antecedents, ensures ample opportunities for students to respond to antecedents, and reinforces or corrects the students' responses without error.

The intensity of the supervision fades gradually (or recourse to dependence on rules from the science fades) as the teacher acquires sound contingency-shaped behaviors and becomes fluent in presenting frequent learn units (i.e., errorless teacher–student interactions) without disruption and with little transition time. Ideally, the supervision, self-observation, and instruction never totally ceases, but with increased skill the nature of the supervisor/teacher interaction becomes increasingly one of professional consultation between individuals with more equal expertise. It is easy for inappropriate repertoires to be learned unwittingly even with the most expert teacher. Flawed skills are likely to develop for even the most advanced teachers, unless supportive and objective monitoring and feedback are regularly provided (Greenwood et al., 1991; Greer, 1991, 1994; Greer, McCorkle, & Williams, 1989; Ingham & Greer, 1992). "Quality control" is as important for teaching as it is for any endeavor.

Thus the contingency-shaped behaviors of the teacher are taught best by a teacher coach who mediates classroom contingencies for the novice teachers. This is done to ensure that the contingency-shaped behaviors that are acquired by the novice are those that are consistent with good teaching practices from the science of behavior, rather than short-term and expedient responses that are detrimental to the long-term benefit of the student. The *advantages* of direct experience are enhanced, while the *disadvantages are avoided* by ensuring that the supervisor mediates classroom contingencies.

There are various levels of sophistication associated with contingency-shaped behaviors. Initially, teachers are able to provide only minimal numbers of learn units during the instructional day (i.e., low numbers per day of learn units) and minimal levels of instructional engagement by students and by teaching assistants. Initial observations of teachers using the teacher performance rate accuracy observation procedure or TPRA (the TPRA will be explained in later chapters) will show high numbers of reinforcement errors and omissions as well as correction errors and omissions (Ingham & Greer, 1992). This teacher inaccuracy will, in turn, result typically in lower rates of correct responding and higher rates of incorrect responding for the student(s) being taught during the observation. With increased supervisor observation and instruction, errors will decline and teachers and students will make fewer errors and produce higher rates of correct responding. Increasingly, teachers will learn to manage classrooms that produce increasing numbers of learn units, and the achievement of higher numbers of learning objectives by students will result. Transition and off-task behavior of the students will decrease, and the classroom will become a more effective and pleasant educational setting. Inappropriate behavior will be eliminated as a result of efficient teaching practices that maximize student reinforcement for academic responding (Kelly, 1994; Kelly & Greer, 1996; Martinez & Greer, 1997).

The rate at which teachers acquire these skills will vary for individual teachers based on:

- The students and subject matter taught in the classroom;
- The learning history of the teacher; and
- The reaction times of the teacher to events that occur in the classroom.

Of course, the skills of the supervisor (or alternately the ability of the teacher to follow written instructions) will play a role in the rate of skill development also.

As in the case of the student, instruction *must continue for the teacher until objectives and subobjectives of mastery and fluency are acquired.* Each teacher is unique and each classroom is unique, resulting in individual rates of teacher progress.

The coaching strategies and tactics for teaching teachers to present accurate learn units at effective rates are provided in the chapter on supervision and teacher consultation. When teachers are effective they engage in contingency-shaped instructional practices that are errorless and accurate regardless of whether instruction is occurring with a group or one on one. If the learn units are to address individual differences errorless contingency-shaped teacher performance is necessary in the teacher's presentation of learn units in written form via handouts, workbook programs, computer-assisted instruction, or one-to-one instruction. They are equally critical when the teacher is conducting total class drills involving many student responses. The teacher with good contingency-shaped behaviors will manage classrooms such that teacher assistants and all pupils are engaged in responding to learn units, while the teacher is presenting and consequating learn units to one or more students simultaneously. If the accurate performance of the teacher with the student is "automatic" and does not require the teacher to "think" about what he or she is to do next, we refer to his or her performance as being contingency shaped. Fluent contingency-shaped repertoires result from (a) the design of the classroom and (b) the second nature (automaticity) with which the teacher responds to and predicts events in the classroom in a manner consistent with the science.

When the teacher has to "think" about how to respond to the student in a manner consistent with the science the repertoires are not yet contingency shaped. The "thinking" is verbal behavior that guides the teacher such that she does not respond with teaching behaviors that are inconsistent with the science. Many of the repertoires that become contingency shaped are initially guided by verbal behavior from the science, in the form of either the teacher remembering (i.e., "oh yes, I must ignore the speaking out") or the supervisor's verbal prompt (i.e., the supervisor signals the teacher to ignore). Automatically accurate teaching practices in the classroom are necessary for the teacher who acquires the other repertoires. That is, it is not unlike typing wherein the typing responses must be automatic when the typist is functioning as an author. The expert teacher performs flawlessly in the classroom while simultaneously analyzing events as they occur. That is, he or she can teach accurately, while *thinking* about other events or analyzing contingencies.

What *thoughts* of analysis occur? They are the verbal characterizations or descriptions of what is occurring in terms of the science and they are verbal behaviors. To do so the teacher must master and be fluent in using the vocabulary of the concepts of the science—really the verbal community of the science. We shall refer to this repertoire as the teacher's *verbal behavior about the science.*

The Vocabulary of the Science and Its Role in a Strategic Science of Teaching

The science of behavior and the related sciences of pedagogy and schooling consist of both a set of research practices for engaging in the science (studying behavior/environment interactions to derive principles and tactics) and a set of research findings concerning effective practices to bring about behavior change under certain stimulus control or environmental conditions. The result is a vocabulary of scientific strategies and tactics. Such a vocabulary is a *community of verbal behavior* (an ever-changing and dynamic community). It is a community because all who know the vocabulary and whose actions are dictated by the vocabulary are members of a common community.

One learns the vocabulary (i.e., verbal behavior about the sciences) in order to:

- Engage in the practices of the science;
- To apply the findings of the science (in this case to teaching); and
- To analyze events scientifically.

Although learning this repertoire is necessary, it is only one component of applying the science to teaching. One can learn a limited number of good classroom practices without knowing much of the vocabulary. One can know the vocabulary but not engage in good classroom practices. "Knowing" the vocabulary and its correspondence to good classroom skills is a necessary step, but it alone is insufficient to achieve the status of strategic scientist. The verbal repertoire is used:

1. To describe behavior and environmental occurrences consistent with the science in lieu of using layperson or prescientific categories;
2. To give rationales to others for the practices that are used (in written form and vocally);
3. To read about the new or continuing practices in the science as they apply to teaching;
4. To communicate events, settings, behaviors, trends, and practices with other scientist-practitioners;
5. To form the basis for verbally mediated or analytic repertoires that are the repertoires of the expert applied scientist of teaching.

The verbal repertoire is typically learned through readings from texts that are technically accurate and taught through tutorials, written quizzes, and vocal quizzes. The responses learned are written and vocal statements that accurately define and describe concepts and tactics of the science per teaching practices. The audience for the student of teaching to display his or her verbal expertise is initially the instructing professional and ultimately other practitioners, experts, educators, boards of education, parents, and most importantly the teacher in his or her capacity to edit and change his or her own teaching behavior.

Teachers who are applied scientists of pedagogy learn to define principles of behavior, operations for implementing them, and related tactics. We have found that these are taught best systematically and simultaneously as the novice teacher learns appropriate contingency-shaped repertoires. Thus, the teacher learns to describe the practices being emitted in the classroom before, after, or simultaneously while engaging in learning them *in the classroom setting* (i.e., *in situ*). Learning the verbal repertoire per se will not necessarily engender the appropriate contingency-shaped repertoire initially.

Those who are responsible for the instruction of the teacher *must teach correspondence between the two repertoires for the teacher*. Teacher educators, i.e., supervisors, consultant teachers, or professors, must use precise terms to describe teaching practices and evoke similar verbal precision from the novice teacher when teaching the contingency-shaped repertoires. Meetings between supervisors, or other teacher trainers, and teachers must be designed to teach and use the *behavioral data language* (Johnston & Pennypacker, 1993) to describe graphs of student and teacher behavior as well as definitions of shaping, reinforcement, extinction, and other principles of behavior and their related tactics.

The precision of the language is critical since the precision must be translated into accurate practice. The precision of practice must, in turn, be related to the precision of description in the process of teaching master teachers. That is, *practices must conform with the verbal description and vice versa*. The operations and levels of skill pertaining to teaching verbal behavior about the science are described in the chapter on supervision. For now, it is important to note that the repertoire must result in accurate, succinct, objective, valid, and fluent descriptions of the practices of the science and its application. The critical relationship that must be forged between the two repertoires (contingency shaped and verbal behavior about the science) becomes apparent in the scientific analytic repertoire of teaching—the verbally mediated repertoire.

The Verbally Mediated Repertoire and Its Role in a Strategic Science of Teaching

The three repertoires of the teacher who functions as a strategic scientist are complex. They encompass sophisticated skills of pedagogy in the classroom setting, sophisticated use of the language of the science to describe teaching practices and the responding of the student, and a sophisticated repertoire involving the operations associated with relating the science to instructional practice. The expert teacher manages the classroom using optimum contingency-shaped pedagogy such that students are receiving optimum numbers of learn units that are individually valid and that result in high rates of correct responding and low rates of incorrect responding relative to the stimulus control and responses being learned (e.g., spelling under various antecedent stimulus conditions, essay writing under specific antecedent stimulus conditions). Next, the expert teacher can describe all of these operations in the classroom using terms that can be translated into practice by another similarly trained teacher with equally good effects.

In an optimally managed classroom, the teacher avoids passive or off-task behavior, long periods of transition, and inappropriate classroom behavior that result from ineffective instructional design and ineffective pedagogy and he or she does so automatically. If the curriculum presentations are designed entirely in an individualized manner and presented quickly and accurately, loss of instruction to transition is nonexistent. Each student moves through learn units designed for that student's ability level, receiving the reinforcer (and schedule of reinforcement) that is right for that student. The teacher dispenses reinforcement, response, and stimulus prompts; records and plots data; and changes pedagogical procedures based on the data per student. Our expert teacher uses procedures anchored to the research literature, reinforces and prompts teacher assistants and peer tutors, and provides corrections for incorrect responses. As the breadth of the tested scripted, or programmed, curriculum increases and the diversity of the teachers' pedagogical skills increases, the frequency of students who experience plateau levels of responding declines. The diversity of contingency-shaped teaching behavior also expands. Moreover, the teacher learns to describe events and ask questions consistent with the science.

However, regardless of how good the teacher's contingency-shaped repertoire is and how individualized the curriculum, the teacher will encounter teaching problems. How the teacher goes about solving these problems lies at the heart of verbally mediated analytic repertoires of teaching. The *verbally mediated repertoire* constitutes the contingency analysis expertise of the science of pedagogy. Contingency-shaped and verbal behavior about the science are perquisite repertoires to the analytic operations.

Prescientific descriptions of teaching would characterize what a teacher undergoes to alleviate instructional problems as *reflective teaching*. However, reflective teaching is simply a description, but one that is useless unless it describes specific and operations that any teacher might use to solve the problem. We use the term *verbal mediation* to indicate the source that a scientifically reflective teacher draws on to solve instructional problems. The verbal community of the science is the source of analytic expertise and because it is a verbal community, the expertise can be passed on to those with the necessary vocabulary and contingency-shaped skills.

For example, the occurrence of a learning plateau would be characterized or described fully in terms of the science by such a teacher as a first step in analyzing the problem (e.g., the data are low or descending and the trend is established by a minimum of three data paths, usually four data points). Data paths are the connecting lines between data points and show the trend; it is the data paths and not the data points that show the trend. These terms, in turn, suggest several operations to a well-trained teacher that can enact to solve the problem (e.g., the student needs an intrastimulus prompt). Each attempt that is unsuccessful is followed by other terms characterizing the situation that, in turn, directs the knowledgeable teacher to still additional tactics (specific teacher actions such as teach a missing component skill) that are likely to work. Eventually, the solution is found and this solution can also be described using scientific and terms that allow the initiated to replicate them. Another professional who is knowledgeable about the science could perform the same operations given a similar problem.

The identified problem is solved through verbal mediation that drives the analysis of the teaching and learning contingencies. That is:

1. A series of verbal statements derived from the science *describe the problem* from the precision of the science.
2. These *series of statements* which characterize the problem *are linked to research-based operations* that also are described with appropriate terms, and
3. *When verbal terms are applied* to analyze the problem *and to change pedagogy*, the student's prognosis improves.

It is important to characterize both the behavior of students in precise scientific terms and the behaviors or operations of instruction. The teaching actions are governed by the terms of the verbal community of the science. Members of that community can perform the same operations given the descriptions of the vocabulary of the science. Prescientific descriptions of students such as "has poor self-motivation" are replaced by accurate descriptions of the rate of the behavior across time and the setting events and contingencies affecting the behavior.

The vocabulary or terminology itself suggests or *prescribes* strategies and tactics (also other terminology) to analyze and solve the instructional problem. Without the science, teachers typically blame the problem on fictional or tautological constructs (i.e., circular reasoning) such as a "decoding" inadequacy. The latter approach frequently results in labeling a student in a way that not only fails to suggest instructional alternatives, but leads to categorizing the student as being somehow inadequate. Characterization of the problem in scientifically valid terms, on the other hand, suggests scientifically tested operations for teaching the deficit responses or repertoires.

For example, a student who has achieved mastery on multiplication facts is having difficulty achieving a fluency criterion (i.e., responds methodically and slowly). The visual display is identified by the teacher as having a scallop pattern and slow rates of responding consistent with fixed rates of reinforcement. That is, after each correct response the student pauses. This pattern of responding suggests that the student's lack of fluency may be a function of the schedule of reinforcement received by the student. These series of verbal descriptions and their relationship to findings from the science suggest a change in the pedagogical procedures (e.g., thin the schedule of reinforcement and place reinforcement on a variable schedule). When the teacher's teaching actions are governed by such statements as "thin the schedule" we say the teacher's behavior is no longer governed by the immediate contingencies but is governed by the scientific verbal directions. That is, rather than reinforce after each correct response, typically a good automatic response by the teacher teaching acquisition of a skill, the teacher reinforces after a number of correct responses that meet a rate or speed-of-correct-responding objective. If the suggested solution is correct the student becomes fluent; that is, he or she performs correctly and quickly without pause. The problem was identified in scientific terms, and solutions were drawn from verbal behavior about the science. This resulted, in turn, in teaching operations or tactics that were applied to the actual instruction. Thus, rather than the automatic performance of standard teaching practices for each learn unit controlling instruction as they should for a student experiencing no difficulties, verbal behavior from the strategies and tactics of science mediates between the early unsuccessful attempts and the later successful attempts. The distinction is much like the difference between a pilot flying by his or her own vision versus knowing when to react to instrumentation and actually flying the airplane governed by the dictates of the instrumentation. Both skills are needed and, in addition, judgment is required for determining which course of action (e.g., contingency shaped or verbally mediated) needs to occur.

In still another case, a preschooler with a history of speech delay has acquired an age-appropriate vocabulary but still continues to use language only when requested to speak. In prescientific terms, she would be said to lack "spontaneous" speech. The research on verbal behavior, however, suggests that *pure tact and mand* repertoires have not been trained. No one has any idea what controls spontaneous speech since spontaneous refers to unknown controlling variables, but the terms *tact* and *mand* are precise descriptions of the controls of certain functions of verbal behavior. Tactics for training non-verbal antecedent control of these two forms of verbal behavior are located in the literature (e.g., stimulus delay, use of establishing operations, and avoiding verbal antecedents). The tactics are then introduced to the child's daily in-struction. As a result, the student acquires developmentally appropriate forms of verbal behavior characterized less precisely by laypersons as spontaneous speech. The initial problem was due to the prescientific description of the behavior. The solution followed from applying the dictates of the science.

The particular learning problem is first reliably characterized from the data collected and visual displays (e.g., charts or graphs) of those data as a trend and not a momentary occurrence. Both the *verbal labels* for what has transpired and the *data collection procedures* are *scientific tacts*—that are part of the verbal community of the science. The scientific verbal description in turn suggests that the existing teaching operations that are occurring are not effective for this student because the problem was not characterized accurately before. The vocabulary of the science that is used to characterize the problem (e.g., the repertoires that are deficient are mands and tacts) also suggests research-based solutions. The scientific verbal repertoire of the teacher *mediates* between the original teaching procedures and the resulting changes in teaching procedures. The teacher in effect edits or corrects his or her instruction as a result of her own verbal identification of the problem. Her verbal identification, in turn, directs her teaching (e.g., he or she has correspondence between her verbal terms and his or her teaching actions). The new procedures are initially under verbal control from the science rather than being directly controlled by the immediate contingencies; thus the new procedures are verbally mediated. They are the thinking behaviors of analytic instruction.

When teachers become proficient in using verbally mediated strategies and tactics from the science to solve problems, the teacher is functioning as a strategic scientist. Acquiring such repertoires is a function of the instruction, supervision, and support provided by supervisors and professors of education who are scientifically sophisticated. Teachers who are strategic scientists have classrooms in which error rates are low, achievement of objectives are high, and plateaus or descending trends are minimized as a result of the teachers' verbally mediated repertoires as well as their contingency-shaped repertoires. Changes are made in the scripted or standard pedagogical procedures in a systematic manner based on strategic scientific analyses resulting in the application of alternative operations that too are based on scientific findings or their extension.

The analytic expertise that is called for in the application of the methods and findings of behavior analysis is the critical core of teaching as applied behavior analysis. It is this repertoire that distinguishes the teacher as a strategic scientist. A strategic scientist of instruction is a professional who can draw on the breadth and specificity of the science to fit research-based tactics to individual students. It is only recently that we have begun to operationalize the contingency analysis repertoires as verbally mediated strategies.

SUMMARY

In this chapter we introduced the three repertoires of teaching. They included the contingency-shaped behaviors of a teacher whose classroom practices conform with teaching practices that correspond to the science of individualized instruction, the concepts of the science implicit in the vocabulary or verbal community of the science, and the verbally mediated repertoires that comprise analytic behaviors of teachers who are strategic scientists. A brief description and some examples for each repertoire were presented in order to provide an overview for the detailed comments that are to follow in subsequent chapters.

We introduced the strategic questions that need to be asked about the various components of the learn unit, setting events, and the student's instructional history. Thus, the information on the teacher repertoires built on the information provided in the prior chapter on the learn unit. A firm grasp of the learn unit and its context is necessary to master the information presented in this chapter. Similarly, the reader will need to be familiar with the repertoire of the teacher in general in order to benefit from the coverage in the subsequent chapters. In effect, the material represents the prerequisites needed by the reader much the same as the prerequisites for any student determines the ease with which he/she grasps new material.

References

Greenwood, C. R., Carta, J. J., Arreaga-Mayer, C., & Rager, A. (1991). The Behavior Analyst Consulting Model: Identifying and validating naturally effective instructional models. *Journal of Behavioral Education, 1*, 165–192.

Greer, R. D. (1991). The teacher as strategic scientist: A solution to our educational crisis? *Behavior and Social Issues, 1*, 25–41.

Greer, R. D. (1994). A systems analysis of the behaviors of schooling. *Journal of Behavioral Education, 4*, 255–264.

Greer, R. D., McCorkle, N. P., & Williams, G. (1989). A sustained analysis of the behaviors of schooling. *Behavioral Residential Treatment, 4*, 113–141.

Ingham, P., & Greer, R. D. (1992). Changes in student and teacher responses in observed and generalized settings as a function of supervisor observations. *Journal of Applied Behavior Analysis, 25*, 153–164.

Johnston, J. M.,& Pennypacker, H .S. (1993). *Strategies and tactics of behavioral research* (2nd ed.). Hillsdale, NJ: Erlbaum.

Kelly, T. M. (1994). *Functional relations between numbers of learn unit presentations and emissions of self-injurious and assaultive behavior*. Unpublished dissertation, Columbia University, New York.

Kelly, T. M., & Greer, R. D. (1996). *Functional relationships between learn units and maladptive behavior*. Manuscript submitted for publication.

Madsen, C. H. J., Becker, W. C., & Thomas, D. R. (1968). Rules, praise, and ignoring: Elements of elementary classroom control. *Journal of Applied Behavior Analysis, 1*, 139–150.

Martinez, R., & Greer, R. D. (1997). *Reducing aberrant behaviors of autistic students through efficient instruction: A case of matching in the single alternative environment*. Unpublished paper, Columbia University, New York.

White, M. A. (1975). Natural rates of approval and disapproval in the classroom. *Journal of Applied Behavior Analysis, 8*, 367–372.

CHAPTER 4

The Strategic Analysis of Instruction and Learning

TERMS AND CONSTRUCTS TO MASTER

- Strategic analysis of instruction
- Learn unit analysis decision
- Tactics from applied behavior analysis
- Selection of the tactic
- History of instruction (strategic questions and tactics)
- Response independence (strategic questions and tactics)
- Setting events (strategic questions and tactics)
- Motivation (strategic questions and tactics)
- Students target S^d (strategic questions and tactics)
- Teacher S^d (strategic questions and tactics)
- Consequences (teacher, student) (strategic questions and tactics)
- Response substitution (strategic questions and tactics)
- Corrections (strategic questions and tactics)

- Reinforcement operations (strategic questions and tactics)
- Establishing operations (strategic questions and tactics)
- Data paths
- Ascending/descending trends
- Functional analysis of the controlling variables
- Perquisite repertoire
- Antecedents
- Postcedents
- Probe sessions
- Data point
- Data path
- Transducing data
- Data language
- Decision tree
- Incidental teaching
- Captured learn units
- Recyling
- Corrections

- Consequation (consequating operation)
- Stimulus prompts

- Establishing operations

- Intact learn unit

TOPICS

Verbally Mediated Repertoires

Strategic Questions

Summary of Applications of the Verbally Mediated Repertoire

VERBALLY MEDIATED REPERTOIRES

Teachers who are expert in the verbally mediated repertoires associated with analyzing the contingencies of instruction can solve instructional difficulties as they are encountered. They achieve instructional objectives with their students in fewer numbers of learn units than teachers who do not have analytic expertise. We have categorized the operations of analysis into two broad categories. We use two terms to describe two repertoires in the analytical process. They are verbally mediated operations associated with the (1) *strategies* and (2) *tactics* of the science of behavior as applied to instructional problems.

These strategies and tactics are needed when learning plateaus are encountered and when descending trends occur—in summary when repeated instructional sessions result in no progress. Indeed they are the same strategies and tactics used by the scientist of behavior to make decisions, but their application to decision making in instruction require new analytic skills. The strategic repertoires and the tactical repertoires are both necessary to maximize the progress of *individual* students. The facile use of these repertoires to solve instructional problems is characteristic of the very best educational programs and teachers (Keohane, 1997).

Analytic expertise is necessary because instructional problems arise for all students and teachers, even when teachers use expert practices from behavior analysis continuously and automatically. We use the term *verbally mediated* to characterize these analytic repertoires consistent with our treatment of all complex behaviors as observable and teachable. We also use the expression "*contingency analysis*" to refer to the focus of the verbally mediated repertoire (i.e., the *verbal mediation* is used for contingency analyses). It is important for the reader to fully understand the construct of verbal mediation, thus we explain the origin of our choice of the expression verbally mediated.

The term *verbally mediated* is a refinement and extension of Skinner's term rule governed behavior for which Vargas suggested the term *verbally governed* (Skinner, 1957; Vargas, 1988). Vargas defined the term *verbally governed* as behavior that is mediated by verbal behavior. It is distinguished from contingency-shaped or *event-governed* behavior. Contingency-shaped behavior (as described earlier and used by Skinner in his original formulation) is behavior that is under the control of the direct events in the environment. As described in the prior chapter, contingency-shaped teaching behavior is teaching behavior controlled directly by events in the teaching environment, particularly the behaviors of the students themselves. The contingency-shaped behavior of the teacher is under the direct control of classroom events. Verbally governed or mediated behavior is behavior in which *verbalization mediates between the behavior of the teacher and the events in the environment*. Verbally mediated analysis involves a series of verbally governed behaviors that we categorize as:

- *Strategic types of mediation* (asking scientific questions about the problem) and
- Mediations associated with *tactics* (choosing scientifically tested tactics).

The verbally mediated repertoire is instructional behavior under the control of forms of verbal behavior, wherein *the instructional efforts of the teacher are guided by verbal behavior from the science*, rather than the behavior of the student directly.

The term *verbally mediated* means that verbal directions from the science control the analytic operations that a teacher applies to a particular learning difficulty experienced by a student. At one level these verbal directions take the form of *strategies* that suggest *questions* to ask about components of instruction and ways to answer the questions. At a second level the answers found by applying the strategies suggest *tested tactics reported in the applied literature*.

While our use of the term *verbally mediated* to describe the analytic repertoire is several steps removed from Skinner's (1957) term *rule-governed behavior*, it is an extension of the same notion. That is, there are several verbal mediations taking place. The original characterization and measurement of the learning problem are verbal tacts by the teacher derived from the teacher's training in the precise use of scientific terms and data collection procedures to describe environment/behavior relations. The occurrence of certain events during instruction acts to *evoke scientific descriptions*—scientific tacts. That is, nonverbal events occurring in the teacher/student interaction are contacted verbally by the teacher; hence, the expression *scientific tact*. Laypersons would tact the events in nonscientific terms—terms that would not lead to solutions to the instructional problems encountered. Table 1 in chapter 1 provides examples of the difference between scientific and layperson tacts.

The student's difficulties are tacted or observed in ways that can *only be characterized using the vocabularies of the basic science and the applied science of pedagogy*. For example, the teacher describes the lack of speech of one or her students not as a lack of spontaneous speech but rather as the lack of a *mand* repertoire. Mands are described in the science as forms of communicative behavior that are under the control of nonverbal antecedents and certain forms of deprivation (momentary or extensive). Additionally, the form of the behavior specifies its reinforcer (i.e., the statement "a pint of bitter, please," or a gesture to a listener who is dispensing the libation results in a pint of bitter for the person emitting the mand). The research literature provides verbal instructions in how to teach the mand repertoire to these individuals who have mastered the vocabulary and whose behavior is governed by that vocabulary.

Verbal descriptions use the vocabulary of the sciences and, in turn, evoke still other scientific terms drawn from the research literature. For example, an occurrence of a single mand operant requires that certain establishing operations be present. Terms that describe tactics from the applied research such as *time delays or momentary deprivations* are two of several tested tactics that can be used as operations to increase the probability that establishing operations are present. These terms prescribe teaching operations which, when used accurately, increase the likelihood that the mand operant will become part of the student's repertoire.

When the teacher uses the time-delay procedure he or she provides an appropriate establishing operation (e.g., momentary deprivation). She initially shows the item under deprivation that evokes the desired response. Then the teacher ensures that the response (and only that response or its successive approximation) results in delivery of the item specified. The problem that the teacher had in teaching the goal behavior is now identified as the absence of teaching operations in which the *antecedent and consequence were used with the necessary establishing operation and antecedent conditions*. The teacher applies the tactics or teaching operations by following the verbal specifications found in the literature. The resulting effects on the student determine the success of the operations. Lack of success with the student results in still another application of the strategies of analysis.

The Decision Tree

Thus, the analytic repertoire consists of a set of questions that result in decisions about particular tactics to apply. *It is a process that might be described as a decision tree.* Our decision tree consists of strategies to locate the problem and tactics to apply based on the branch of the decision tree selected as a result of the strategic analysis. A decision tree is a task analysis of various operations that one might go through to solve a particular problem. For example, when we follow a set of steps to solve a problem with a software program in our computers we follow a set of *if–then* procedures. Our behavior is governed by the written instructions. Thus, our decision tree for doing contingency analyses for instructional purposes is a set of *if–then verbal instructions*, and the verbal forms are part of the community of verbal behavior that constitute our science.

We use the terms *strategies* and *tactics* to differentiate operations in the decision tree process. The term *strategies* include:

(a) The kinds of *questions* that one might ask about a particular problem (e.g., the possible source of the problem);

(b) The relationship of the problem to particular *principles of behavior* (e.g., reinforcement, discriminative stimulus); and

(c) the kinds of questions one can derive from the perspective of behavior selection (i.e., *the epistemology of the science of the behavior of the individual*).

The term *tactics* refers to the specific procedure from the applied research literature suggested by the result of strategic analysis. These include tactics such as incidental teaching, intrinsic stimulus prompts, and self-monitoring, to name a few of the hundreds of tactics available in the literature. In this and subsequent chapters we provide lists of those tactics and exercises to help the reader learn *to fit tactics to the answers to strategic questions*. In almost all cases the tactics are directly related to the *principles of behavior*.

Strategies from the science prescribe *where to look and how to*:

(a) *Describe the problem in scientific terms* (e.g., relate the problem to the principles of the science of the behavior of the individual or identify a possible controlling variable in the environment);

(b) *Isolate components of the learn unit and its setting shown* in Figure 1 that are potentially at fault (e.g., is the problem in some part of the student's *nucleus operant* or in the setting events); and

(c) *Select tactics from the applied or basic science* literature that are likely to be successful.

Strategies incorporate the nature and type of questions that *focus the analyst's search on instructional variables* rather than on hypothetical constructs about the student. Table 1 in chapter 1 contrasts postscientific strategies with prescientific strategies. The postscientific strategies listed in Table 1 are generic questions that lead in turn to still other strategies specific to learn unit components or the prerequisites for learn unit components. These generic questions and their subquestions are called strategies because they identify options for where the problem is likely to be found. The strategic questions that are selected, in turn, suggest still other strategies to find the source of the problem. The question is never "can the student learn"; rather, the question is "what does the student require in the way of instructional operations."

The steps that the teacher goes through consist of the following.

1. The teacher looks for the source of the problem in the learn unit context field (Figure 1).

2. A series of questions suggests a likely source (a logical contingency analysis).

3. The problem is related to a principle of behavior (e.g., the problem is inadequate S^d control).

4. Tactical procedures from the applied literature are weighted as solutions and one of the choices is selected. For example, one possible cause

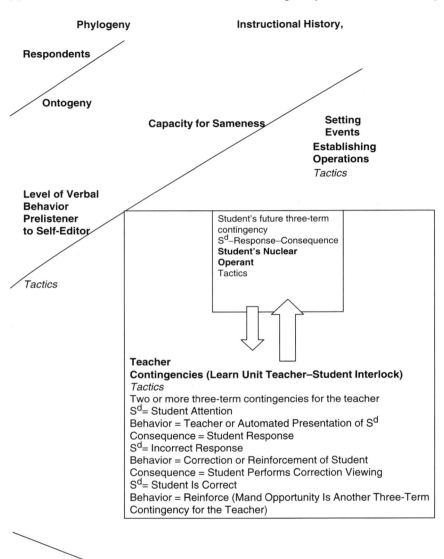

FIGURE 1

might be that the student is focusing on the wrong attribute of the stimulus and the tactic of general case instruction is selected as a teaching intervention.

5. The teacher applies the tactic for the first time. That is the application is under the control of the verbal behavior of the science. At the point in which the teacher applies the tactic *under the control of student behavior rather than the verbal directions*, the repertoire becomes contingency shaped. The verbal directions no longer govern the teacher's behavior.

6. The tactic is successful and the student progresses; alternately, the tactic is unsuccessful. In the latter case another tactic is chosen.

Figure 1 shows the learn unit and the context for the learn unit. This figure represents the factors affecting the instructional process associated with the learn unit and the learning of the three-term contingency by the student. Within the learn unit is the future three-term contingency the student is to learn (the target three-term contingency). This is surrounded by the three-term contingencies of the teacher. Together, the three-term contingencies of the teacher and the target contingencies for the student constitute the learn unit. Surrounding the learn unit is the context of (a) the setting events (establishing operations) for teacher and student, (b) the instructional history or ontogeny of the teacher and the student, and (c) the phylogeny of the teacher and student. Tactics from the literature can be identified that may serve as solutions for learning problems associated with each of the components shown in the diagram.

Presume that the nature of the learning problem suggests that the possible source of the difficulty is found in the way in which the target antecedent is presented (e.g., the student confuses square and rectangle). In this case, the teacher identifies the source of the problem by systematically experimenting with components of the antecedent or by a logical contingency analysis based on the science. By the target antecedent, we mean what the teacher seeks for the student to see, hear, or feel prior to when the student is to respond (i.e., the target S^d). The particular problem is tacted or characterized scientifically as lack of stimulus control for shapes presented at the level of distinction between rectangle and square. Given this difficulty and the elimination of teaching operations or the reinforcement operations as a possible problem, the literature (i.e., the vocabulary of the science) suggests that the discrimination may be taught by using a series of *stimulus prompts* or alternatively *general case* teaching procedures. These choices are *tactics* from the research literature in applied behavior analysis. The teacher then applies the verbal directions from the literature on how to use the stimulus prompts or general case tactic to direct her specific instructional actions. Subsequently the student masters the distinction between rectangle and square.

Strategic Questions

1. Does the student have the instructional history (*prerequisite repertoires*) and related stimulus control to enable him or her to respond to the learn units for the new target objective? Is the problem a function of inadequate or incomplete prior instruction? How does one go about determining the answers to these questions, and after doing so, how does one fix the problem?

2. Is the source of the student's difficulty found in the *motivational context* (e.g., setting and setting events) of the presentation of learn units for the target objective? Are there environmental factors in the sequence, setting, or prior or current stimuli which interfere with or mask the reinforcement value and the antecedent stimulus control for the learn unit presentations? If so, where are these problems and how can they be compensated for by instructional design?

3. Does the source of the student's difficulty lie in the manner in which *the teacher presents the antecedent* or the configuration of the target stimulus and foils (e.g., negative exemplars) to the target stimulus? Are the difficulties located in the teacher's way of presenting the antecedent? Are the difficulties located in the presentation of the target stimulus alone or in rotation with nonexemplar foils? Is the antecedent the correct one for the repertoire being taught?

4. Are there *phylogenetic, anatomical, or physiological differences* that suggest restructure of the instruction or analysis of the source of the problem?

5. Can the student emit the *response* under nontarget instructional conditions and are the responses adequate to the repertoire being taught? Does the response not yet belong with the instructional antecedent, setting, and consequence or is the response not presently in the student's repertoire under any antecedent/consequence arrangement?

6. Is the source of the difficulty located in the *postcedent or consequence* for the student's response (Vargas, 1988)? Is the problem one of prosthetic versus natural reinforcement?

The source of the student's learning difficulty may be located either in the *current way* in which the student is being taught (i.e., the manner in which learn units are currently presented or the way in which the three-term contingencies are operating on the teacher's behavior) or in the *history of the student's instruction* (i.e., the *existing skills* in the child's repertoire). Phylogenetic, anatomical, and physiological factors affect learning in powerful ways also, but if the instruction is consistent with the science of behavior, the student's behavior will determine how the instruction is adapted to compensate for these individual differences. These three broad sources for problems are:

1. *The way in which the student is currently taught* or components of the learn unit itself (i.e., materials, pedagogy);
2. The *setting events or events preceding the learn unit* presentation; and
3. The *way in which the student was taught in the past* or his instructional history.

The sources for the search for instructional tools are similar whether the student is a 2-year-old learning a mand repertoire or a 15-year-old learning calculus. There are differences in tactics and instructional histories that are applicable, however, and these differences are the subject matter of future chapters.

The remainder of this chapter is devoted to describing the strategies associated with verbally mediated solutions to instructional problems. The four sources of difficulty

1. the *way in which the student is currently taught* or components of the learn unit itself (i.e., materials, pedagogy);
2. the *setting events or events preceding the learn unit* presentation;
3. the *way in which the student was taught in the past* or his history of instruction (e.g., prior learn units *or experiences*); and
4. phylogenetic and physiological factors

are presented according to learn unit components in Figure 1.

The search for verbally mediated solutions involves:

(a) The prerequisite repertoire of the student;
(b) The antecedent(s) or target S^d for the student in the learn unit;
(c) The student's response component in the learn unit;
(d) The consequence for the student used in the learn unit;
(e) The motivational setting in which the learn unit is presented; and
(f) Physiological and anatomical conditions affecting the control of any part of the learn unit and its context.

Figure 1 provides a visual representation of the learn unit and the factors surrounding the learn unit and the components within the learn unit. The flow chart provides a visual display of the sources for analyzing the locus of the problem. The reader needs to refer to this chart frequently because it is the source for all of the strategic questions to be asked and, in turn, the source for the tactics that might solve the problem,

Before we can engage in the process of solving instructional problems, it is necessary to describe how and when the problem is detected.

- How do we know that there is a problem?
- When should we change our current strategy?
- What is the source of our decisions?

These questions are also is part of the vocabulary of the science, but it is a special part and consists of *the data language of the science of behavior* and our emerging science of individualized pedagogy.

Identifying Instructional Problems Using Visual Displays of Student Response to Instruction

We shall describe the process for:

(a) making decisions about when there is a problem in the instruction and
(b) at the same time describe how we measure the teacher's analytic responses as part of the process of teaching the analytic repertoire.

The measurement of teachers' analytic repertoires is as important to the development of teacher expertise in the analysis of instruction and related decisions as the teacher performance rate/accuracy measure is to the development of fluent contingency-shaped teaching repertoires (Keohane, 1997).

Decision Opportunities

A count of three *data paths* is the earliest opportunity for a decision about the progress of the student. The first *data point* is a point of origin. Thus, from the first data point of a beginning or new instruction in a phase to the second data point is 1, to the third data point is 2, and to the fourth data point is 3. At this point it can be determined whether the trend is descending, stable, or ascending. If a trend, or achievement criterion, is present the decision is made whether to continue, change tactics, or move to a new objective. If the direction changes for one or more data paths, a judgment of trend across five data paths should be made. If no trend is apparent at five data paths, a decision opportunity is at hand. Of course the achievement of criterion involves only the number of consecutive sessions and data points per the program under review. If the criterion for your school or curriculum is two consecutive sessions with a program or lesson at 90% correct then two data points at 90% constitutes the decision point. A broken vertical line should appear after each decision point and a brief written statement of the tactic chosen should appear on the graph. New objectives should be labeled following the achievement of an objective at the criterion level.

Trend Determination

Ascending trends should result in *continuation* in most cases. A descending or no trend should result in a decision to *change tactics* and a broken line should be present. If criterion was achieved the broken line should appear, also. Each decision opportunity that is consistent with these definitions is a correct decision for the teacher, while a decision that is not correct or the lack of a decision is an error in the teacher's analytic responding to the visual display. *Each session run without decision change is another error.* The rationale for this is that every instructional session in which the student is subjected to inappropriate pedagogy wastes educational time and may indeed act to compound the student's difficulty with the instruction in progress. When a decision is made after a decision error it is not counted as a correct decision because it is, at this point, a correction. The astute reader will note the presence of the implicit presence of learn units for the teacher in our analysis.

Learn Unit Analysis Decision

Once a decision for change is made, the next step is to perform a series of strategic analyses of the probable source of the referring problem to the components shown in Figure 1 and the steps described above. If the description of the potential problem includes *a reasonably viable reference* to the learn unit, the three-term contingency, the instructional history, or the setting events, this strategic choice is counted as a correct decision. If the analysis of the potential problem draws on prescientific rationales or hunches not grounded in the science, the strategy analysis is counted as an error. At this point, another correct decision might be to do a correlation (AB design) or functional analysis drawing on the above components as the basis for the analysis. Of course, each correct or incorrect decision for doing the experiment constitutes a decision point, for which correct or incorrect tallies may be made.

Selection of the Tactic

If the tactic selected for the solution to the student's problem is derived from the literature and is related to a principle of behavior or component of the learn unit analysis, another correct decision is counted. If the tactic is not from the literature then the decision is counted as an error. The teacher may skip the experimental strategic analysis and jump directly to the tactic decision, without an error or an incorrect count if the evidence for the need of the tactic is compelling and the teacher has prior experience with this particular type of decision. This usually occurs in those instances when a teacher is experienced with the problem at hand or the student's history and immediately recognizes a needed tactic.

Implementation of the Tactic

If the tactic is implemented correctly, the teacher receives a count of correct. If the tactic is not implemented accurately, the count is an error. Following the implementation, we once again return to the steps in the decision tree spelled out in the preceding section.

Thus, we have described the steps for deciding when there is a problem and how the decision is conveyed on the graph. At the same time we have described the measurement of the teacher's analytic expertise. The operational definitions prescribe the steps that the teacher is to follow in analyzing the visual display of instruction. The measurement provides the means for the teacher or her coach to use in monitoring the decisions and therefore provides the means to improve the repertoire.

We are now ready to describe the strategic analysis in more detail. We shall describe the strategies associated with analysis of the instructional problem consistent with the learn unit and its context outlined in the flow chart found in Figure 1.

Strategic Questions to Ask about Prerequisite Repertoires

One of the possible sources of a student's difficulties with a particular curriculum concern the prerequisite stimulus control of the student's behavior and the possible relationship of this *history of instruction* to his current difficulties. The following outlines a few of the strategic questions that provide directions for locating a likely solution.

1. Prerequisite stimulus control. Are the prerequisite stimulus control and corresponding responses presently in the student's repertoire such that she can respond adequately to the current learn unit presentations? Comparing various antecedent or setting event controls and observing differences in responding can isolate this.

Are the prerequisites scientifically identified; that is, is there evidence suggesting that the presumed prerequisites are necessary or are the prerequisites assumed on a logical basis?

If the necessary prerequisites for the new response are known, what is the student's history with regard to the prerequisites (e.g., has she mastered the perquisites)?

1. How recently were the repertoires found to be present?
2. Did the student achieve both mastery and fluency?
3. What was (were) the setting(s) in which the objectives were achieved and how similar or different are the current setting(s)?
4. What reinforcers or corrections were used during the prior instruction for the prerequisite repertoire and how were they presented?
5. Was the recording of the achievement of the objective reliably done by a trained teacher?
6. What was the scripted or programmed instructional methodology and how reliably was it implemented?
7. In what ways were the instructional operations similar or different from the operations used to teach the new objective?

If the above answers to the above questions suggest that it is necessary to teach prerequisite repertoires, which component(s) of the prerequisites(s) need(s) training (or retraining) in order to ease the student into the new target objective?

1. Were there requisite responses (or stimulus controls) that were not considered when assigning the student to the new target objective?
2. Is there a chaining sequence of behaviors necessary for the new target instruction that was presumed but not actually tested or trained?
3. Is a more detailed task analysis that includes instruction in the additional prerequisite steps needed prior to reintroducing the student to the new objective?
4. Are the formerly learned stimulus controls consistent with the new ones introduced (e.g., "point to" vs. "touch the" vs. "give me the")? Is there some intervening stimulus control that needs to be taught (e.g., before training vowel sounds, can the student discriminate between word or letter formations and nonwords on pages)?
5. Was the prerequisite stimulus control adequately trained (e.g., is the student under adequate instructional control)? For example, were the procedures of general case training (Engelmann & Carmine, 1982) used such that the ranges of exemplars of the target S^d were programmed adequately?
6. Were irrelevant components of the stimulus rotated across the exemplar such that the response came under the control of relevant stimuli rather than irrelevant stimuli?

New prerequisites may need to be identified or in some cases a set of prerequisites may need to be identified. It is important to keep in mind the fact that behaviors *are often independent, even though they may be associated with the same stimuli.* Assumptions that a central concept (e.g., color concepts) has been learned, simply because a student can point to colors when asked to do so, may result in no instruction in naming the color. An assumption was made that the two were equivalent when, in fact, the two responses were independent repertoires that could be related only through instruction to be related. The first response of *pointing* was selection based responding, while the latter involved *production* of verbal behavior. Thus both repertoires must be taught.

   2. The student and instructional control.   By instructional control, we mean the behavior of the student's attention is guided reliably by the behavior of the teacher (Skinner, 1957). When the student is instructed to do something by the teacher, can the teacher expect that the student will follow the instruction immediately and reliably (e.g., the child is looking elsewhere, the teacher says "look at me," the student immediately looks at the teacher and is ready for a learn unit presentation)? Is the difficulty that the student is having with a particular program really due to incomplete or weak instructional control by the teacher for the student? Is the instructional control problem based on the teacher's inadequate or incomplete teaching skills; is the teacher using and reinforcing the incorrect attention signals? This question must be resolved before proceeding with subsequent analyses by using the TPRA observation procedure. If the student is not compliant and attentive, additional questions cannot be answered.

These questions and the questions associated with the sources of difficulty (e.g., antecedent, response, consequence, and motivational settings) are answered by reviewing the existing data for the student and by systematically observing the student under controlled instructional conditions. If the existing data record helps eliminate many or some sources of difficulty, much time is saved. Some of the possible difficulties may need to be identified by experimentally manipulating the variable(s) associated with the potential problem. The strategies of applying the experimental method (e.g., see the description of the Mill's method of disagreement in the glossary) through single case experimental design (i.e., reversal, multiple baseline, multitreatment designs) may be done rapidly to isolate the locus of the problem. Typically, this is done through *probe sessions* rotated with current instructional procedures isolating one variable at a time (see the design chapters in tests concerned with introductions to applied behavior analysis). However, the tactics already identified in the literature provide likely and immediate solutions that *obviate the need for prolonged experimentation in most cases*. This fact emphasizes the necessity for continuous expansion of the teacher's repertoire of verbal behavior about the science under conditions that reinforce verbally mediated tests with students.

A probe session is a teaching session in which one tests for the presence or absence of the prerequisite response/stimulus control under various environmental changes and may involve the range of controlling variables including the (a) setting (motivational conditions), (b) the response, (c) the consequence for the prerequisite behavior, and (d) the antecedents for the response (both teacher and target stimulus). If the response was there before, then one of these conditions may have been different in the student's instructional history. However, if the prerequisite is emitted under the same conditions as used in the present instruction (e.g., the same contingencies and motivational settings), the prerequisite is not necessary for this student. The prerequisite may actually belong to another repertoire (e.g., when *pointing to* is not a prerequisite for naming but actually a different repertoire). Simply because someone can identify an exemplar in a multiple choice setting is not equivalent to naming or constructing an exemplar.

The result of the above analysis will be the identification of deficits in the student repertoire or deficits in the teacher's presentation vis-à-vis the prior instructional history of the student. Thus, the necessary prerequisites can be taught or the instructional presentation corrected before resuming the new instructional objective. When the instruction is adequate per learn unit presentations (i.e., the learn unit is intact), consistent with the student's history, and the student has the necessary prerequisites, *the student will learn*.

Strategic Questions to Ask about Motivational Conditions and Settings

In the science of the behavior of the individual, motivation is viewed as the current and prior environmental events (setting and setting events) which act to affect the reinforcing and punishing effects of the existing consequences of behavior and, therefore, the antecedent control (Michael, 1993).That is, the motivational setting affects the rate and strength of the behavior associated with the postcedent value of reinforcement or punishment, which in turn affects the strength of the antecedent control. Establishing operations are settings that affect the three-term contingency on a momentary basis and are the results of deprivation or satiation of food, water, attention, and activity, as well as competing activities and their reinforcers/punishers.

Teachers need to recognize the control of these variables and design or redesign classroom settings or sequences of experiences to enhance instruction rather than to allow them to compete with instruction. There are a few tactics from the applied literature that suggest ways to use establishing operations to enhance instructional effectiveness or to compensate for ways in which motivational settings interfere with instruction.

Reinforcements and the preference of one activity over another are relative and change from moment to moment. Whether or not an event, activity, or item functions as a reinforcer is affected by prior, simultaneous, and subsequent events (i.e., setting events, see Figure 1). Good strategic analyses will take these conditions into account as potential variables that may affect instruction. While much more needs to be known about how to compensate for or ameliorate deprivation or satiation conditions through momentary operations in the classroom, much can be done by using what is known or seeking out new tactics through *in situ* research.

A few of the tactics from the existing research findings include:

- Setting up situations that deprive a stimulus in a chain of operations (e.g., withhold a needed item to enhance its reinforcement effect);
- Alternating learn units between peer and target students;
- Changing the sequence of activities for individuals in a flexible manner contingent on student behavior;
- Using slightly preferred activities to reinforce low preference activities;
- Alternating instruction requiring much movement with activities requiring little movement or providing brief quiet periods prior to instructional presentations;
- Eliminating or *using* competing stimuli, briefly eliminating access to an object associated with low preference activity or instruction. For example, use the competing stimuli as brief reinforcers for the new response/ stimulus control.

Of course, one can simply wait until the competing stimuli or motivational conditions are ideal. Certainly, these ideal moments should result in *captured learn units* by the teacher (e.g., incidental teaching). But delays between natural moments result in drastic reductions in learn units and an unwitting increase in inter-learn unit latency time. It is best to *both manipulate the natural conditions that will result in the immediate resumption of successful learn units and capture instructional earn units when the opportunities are there* (Schwartz, 1994). A short enforced quiet period with no activity, or the teaching of learn units for sitting still or eye contact for students functioning at this level, may be all that is needed. Alternately, teaching learn units for a preferred activity creates a momentum (reinforcement schedule) that will allow the teacher to insert the new program learn unit with success.

When students are at the self-management stage and proceed through individual assignments at their own pace, simply rearranging the order of these tasks can set up the appropriate motivation. Tasks (and their reinforcers) that are marginally performed or low in reinforcement value may be assigned under conditions in which there are few competing stimuli (antecedent or postcedent), while momentarily effective reinforcers (ones with more extended deprivation) and responses on the brink of mastery or fluency may be used only at moments in the day when strong competing stimuli are likely to occur.

If children (or adults for that matter) are hungry (food deprived), ill, sleep-deprived, or uncomfortable, it is necessary to find resources to alleviate the problem directly. Also, by providing reinforcers to overcome more minor discomforts, one can make some progress while finding ways to address the problem more directly (e.g., allowing the child to sleep, sending the child to the nurse, providing nutritious food, finding appropriate services). Students returning from vigorous activity respond differently than they would have if a quiet period had preceded the instructional session. Teaching a program (or assign one) in which the student is currently having success (almost achieving criterion) before returning to the new program is one tactic to use to rearrange the motivational conditions.

If there are strong competing contingencies (e.g., peer interactions occurring), change those contingencies or use them to teach the new program (e.g., use peer tutors, alternate learn units between pairs of students, use peer models). If the classroom for students with reader/writer repertoires is arranged such that individual desks have carrels, many competing stimuli can be alleviated or modified. Once students come under the control of recording and graphing their own data in partially isolated carrels, a new set of classroom contingencies can be invoked to alleviate competing stimuli. If competing stimuli are the problem, it is possible that instructional control is not present. This can be reactivated by invoking new reinforcers, reinforcing the behavior of others for similar target responses, and reshaping target control with the target student before proceeding with the new program.

A major problem in many classrooms is that there is *not enough prosthetic reinforcement for academic responding compared to peer reinforcement.* Simply running intact learn units will alleviate this problem in most cases.

Antecedent Control

Some sources for antecedent problems include answers to the following questions.

1. Does the teacher have antecedent or instructional control of the student's attention and is the teacher's antecedent unambiguous to the individual student?
2. If not, are the problems due to the teacher's inaccurate presentation of the scripted or programmed antecedent vis-à-vis the student's history?
3. Is the sequence of presentation of learn units programmed to teach the essential control by the target stimulus (e.g., shape or color or texture across irrelevant settings) by contrasting the range of needed exemplars?

Two prominent components of antecedents are potentially at fault. These are:

1. *The teacher's antecedent* (preferably scripted or programmed to avoid this problem) or
2. *The sensory (auditory or visual) display* of the target controlling stimuli. Both of these are related to the instructional or teacher antecedent presentation in the learn unit.

Much academic instruction involves bringing the student's behavior under multiple antecedent control involving both the teacher, or text, instructions *and* the target stimulus (e.g., "add these columns to obtain a sum," along with the display of numbers). We call *this multiple control* after Skinner (1957) because there are two or more antecedents for the response. Multiple control is also termed conditional stimulus control in some cases. Any of the "multiple controls" is a potential source of difficulty for the student and a potential source for an instructional error by the teacher.

The first and most common problem is the lack of attention to either the teacher or the target stimulus or indeed both as were described in the section on prerequisite stimulus control. When the problem is momentary rather than a problem of prerequisite stimulus control, one may satisfactorily resolve the problem by delaying either one or both antecedent presentations until attention is obtained (i.e., stimulus delay). If the delay procedures have to be continually used, or extended to obtain correct responses, achieve mastery with the stimulus delay conditions and then add a new objective for speed (fluency) and mastery. If this is done, subsequent programs (e.g., new or more advanced instruction) can be introduced with *both* mastery and fluency criteria (Johnson & Layng, 1992).

A second common problem is that the teacher emits too many variations in the antecedent at a stage before the student has been taught a particular antecedent. For example, substituting a statement such as "Do you know this color?" for the prescribed antecedent "What color is this?" or "be still" for the prescribed antecedent "sit still" results in potential confusion for the student. Of course, once the student's behavior comes under the control (e.g., has mastered) of the target stimulus, the teacher may wish to *systematically* program variations in teacher antecedents. Good programs are designed to maintain consistency in antecedent presentations. As a first course of action, provide consistently unambiguous teacher antecedents.

As the student gains increasing independence from the necessity for direct vocal teacher antecedents, teacher antecedents are replaced with written or textual instructions. The student with independent reader status can then move at his or her own pace through modules of instruction following written instructions. Such behavior is more closely related to repertoires needed in life.

The Target Antecedent Stimulus Control

The configuration of the target stimulus must be accurately portrayed with no variation in the essential component (e.g., squares need to be perfectly square). The use of sloppily constructed stimuli may introduce unnecessary ambiguity. Numbers, letters, words, or questions that are less than crystal clear create problems. Sometimes the stimulus is diluted with unnecessary additions (e.g., an alphabet card for the letter *g* presented with the picture of a goat). When the irrelevant components of the stimuli are varied across presentations containing unambiguous features of the essential stimulus configuration, the student is taught the essential stimulus control. That is, when teaching shape discrimination, vary size, color, and texture of the target shapes. At the same time vary the nontarget negative exemplars. It is important also to vary the order and location of the presentation of the target stimulus.

If all of the above factors are in effect and the student still needs help, the teacher may use stimulus prompts such as color highlights, room orientation, or stimulus accents or highlights and gradually fade the extra stimulus prompts across learn units. For example, the systematic fading of the number of dots follows providing dotted lines prompt the construction of letters initially until letters are formed with no prompts. Backward chaining procedures are another example of a tactic to use with target stimulus presentations.

When the response calls for identifying exemplars (e.g., point to blue or circle the correct answer), rotate unambiguous exemplar presentations (systematically varying the irrelevant stimuli) beginning with the least remote comparative stimuli that are discriminable to the student. If difficulty results, increase the disparity until correct responding occurs with the narrow disparity. The range of discriminable difference beginning with narrow differences avoids prolonged programming for students who immediately discriminate. However, prolonging presentations that are not discriminated increases the difficulty for students who have problems (see Engelmann and Carnine, 1982, for an extensive treatment of general case teaching procedures).

Each of the tactics described above are derived from the applied research and each tactic is a response to strategic questions. As in the case of searches for the locus of difficulty with other components of the learn unit, experimental analyses using probe sessions will allow the teacher to locate the source of difficulty. Once the source is located, existing tactics from the science (e.g., stimulus prompts with fading procedures) may be used to correct the problem.

Response Parameters

There are several questions that need to be asked to assess the role of the student's response in the instructional problem encountered. There are six issues that one can address:

1. Response presence;
2. Response independence;
3. Response definition;
4. Response substitution;
5. Accuracy of observation/data collection; and
6. Relationship of response and motivation.

1. Is the response in the repertoire?   If the response is in the student's repertoire, to what other repertoire does it belong? For example, if the new objective requires using addition derived from story problems, the student should have achieved mastery and fluency with one or two digit addition and needs to be a fluent reader. Are the responses that are to be taught in the context of the new antecedent control within the repertoire of the student?

2. Is the response independent from what is being taught?   Are different responses presumed to be interchangeable (automatically transferable) when in fact they are not? For example, is the identification of exemplars (e.g., point to the circle, identify the correct description of photosynthesis, circle the correct spelling) presumed to be the same as the production or construction of exemplars (e.g., construct a circle, perform an experiment demonstrating the process of photosynthesis, construct the word based on the phonetic sounds) even though they are not interchangeable responses? The assumption that because one recognizes an exemplar, that one can construct the exemplar, or that one can even apply it to an operation (e.g., calculate square footage for a construction project) is a common fallacy. Perhaps the source of the fallacy lies in the presumption that one teaches understanding (e.g., concepts) independent of specifics. This has never been documented to be the case, yet it is an a priori assumption that is often accepted without reflection. The result of such assumptions is to allow the blame to be placed on the student. Until such time as the science of behavior is able to identify natural relationships between such responses, it is best to assume that they are different and require systematic instruction. Thus, *the response is to be taught in all of the learn unit settings in which the student will be called on to use it.*

Another common obstacle for the student results from the so-called "multi-sensory approach." In this theory the assumption is made that by having the tactile experience of tracing a raised letter, the child learns to visually identify the letter. The *child may actually have learned to attend* to the essential parameters of the shape visually as a result of the stimulus prompt produced by the tactile tracing. In such case a useful and effective extra stimulus *visual prompt* was developed. This is not the same as presuming that a tactile sensory experience resulted in a central processing effect (e.g., in the brain) that allowed the child to codify the letter resulting in another manifestation (visually identifying the latter) of the central process.

A related misconception occurs in the so-called perceptual approach to instruction. That is, the belief that when one teaches a single or several tasks involving "fine motor" control, one is teaching the general response class of fine motor control. Rather, one should teach responses having a particular effect and other responses classified as fine motor are to be taught as part of the separate effects that result. For example, tracing within the confines of a stylus does not result in better drawing of letters. Teach the drawing of letters or the manipulation of a button relative to an environmental effect. In a related vein, the conditioning of muscles in a simulated setting is never quite the same as doing the activity per se.

Translating a language from an oral rendition differs from translating the same language from a written text. Translating (e.g., transcriptive behavior) a language is not the same as using it (e.g., speaker or writer behavior). Nor is the production of language vis-à-vis contrived contingencies (rule-governed re-sponding) in academia the same as using it in the native setting (contin-gency-shaped responding), although the former may represent a necessary approximation for someone not living in the culture of the language.

The science of behavior is beginning to identify stimulus equivalence classes. This area of research has the *potential* to determine the conditions of instruction which result in the production of several antecedent–response–postcedent relationships that result from the instruction of a only a few antecedent–response–postcedent relationships (Sidman & Wilson-Morris, 1974). Nevertheless, until the research has identified these conditions, it is best to assume that responses do not belong to the same classes unless they are taught across at least several classes. Spelling a word, reading a word, identifying a word, pairing a word with its meaning, using the word in a sentence, and reading the word in a sentence are each different, and unless the student's behavior tells you differently, each should be taught to the student. However, you should be alert to the possibility that the student may not need the instruction. Continual measurement of responding will allow you to identify the needs of the student from moment to moment.

3. Is the definition of the response adequate?   Does class define the responses such that the effect on the environment is part of the definition? If that is the case, is the necessary range of topographies delineated in the definition of the response? Are those collecting data on student responses (e.g., teaching personnel, peer tutors, or the student involved) *accurate transducers* of the response consistent with the definitions? A complete response definition is done in order to specify the objectives of the instruction and to guide the teacher transducing the data and consequating the response accurately. The accuracy of the teacher's data collection and reinforcement operations or corrections are linked to the completeness, clarity, and validity (effect of response on the environment) of the definition of the response.

4. Is response substitution needed?   If a behavior needs to be substituted for one that it is not possible for the student to do, can the substituted behavior have the same potential function as the original behavior? One can draw with one's foot, for example, as a substitute for inability to draw with one's hands. One may add by overt rule-governed operations by making tick marks, by counting one's digits, or by automatic responses (e.g., "rote" responses to number combinations). If possible, the accuracy of the response should be defined, at least in part, by the effect of the response on the environment. Improvisational responses in jazz, for example, should be defined by the effect on the audience as well as whether or not the response is within common practice vis-à-vis the harmony (Greer, 1980). Similarly, the correct response to a math problem (or its use to construct an object or define an operation) or the following of written instructions or retelling of the events in a story (e.g., "reading comprehension") all prescribe the effects of the response on the environment. In addition, the range of topographies allowable to produce the effect should be specified (e.g., count on fingers or with a calculator, underline events in a story). When defining the effects of the emission of verbal behavior (e.g., a mand) one needs to (a) define the effect (student obtains milk) and (b) the topographies allowable (the verbal behavior may take written, vocal, or gestural forms).

Alternately, the student may simply have to "think" to form the numbers or labor intently to produce the shapes (Johnson & Layng, 1992; Singer, 2000). In this case the formation of numbers is not fluent. The response of writing the numbers needs to be taught such that the student can write quickly. Determine how many numbers an adult can write in a minute. Teach the student until the student can write the numbers as quickly. Then, return to the initial instruction to determine if lack of response fluency was the problem. See *rate criterion* and *fluency* in the glossary for ways to set speed and accuracy criteria.

5. Is the data collection accurate?   A critical question concerns whether or not the data collector (teacher) is accurate vis-à-vis the valid response. Since the effectiveness of instruction presumes that the teacher reinforces or corrects accurately, the accuracy of the teacher's observation (revealed by data collection) is a necessary condition of the accomplishment of learn units leading to effective performance by the student. The importance of data collection and response class definitions have been treated at length as they pertain to human behavioral research by Johnston and Pennypacker (1992). However, *they are no less important for accurate and effective teaching* than are accurate blood tests critical to *the practice of medicine*. It is increasingly apparent that effective instruction, at least at the learn unit level, presumes the same types of operations that are characteristic of good science in the laboratory or applied research settings.

6. Are the responses considered independently of the motivational setting? Responses, as components of instructional objectives, cannot be considered independently of the motivational setting (setting events and setting stimuli), the antecedent, and the postcedent (reinforcement, correction, or punishment). These are, of course, the components of the operant and the learn unit when taught by a teaching agent. Once it is known that the response can be imitated (e.g., is within the student's repertoire), the task of instruction is to place the response for the student into the context of the contingencies being taught. This is in fact what instruction is about. Thus, if the response is in the repertoire, appropriately defined, accurately observed and recorded, appropriately substituted if necessary, and treated as independent, then one can address the motivational setting issues.

Strategic Questions about
Postcedents (Consequences)

Reinforcement Questions

Are the postcedents for the behavior (e.g., the reinforcement operations) sufficiently reinforcing? Is the moment-to-moment relativity of reinforcement considered? Are *prosthetic* reinforcers needed? Are the "natural" (e.g., conditioned and unconditioned) consequences reinforcing and, if not, does the teaching process condition the natural consequences as conditioned reinforcers? Is the learn unit present at the right moment in the teaching process (e.g., is the reinforcement present at the right place in the instructional sequence? Is the issue of generalization part of the reinforcement choice? These issues may be summarized as:

1. The presence of reinforcement.
2. Reinforcement relativity.
3. Prosthetic versus natural reinforcement.

4. Need for prosthetic reinforcement.
5. Learn units and reinforcement.
6. Reinforcement and generalization.

1. Is reinforcement present?   Lists of potential reinforcers are available in several texts (Cooper, Heron, & Heward, 1987; Sulzer-Azaroff & Mayer, 1986). Each student has a community of reinforcers and these vary from moment to moment based on motivational conditions. Instructional analyses must continually include close observation for reinforcement effects. Indeed, the first source of difficulty is often the reinforcement operation.

2. Is reinforcement relativity considered in the analysis?   Reinforcers change with the settings and setting events present, as described in the section on motivation. The importance of selecting and using reinforcing operations that function to reinforce is a critical component of effective instruction. Usually, the isolation of other problems within the learn unit is critically tied to a reasonable certainty that the reinforcing operations used with the students are, indeed, reinforcing. Thus, ensuring that reinforcement is present is frequently the first place to look when instruction is not effective. The fact that the discussion of the postcedent occurred last in the treatment of verbally mediated strategies does not mean that it is the last place to look. Rather, it is the first and necessarily a prerequisite that must be eliminate as a source of difficulty before isolating other components of the learn unit that may be faulty.

3. Prosthetic and natural reinforcers.   Similarly, responses that are mastered and fluent and, under the control of prosthetic reinforcers, are not *adequately taught* until they come *under the control of the natural reinforcing effects of the behavior*. While prosthetic reinforcers are frequently necessary at the outset and also periodically to maintain control of the natural effects, the responses are not truly mastered until they are controlled by the natural consequences. For many students, learning under both types of reinforcement is a necessary stage of instruction.

Running repeated probe sessions under prosthetic and natural consequences, or naturally occurring generalized reinforcers such as obtaining the correct response or praise, can test the control of the natural or prosthetic contingencies. The natural reinforcer can be taught by gradually thinning out the prosthetic reinforcers, until the response is under the control of the natural reinforcer (e.g., the effect on the environment or generalized reinforcers). In addition, measures of the frequency and duration in which the student selects and attends to the response and natural reinforcing conditions in free time (learner-controlled responding) is still another measure of the attainment of natural reinforcement control. One critical condition associated with the development of new conditioned reinforcement is the use of praise and unconditioned reinforcers in the teaching process (Greer, 1980).

4. Is the need for prosthetic reinforcers considered?   Ideally, obtaining correct responses should reinforce academic responding. In reality, prosthetic reinforcers in the form of praise, tokens, and edible food items are often needed for the types of instruction done by schools before the natural effects that the behavior has on the environment comes to reinforce the behavior. One of the *major problems that affect poor instruction is the assumption that learning should be intrinsically rewarding, wherein intrinsic is interpreted as intrinsic to the person.* Simply because a student reads independently during free time does not mean that the student is *intrinsically motivated.* Rather, the *reinforcement has to do with the textural stimulus control* and the setting events that determine the *reinforcement value.* Reading written directions in how to assemble a contraption is under the control of the need for the assembly and thus the resulting reinforcement of the assembly. Written responses are under the control of the reinforcement of a reader, even if the reader is also the writer himself (Lodhi & Greer, 1989).

It is important to use prosthetic reinforcers when they are needed because of the importance of conditioning school-related stimuli as positive reinforcers. Textual stimuli such as words, numbers, maps, blueprints, and music symbols are neutral or even nonpreferred stimuli (e.g., they do not exert reinforcing control), and they need to be taught such that they become *attention attracting.*

The existing high levels of school truancy are a function of inadequate reinforcers in classrooms for the truant students' behaviors (Mayer, Butterworth, Nafpaktitis, & Sulzer-Azaroff, 1983). Approvals or social reinforcements are naturally conditioned reinforcers but they are not the natural reinforcers for the needed operant. Praise and social reinforcement need to be faded and replaced with the conditioned reinforcement effects of the students' behavior on solving the problem at hand if the student is to be independent. This too will need to be taught such that the effects of the natural reinforcers (that result from the impact of the behavior on the student's environment) control responding.

5. Are the need for learn units considered relative to the student's existing operants?   At this point, we need to consider the *learn unit.* Is the learn unit appropriate for the student. That is, should the learn unit occur at a different point? Has the student already mastered correctness and is now ready for fluency? Or, does the student need to have a smaller component of the instruction serve as a learn unit? In this latter process, we might break the instruction into a smaller component and insert a learn unit at a different point in the instruction. If the student has mastered the learn unit at hand then the unit should constitute a learned operant. Repeated instruction at this level is not necessary and may even be detrimental. Independent responding should be reinforced and learn units inserted at the point where the stimulus control is not an operant. The reader is referred to chapter 2 on the learn unit for an in depth treatment of the strategic analysis of learn units.

6. Is reinforcement considered relative to generalization? It is important that the student use the newly learned response in all of the *settings* in which it is required. The concern about generalization in education and applied behavior analysis (Stokes & Osnes, 1989; Stokes & Baer, 1977) is an important concern. But perhaps the concern is really with the antecedent/behavior/consequence under the motivational conditions that the behavior will be needed. If those conditions are taken into account in the presentation of the learn unit and its setting, the process of programming instruction across different teachers and settings will be needed less often. See the chapter on the learn unit for the relationship between learn unit presentations and generalized stimulus control.

Corrections

The consequence to an incorrect response is as important as is reinforcement to a correct response. The answers to the questions that need to be asked about corrections are critical to the outcomes of instruction. Some of the questions that need to be asked are outlined in the following box.

- Are the consequences to incorrect responses functioning to *prompt* correct responses to subsequent learn unit presentations?
- Or are the corrections functioning as *punishment* that has a detrimental effect on responding?
- Are the incorrect responses dealt with such that differential reinforcement results (e.g., does the student learn discriminative control)?
- Does the student attend to the critical components of the target antecedent during the response correction?
- Does the student emit the correct response as part of the correction operation?
- Do the *learn units move quickly enough* to use the correction as a prompt for subsequent learn units and is the student attending to the target stimulus during the correction?
- Is the response one that calls for planned ignoring rather than a correction? Is the student's difficulty a function of competing stimuli (e.g., maladaptive behavior) that is best dealt with by a brief negative reinforcement procedure (e.g., student redoes written response to criterion) or a punishment procedure—does the punishment correct the response or does it negatively reinforce postponement of appropriate responding? Is the reinforcer for correct responding functioning as a reinforcer at the moment of correction?

Correction procedures appear to function, when they are effective, as prompts for correct responses during subsequent units. They may also function for some students on some occasions as punishment. The term punishment refers to operations that function to decrease behavior. It is important to use the correction as a prompt and to have it serve that function. At *early stages* in developing stimulus control for a student who is having some difficulty, the teacher may need to reinforce the correction response by the student. If this serves to keep the student attending at a critical juncture in shaping the response and ameliorates escape responses that may result from teaching operations that unwittingly result in escape or avoid responses (often by-products of punishment), then the reinforcement of corrections is appropriate. Rapid presentation of correction units with ones that result in correct responding may have the same useful result. If reinforcement of correction responses by the student is used, be certain to quickly fade their use as soon as the student begins to emit correct responses to the learn unit without corrections. Continuing to reinforce correction responses will result in loss of the differential reinforcement effect for correct responses. This result may occur with children first learning to speak or college students in a PSI program who are recycling too frequently.

Recycling procedures for written response (or complex vocal recitation responses) are also included as correction procedures. Recycling is a term for *restudying* and convincing a tutor or instructor that one has learned a portion of the subject matter that previously was deficit. With students who have the necessary self-study skills to avoid future recycling efforts by more careful preparation, the recycling procedure should require a punitive component. For example, requiring the student to retype the question and answer, locate the page of the answer in the text, and provide an elaboration are useful requirements to put in effect. Also requiring that the student restate a related question may be used together with the overcorrection.

However, if the student truly needs a compensatory program of instruction, it can be provided through tutoring followed by a retake of related question(s). In the latter case, the problem is lack of a sufficient repertoire, not a problem of inadequate study time by the student or loose stimulus control by the teacher/ professor as in the former case. As in the case with all corrections, the shorter the delay between the emission of corrected response and a correct response *without corrective prompts*, the more likely the student is to learn.

In some cases, reverting to part or all of a stimulus or response prompt procedure with which the student has achieved prior success may do the correction procedure. If this is done, the student is not given a correct response, but instead is given a prompt (full or partial) based on previous responding. Subsequently, the prompt should be easily faded. A similar operation occurs also with reinforcement; if the target of instruction is to bring the student under the control of a new conditioned reinforcer, then the newly conditioned reinforcer is used on a carefully programmed schedule that obtains correct responding while appropriately thinning the former reinforcement.

There is much that remains to be learned about correction vis-à-vis the *importance* of errorless learning. Presumably, if shaping or programming realizes the principle of successive approximation perfectly, few if any errors occur. However, in order for the shaping program to result in errorless learning, the steps in the program need to exhaust all potential errors by an exhaustive set of steps that sample the complete range of exemplars and nonexemplars. This is often not feasible (e.g., it is too time-consuming and labor-intensive to develop such a program); thus, the correction procedure is an attempt to compensate for thoroughly programmed material. The student's instructional history as well as the responses to the current instructional contingencies must be the guide. The expert teacher learns to observe the contingencies in effect and their relation to the instructional history of the student.

SUMMARY OF APPLICATIONS OF THE VERBALLY MEDIATED REPERTOIRE

The strategies for locating factors that impede student learning involve at least three components.

One must know:

- where to look,
- what questions to ask, and
- how to characterize the situation in scientific terms.

The prior sections concerning the strategies of locating instructional problems within the learn unit and its context suggest where to look and what types of questions to ask.

One must apply strategies from the scientific method to locate the source of the problem. The strategies for such a role place the teacher at a level of professional sophistication unparalleled in the history of education. There are similar parallels in medicine, engineering, pharmacology, and behavior therapy. The operations for doing instructional analyses described in this chapter suggest that teaching has finally become a profession—one with a true science of pedagogy.

If education is to truly teach rather than select, the task of pedagogy is to locate the source of the student's difficulty in the practices of instruction, not in the student per se. Such a step is a necessary, if not sufficient, means to reform our schools and to save our children. It is a necessary repertoire now for special educators who must be strategic scientists of instruction if there interventions are to be effective.

The strategies and tactics suggested in this chapter are only a beginning and are not meant to be exhaustive. They are, however, first steps. The next two chapters deal in more detail with the particular teaching repertoires needed by teachers working with particular types of students.

References

Cooper, J., Heron, T., & Heward, W. (1987). *Applied behavior analysis.* Columbus, OH: Merrill.

Engelmann, S., & Carnine, D. (1982). *Theory of instruction: Principles and applications.* New York: Irvington.

Greer, R. D. (1980). *Design for music learning.* New York: Teachers College Press.

Johnson, K. R., & Layng, T. V. J. (1992). Breaking the structuralist barrier: Literacy and numeracy with fluency. *American Psychologist, 47*(11), 1475–1490.

Johnston, J., & Pennypacker, H. (1992). *Strategies and tactics of human behavioral research* (2nd ed.) Hillsdale, NJ: Erlbaum.

Keohane, D. (1997). *A functional relationship between teachers use of scientific rule governed strategies and student learning.* Unpublished Ph.D. dissertation, Columbia University, New York.

Lodhi, S., & Greer, R. D. (1989). The speaker as listener. *Journal of the Experimental Analysis of Behavior, 51*, 353–359.

Mayer, G. R., Butterworth, T., Nafpaktitis, M., & Sulzer-Azaroff, B. (1983). Preventing school vandalism and improving discipline: A three-year study. *Journal of Applied Behavior Analysis, 16*, 335–369.

McCorkle, N. P., & Greer, R. D. (1995). *Motivational functions of yoked and competitive peer contingencies.* Paper presented at the 15th annual meeting of the International Conference of the Association for Behavior Analysis.

Michael, J. (1993). Establishing operations. *The Behavior Analyst, 16*, 191–206.

Schwartz, B. (1994). *A comparison of establishing operations for teaching mands.* Unpublished doctoral dissertation, Columbia University, New York.

Sidman, M., & Wilson-Morris, M. L. (1974). Testing for reading comprehension: A brief report of stimulus control. *Journal of Applied Behavior Analysis, 7*, 327–332.

Singer, J. (2000). *Rate and contingency-shaped responding versus verbally governed responding and subsequent acquisition of complex repertoires.* Unpublished doctoral dissertation, Columbia University, New York.

Skinner, B. F. (1957). *Verbal behavior.* Cambridge, MA: B. F. Skinner Foundation.

Stokes, F., & Osnes, P. G. (1989). An operant pursuit of generalization. *Behavior Therapy, 20*, 337–355.

Stokes, T. F., & Baer, D. M. (1977). An implicit technology of generalization. *Journal of Applied Behavior Analysis, 10*, 349–367.

Sulzer-Azaroff, B., & Mayer, G. R. (1986). *Achieving educational excellence using behavioral strategies.* New York: Holt, Rinehart & Winston.

Vargas, E. A. (1988). Event governed and verbally governed behavior. *The Analysis of Verbal Behavior, 6*, 11–22.

CHAPTER

5

Teacher Repertoires for Students from Prelistener to Early Reader Status

TERMS AND CONSTRUCTS TO MASTER

- Prelistener
- Listener
- Speaker
- Conversational stage
- Conversational units
- Speaker as own listener
- Operations for teaching independent play or seat work
- Managing individualized instruction in a classroom of students who are pre-self-editors
- Verbal behavior repertoires and independence
- Evolution of verbal behavior
- Self-talk

- Intraverbal
- Autoclitics
- Tacts
- Mands
- Selection behavior
- Production behavior
- Scientific tacts of classroom events
- Correspondence instruction
- Saying and doing
- Equivalencies between stimuli
- Emergence of stimulus equivalence
- List of tactics in this chapter

TOPICS

The Target Instructional Stages That Determine Teaching Repertoires

Sample Instructional Goals for the Learning Stages for Foundational Communication

Why Learn Units Are So Teacher-Intensive

Teaching Operations Needed

Managing Individualized Instruction in a Classroom Setting: What to Do with the Other Students

Individualized Interactions between Teacher and Student

Tacting the Events in the Classroom as a Scientist: Toward Analytic Expertise

Examples of Tactic Selection Operations

Summary

We have found that the communicative repertoire of students is a reliable predictor of the types of strategies that their teachers will need. There are two broad categories of teacher competence based on the communicative repertoires of their students. They are (1) the basic repertoires of communication and (2) reader and writer repertoires. Those strategies and tactics associated with *the basic repertoires of communication* are the subject of this chapter. The basic communication category includes *prelistener behavior, listener behavior, speaker behavior, speaker/listener exchanges, the speaker as her own listener, and prereading repertoires* as well as the intervening stages.

The students who need the teaching operations described in this chapter include the following:

- Students who cannot follow instructions;
- Students without the speaker responses to tell others what they need;
- Students who cannot yet function in speaker/listener exchanges with others;
- Students who have not acquired appropriate self-talk repertoires; and
- Students who have mastered some, or all, of the above and who are beginning to read.

The strategies and tactics involved in teaching students these repertoires are different than those needed to teach students who already have basic communication repertoires and who are under the control of written forms of verbal behavior. While the basic principles of behavior are not different, the tactic, analytic, and the contingency-shaped repertoires needed to implement those principles are so different as to require separate treatments.

Those *students who have not incidentally picked up the basic communication skills at an early age require* teachers who have sophisticated repertoires for occasioning first instances of speaker and listener behaviors. Students with serious developmental delays simply will not learn to communicate unless their teachers have the necessary expertise. The process of teaching students with serious delays to use basic communicative repertoires requires complex and sophisticated teaching actions and instructional analyses. Prior to the development of a science of individualized instruction (Skinner, 1968), it simply was not possible to teach listener and speaker repertoires to some students. It was also not possible to teach a host of self-care and independent skills.

Young children who are not categorized but who have language deficits will need similar instruction (Hart & Risley, 1996). This latter group consists of students without the requisite numbers of language interactions that result in the range of vocabulary that they will require to be successful in school. However, when they receive frequent and valid learn units to compensate for their lack of verbal experiences, these students move rapidly through the verbal repertoires. Nevertheless, they too need teachers with expertise in teaching communication.

When children learn fluent speaker/listeners repertoires, they have the component or "tool skills" (Hart & Risley, 1996) that they will need for advancement to more complex repertoires. Often an individual student's difficulties with problem solving, self-management, or reading comprehension, among other repertoires, are related to the students' lack of fluency with the foundation verbal repertoires. By fluent, we mean that the students:

1. Use an extensive vocabulary that they maintain over time;
2. Use the vocabulary in all settings in which it is needed;
3. Use their existing vocabulary to extrapolate new vocabulary, and
4. Are reinforced naturally by communicative behavior (e.g., they ask questions, engage in self-talk, and repeat newly learned tacts). (Hart & Risley, 1996.)

We view the communicative repertoires that students do, or do not have, as *instructional stages*, although "developmental" stages are not necessarily involved. The stages simply represent existing learning repertoires that determine the pedagogical operations needed. The stages of instruction are sequential to some degree, but there is often overlap.

We refer to the communicative repertoires collectively as *repertoires of verbal behavior*. However, the word *verbal* refers to vocal, gestural, and later textural forms of communication. In other words, speaker behavior may be either *vocal or gestural* verbal behavior. Listener behavior may also refer to responses of the student to any of the modes of communication (vocal or gestural). Later we treat reader behavior as an extension of listener behavior and writer behavior as an extension of speaker behavior. The categorical stages determine the communication objectives to be taught and they specify the range of teacher actions that are possible and necessary (Greer, 1986, 1990).

THE TARGET INSTRUCTIONAL STAGES THAT
DETERMINE TEACHING REPERTOIRES

1. *Students who are at the prelistener stage* are not under the instructional control of others. They do not have the listener skills to follow the spoken instructions of the teacher. Some psychological and linguistic theories use the computer metaphor "receptive" as an adjective (e.g., receptive speech) to characterize the absence of responding to speakers. Some individuals are said to lack "receptive speech." We avoid this term because it does not specify clearly what a student is to learn and how she is to be taught. The term *listener* specifies a range of actions to events that we can identify, count, and hence teach.

2. Students who respond reliably to at least some spoken instruction from the teacher (i.e., they have some dependable listener repertoires) but who do not use gestural, assisted, or vocal means to communicate with a listener (lack of speaker behavior) are described as *students who are at the prespeaker stage of learning*. Again we avoid the metaphorical term "expressive" because it is not pedagogical or scientifically useful. The term *speaker* refers to behavior/environment relationships that are observable, countable, and teachable. When students can specify what it is they want a listener to do for them, or when they can communicate to a listener that they are identifying certain features in their environment (e.g., they make verbal contact with their nonverbal behavior), we say that they have speaker behaviors and are acquiring speaker repertoires.

3. *Students at the preconversational stage of instruction (prespeaker\listener stage)* are under good instructional control (e.g., they respond reliably to spoken directions) and they communicate with others such that they can specify or command others to supply what they are lacking at the moment (e.g., they *mand* reliably). Also they make verbal *contact* with nonverbal events in their environment (e.g., they *tact* common objects, events, or activities). However, they may not rotate or exchange speaker/listener roles with another individual (i.e., they are not fluent in verbal exchanges with others); thus, they require instruction in conversational exchanges.

4. Students who have speaker/listener repertoires with others, but who do not engage in those speaker/listener exchanges with themselves: We shall refer to these as students who are at the *prespeaker as own listener* (i.e., they do not engage in self-talk). We observe students *who do have these repertoires* talking to themselves when they play with toys or engage in imaginary play. This repertoire is referred to in some literatures as self-talk. It is an early form of "thinking" that is observable and may be a necessary component skill to other forms of thinking. With normally developing children, audible self-talk becomes silent as children become more proficient because the children are censored by parents for speaking aloud at certain stages in their development. One learns, so to speak, to

keep one's fantasies to oneself. This repertoire is the basis of thinking as behavior from the verbal behavior perspective (Skinner, 1957).

Finally, students who are acquiring the first steps in reading are referred to as early readers or prereaders—they are in the first stages of coming under the control of textual stimuli. They have speaker repertoires and are reliable listeners. They should also have acquired conditioned reinforcement for textual displays (i.e., children's books are objects that they seek out for some portion of the time in free play settings). Table 1 summarizes the repertoires and their effects on the world.

TABLE 1
Evolution of Verbal Repertoires and Their Effects

Verbal repertoires	Effects in the evolved verbal world
1. Prelistener status	Individuals without listener repertoires are entirely dependent on others for their lives. Total dependency.
2. Listener status	Individuals can perform verbally governed behavior (e.g., come here, stop, eat). Individuals can comply with instructions, track tasks (e.g., do this, now do this), and avoid deleterious consequences while gaining habilitative responses. The individual is still dependent, but direct dependent physical contact can be replaced somewhat by indirect verbal governance. Contributions to the well-being of society become possible since some interdependency is feasible.
3. Speaker status	Speakers, when in the presence of a listener, can govern consequences in their environment by using another to mediate the contingencies (e.g., eat now, toilet, coat, help). This is a significant step toward the individual controlling events that, in turn, control his or her behavior. The community of listeners benefits proportionately too.
4. Speaker as own listener status	Individuals can function as a listener to their own verbal behavior (e.g., first I do this, then I do this), reconstructing the verbal behavior given by another, eventually constructing verbal speaker–listener behavior. At this stage, the person achieves significant independence. The level of independence is dependent on the level of the person's listener sophistication. What level of sophistication does the speaker have as a listener? Can the speaker function as a sophisticated listener? The function is a self-editing one. One way that young children use this repertoire occurs when they play with toys (Lodhi & Greer, 1990).

(continues)

TABLE 1 (continued)

5. Reader status	Individuals who have reading repertoires can supply useful, entertaining, and necessary responses to setting events and environmental contingencies that are obtainable by seeking out and following the directions of written text. The reader may select the verbal material without the time constraints controlling the speaker–listener relationship. The advice of the writer is under greater reader control than the advice of a speaker.

The pedagogical tactics, or teaching operations, and analytic tools (i.e., strategic questions to ask about learning obstacles and ways to answer the questions) that are associated with teaching these five repertoires are the subject matter of this chapter. Those students who meet these broad characteristics specified above vary in age and size; but, from an instructional point of view, their limited communication skills, or as we shall say their limited *verbal repertoires*, specify what kinds of teacher/student interactions are possible and what kinds of teaching interactions are necessary if the student is to learn not only the communicative repertoire but all curricular objectives. For example, these students are taught matching responses (e.g., "place same with same," "put red with red"), responses to one- and multistep commands, listener behavior, independent play, and social interaction. But how they are taught or the nature of the learn unit presentations is determined by their existing verbal repertoires. Instruction that results in changes in verbal behavior is critical to their progress and prognosis.

Of course, there are developmental and age-related issues associated with these students, but we will not consider those in any detail in the discussions in this chapter for brevity's sake. These may be learned from other literatures (i.e., developmental psychology or studies of the characteristics of groups) and are learned also by paying attention to the behavior of each individual student. It is common practice in education courses devoted to methods of teaching that are not built on a science of behavior or pedagogy to present "teaching methods" according to age levels or developmental milestones. We avoid repeating that information. Instead, we supplement those texts by providing the scientific pedagogical expertise that is unavailable in treatments devoted to the normative practices of teaching.

There are many gradations within each of the broad instructional stages outlined above and there is much that we do not know. Some students will have a few dependable listener responses but to only a few adults, while others will respond to most speakers. Some students will acquire speaker behaviors before they have good listener behaviors. We are often tempted to describe the latter children as "willfully disobedient" but all we know for a fact is that they do not have sufficient listener repertoires.

Some students will have inappropriate echoic behaviors that are described as echolalia (i.e., the teacher says "what's your name?" and the student repeats "what's your name?"). In the latter case the *intraverbal* responses of the student are faulty; that is, a verbal antecedent evokes an echoic response even though the echoic response is not appropriate or necessary. Rather than responding with vocal-verbal behavior that has *point-to-point correspondence* to the vocal-verbal behavior of another, a person with good intraverbal behavior provides responses that are responsive to the others verbal behavior using different forms of verbal behavior. Some of those with echolalic problems have good speaker-as-own-listener repertoires, but they may not have adequate listener repertoires or they may have faulty intraverbal repertoires. Still other students have the basic communication and advanced reader writer behavior, but they engage in speaker as own listener behaviors that are inappropriate. Some characterize such students as having "delusional speech;" but, from an instructional viewpoint what is observed is audible speaker as own-listener responses that should not be audible (e.g., inappropriate self-talk).

The following list shows the instructional stages associated with teaching students who are involved in learning, or are ready to learn, the basic communication repertoires. The list provides some illustrative instructional subobjectives for students at each stage. Illustrations are used at this point to clarify the strategies and tactics needed to teach these students, not to outline the curriculum for teaching those repertoires in detail. The information provides the rationale for the types of learn units and the classroom arrangements that make such instruction feasible. This chapter is concerned with only a subset of the learning stages; the full set is listed in the chapter on curriculum. The procedures associated with the stages not covered in this chapter are the subject matter of the next chapter. The subobjectives listed below are only a few of the instructional programs associated with mastering the various communicative repertoires. We list them to provide examples of the curricular settings for the teaching operations and analyses that follow.

SAMPLE INSTRUCTIONAL GOALS FOR THE LEARNING STAGES FOR FOUNDATIONAL COMMUNICATION

The five stages are the prelistener stage, the listener stage, the speaker stage, the conversational exchange stage, the self-talk stage, and the early reader stage.

Listener and Instructional Control Stage

The following are some of the instructional objectives that these students at this stage will require.

1. The student responds to "sit still" and "look at me" for up to 20 s of continuous eye contact per learn unit for 20 learn unit sessions at 100% for two successive sessions.
2. The student responds to one-step commands (e.g., "come here," "stand up," "get your coat," "clean up," "get in line," "stop," "no," and related commands) for 20 learn unit sessions at 100% for two successive sessions.
3. The student learns to respond with various verbal repertoires through a sequence of consequences as reinforcers as the following scenario illustrates. The student initially responds as a listener to prosthetic reinforcers and prompts, then for praise, and then with infrequent praise and increased reinforcement from the natural environmental effects of the behavior.
4. The student goes to time-out, when told to do so, immediately and without resistance and no gestural or physical prompts. A more advanced level would involve the student putting herself in time-out, and as such this learning objective is also part of the early stages of self-management. This student's level of listener expertise is closely tied to what we commonly refer to as self-discipline (e.g., they have verbally governed behavior).

The students must learn to respond immediately to the instructional commands of the teacher as a prerequisite to the teacher's subsequent success with the student. This does not mean that the student will not learn the other repertoires if this one is not first mastered, but until it is, subsequent instructional difficulties will frequently come about as a result of the student not being under adequate instructional control. Thus, students with advanced speaker repertoires may still require instruction to achieve listener fluency. Students with speaker behaviors or even early reader responses can receive both speaker and listener instruction simultaneously.

Speaker Behavior (Speaker Stage)

The following descriptions of curricula are characteristic of the curricula that students at this stage will need.

1. Initially the student learns to imitate the *form* or topography of the word or words, we refer to this as *echoic* behavior. The form then needs to be learned by functions such as the mand function or the tact function. Even though the form may be the same, the function must be taught just like any other behavior. The function that is learned is the emission of the

behavior for different settings, antecedents, and consequences. Echoics ''are graduated'' to mands (e.g., the student is taught to use the word(s) in the *function* of the mand), while the same and other echoics are graduated to tacts. Mands are emitted with and without the presence of the reinforcer manded and without verbal antecedents. They specify their reinforcer. True tacts are responses without verbal antecedents and are reinforced by generalized reinforcers (see the Glossary for a definition and examples of tacts and mands).

2. The student learns to apply autoclitics with tacts and mands and as intraverbals (e.g., that is the . . . , this is the . . . , please . . . , the one on/under/beside, higher, lower, and other comparatives) that are used also in conjunction with the initial introduction to learning to discriminate between concepts or stimuli as part of general case instructional tactics. Autoclitics are verbal behaviors that quantify, qualify, specify, and negate or affirm mands (e.g., ''I want two of the big blue Legos'') and tacts (e.g., ''that is the square'').

3. Examples of intraverbal responses between the teacher and the student (student responses made to teacher verbal antecedents) to academic instruction are responses to ''what color'' or textural stimuli such as flash cards or textural words, point to the picture of———, put same with same, which one is different, and find the sum, followed by verbal statements by the student. Remember, if you require complete sentences of the student when she is ready to do so (e.g., ''that is the color red'') you will be teaching a repertoire that will expand your students' verbal behavior as a discrimination tool!

The Conversational Exchange Stage

The following instructional objectives are characteristic of the curricula that students at this stage might receive.

1. The student emits one conversational unit (i.e., one full exchange in which the student emits both a speaker response that is reinforced by the verbal behavior of another person responding as a listener and in turn a listener response by the initial speaker). See the Glossary for further elaboration on the nature of the conversational unit. An example of such an exchange follows.

   The target student, Mike, asks John for a Lego. John says ''Which Lego?'' Mike says, ''big one.'' John presents the big one. Mike responded to his environment with a mand to someone who could deliver the item. John was under the control of Mike's mand and responded in turn with a question about specificity. Mike responded to John as a speaker showing that he had listened. Thus for Mike, John and the setting functioned to *evoke* verbal behavior. The verbal behavior for Mike was reinforced by

John's verbal behavior as ascertained by Mike's listener/speaker response. Mike functioned as both speaker and listener as did John; thus, Mike was under the control of the verbal and nonverbal contingencies and he reinforced John's verbal behavior. Because both parties were part of interlocking three-term contingencies that were verbal, their responding was a *true conversational unit* as defined by the current literature (Chu, 1998; Donley & Greer, 1993; Lodhi & Greer, 1989,).

2. The student emits a minimum of four conversational units on two successive occasions with one or more peers during a period in which social interaction typically occurs. Obviously this last objective would be an advanced level of what was achieved in No. 1. The following instruction describes what the student may have gone through to achieve her first conversational unit.

3. The student successfully mands a peer to do something for her (i.e., toy please) or tacts an activity of another. This is an example of part of a conversational unit—in this case, the speaker part of a successive approximation of the full conversational unit.

4. The student responds to a mand of another student as requested (i.e., "I want that Lego," followed by the target student delivering the Lego). This is still another part of the conversational unit—in this case, the listener half.

Self-Talk Stage (Speaker as Own Listener)

The following instructional objectives are characteristic of the curricula that students at this stage might receive.

1. During instruction to point to a color, number, word, or other discrimination task, the student begins to tact the item to which he points. "Point to the square.. The student points to the square and begins to say" "square" each time he points. An alert teacher captures this response and reinforces extemporaneous tacts and records their occurrence. Subsequently the tacting of actions by the student becomes a subobjective for self-editing behavior (the component of tacting one's own behavior).

2. After a student mands, the teacher asks the student to specify which color, size, or other parameter of the item manded in order to shape longer exchanges between the student and the teacher. The conversational unit is prompted.

3. The student is taught to teach a doll, puppet, or other anthropomorphic toy (e.g., teddy bear) to do something within the student's repertoire. If necessary, the teacher prompts the student to take on the role of the toy responding as well as the student pretending to be the teacher. The student's responses can then be treated as learn units themselves. Both

speaker and listener roles are shaped with in the same learn unit in which the student functions as her own speaker and listener.

4. After meeting criterion on the other tasks, the student is asked to describe what she is doing as she looks at pictures in a book, plays with toys, or does artwork. Each description or series of descriptions becomes a learn unit as described in the chapter on the learn unit.

5. The student is asked to respond as both teacher and student to describe a part of the typical school day. ''Mary, tell us about getting ready to go home.'' ''Teacher says time to go home. I say 'OK.' I go get my coat'' (gets coat).'' ''I get in line'' (gets in line). ''Teacher says, 'let's go.' We go to bus.''

Of course, at the same time that the student is learning the basic communication repertoires the student is learning many objectives that are not typically viewed as language, also. These include learning to match or place colors, objects, or shapes with matching stimuli; point and tact colors, individuals, and pictures; engage in free play; manipulate objectives for functional purposes; and engage in cooperative play. But as the above outline shows, the communicative repertoire is at the heart of all of the student's instruction. Hence, the problem needs to be addressed as one of verbal repertoires. Using this outline of the curriculum we can now turn to the pedagogical strategies and tactics for teaching these repertoires and for managing classrooms such that comprehensively individualized instruction is possible.

The Early Reader Stage: First Steps to Textural Control

The student may be taught to read when he or she has speaker repertoires, has listener repertoires, looks at books in free time, can match pictures and objects, and can point to shapes and objects when instructed to do so. Typically, it is a good idea to condition books as reinforcers for free time first. One way is to use the Acorn reader series employing learn units for each of the questions in the reading series. If this is not accessible, use storybooks and design learn unit interactions for the events on the page. When the student is performing well at this task, introduce the Edmark series (Edmark Corp., 1992). This programmed instruction introduces the student systematically to textual responding. Convert each of the scripted teacher/student interactions into learn units and graph as rate correct/incorrect.

Once the student has achieved the stage of reading words, you will want to introduce the *SRA Reading Mastery Series* (Engelmann, 1982, p. 38; Becker, 1992, 1977; Becker & Carnine, 1981). Convert each student/teacher interaction into learn units and graph as rate per minute correct/incorrect. Do the instruction individually or in small groups such that you can provide intact learn units and collect the data on the students' individual responses. Alternately you may use the Direct Instruction procedures spelled out in the series (e.g., *Reading Mastery*) for teaching groups and graph their tests. Maintain a token economy for correct responses for those students reinforced by tokens; use praise when delivering tokens. Alternately use preferred items or activities with praise to reinforce correct responding. Prepare your script in advance to avoid slow rates of presentation. (Please note that the capital Direct Instruction refers to the tested materials developed by Engelmann and his colleagues, not direct instruction which is the generic term for all teaching that is direct as opposed to indirect.)

Continue with the Edmark program at the same time that you are doing the *Reading Mastery*. Do both. As the student acquires mastery, begin timed readings with appropriate level reading material (number correct/incorrect in minutes). Ask 10 comprehension questions for each reading passage they do aloud. Increase the rate of reading until the student reads at a "normed rate of speed." Determine the rate by using adult reading speeds as the long-term goal, with subobjectives as slower rates. Make pages with the new words for students to practice until they achieve a normative rate of reading. Then, have the students read in prose form from the storybooks you are using. You will need a graph for rate of reading aloud and correct responses in percentages to comprehension questions on the reading passages. Follow the *Morningside Generative Curriculum Guide* for objectives for fluency and sequence (Johnson & Layng, 1994). When students have difficulties, apply the learn unit analyses and attach appropriate tactics from the science to fix the problem (Keohane, 1997). Be guided by your data. See the analytic examples and the list of tactics at the end of the chapter.

WHY LEARN UNITS ARE SO TEACHER-INTENSIVE

Learn units cannot be presented to these students in written form, because the students are not under the control of textural stimuli. As a result, learn teachers or teacher assistants present units to them. Learn units for these students are, by nature, teacher-intensive, because the learner does not have the prerequisite for receiving written forms of learn units. Thus, individualized instruction requires teacher-intensive one-to-one instructional tactics. The faster that these students learn to master and become fluent with these early stages of verbal behavior the more independent they will become as learners and the less teacher-intensive instruction they will require (see the chapters on curriculum). Individualized instruction can then be presented in written forms and the student can carry out several steps independently after the teacher has presented the multistep antecedent in written, signed, or vocal form.

Students who acquire listener repertoires respond to teacher instruction and they do not require the degree of continuous, close surveillance that students whose behavior is not under instructional control require. As the student's behaviors come increasingly under the instructional control of the teacher and teacher assistants, more learn units can be presented to more of the individual students in the class. The ability to teach more individual learn units expands even more with the increase in these students' speaker behaviors.

Thus, the advancement in verbal behavior as communicative repertoires is closely tied to increased learner independence. As learner independence increases, the student learns more and more by verbal instruction. The student can be supervised by verbal instructions and can specify needs and recruit reinforcement for verbal contact with the nonverbal environment. Learning becomes less directly tied to teacher initiations, because the student can initiate learn unit opportunities. Other stimuli evoke learning responses. Throughout the student's education, the various levels of verbal behavior are tied to learner independence.

TEACHING OPERATIONS NEEDED

The teacher will need to have the students' behaviors under instructional control using positive reinforcement operations rather than aversive or punitive operations. Each learn unit presentation requires that the student is attending and that the prerequisites are in place. The antecedent that is to become the S^d for the child's response (i.e., the target stimulus) must be consistent, unambiguous, and salient. The child must have the opportunity to respond without distractions. The teacher must respond contingently with a reinforcing operation that functions to reinforce or a correction that results in the student correcting her response under the control of the target antecedent stimulus.

The time between learn unit presentations must be minuscule, so that learn units are presented in rapid fashion. Presentations are always intact learn units. That is, if the child is distracted, her attention is regained and the learn unit is presented in its entirety. The presentation resulting in distraction is not regarded as a learn unit. When conditions occasion the opportunity to capture a naturally occurring learn unit, the teacher responds to the conditions with intact learn units.

Given that the teacher is applying the above operations fluently and using appropriate curricula for the student who is instructed, the teacher will next need to be responsive to the data points representing the instructional sessions in order to determine if the adequacy of the operations used are consistent with the analytical decision tree described in chapter 4. The choice of tactics from the research literature is tied to the verbal repertoires of the student. A list of many of those tactics is found at the end of this chapter.

Thus far we have described the individual one-on-one procedures needed for instruction. However, the teacher will also need to draw on other teaching operations and tactics to teach and manage all of the children in the class at the same time.

MANAGING INDIVIDUALIZED INSTRUCTION IN A CLASSROOM SETTING: WHAT TO DO WITH THE OTHER STUDENTS

Even with very small student/teacher ratios, the students must still learn to engage in independent play or individual seat work in order for the teacher to carry out individualized instruction on a one-to-one basis with each of the students in the class. The development of independence repertoires is also an important objective of instruction for all of the students. The teacher must learn to teach independent and nondisruptive play and appropriate self-entertainment while simultaneously: (1) reinforcing teaching assistants for reinforcing independent play by the student (e.g., four times per minute) and (2) presenting individual learn units to one or more students. The teacher presents individualized learn units while keeping an eye on what's happening in the classroom. As the teaching assistant becomes more proficient, she will keep the students who are not receiving individualized instruction engaged in independent play (i.e., puzzles, play dough, manipulative toys), cooperative play, or the performance of early forms of independent academic work (i.e., sorting activities, matching stimuli by circling comparative exemplars). As the students become increasingly independent, the teaching assistant can also run individualized learn units, while both the teacher and the teaching assistant reinforce independent engagement from a distance.

It is important to establish individual play and what will later be *independent learning skills* (e.g., looking at books and early responses to worksheets). Physical areas for those activities are established for younger children as is a standard seating arrangement for older students. The students are taught to stay in the area or seats and to engage in independent play or work. This is a critical repertoire for all students. It is the beginning of the self-management curricula. As the students learn to be engaged, when not receiving one-to-one instruction, their engagement creates the necessary environment to provide more and more individualized instruction and sets the occasion to teach the students to "learn to learn" independently.

It is a common fallacy to assume that because a teacher is attempting to present "instruction" to a group of students that more learning is taking place. In fact, fewer learn students receive units when so-called group instruction occurs (Greer, 1994). The teacher is the one who is behaving and those who are not expert are likely to presume learning is taking place. Like the college teacher who races to cover the material in his lecture notes, the teacher is under the false impression that the students are learning. Most teachers are locked into group instruction because they do not have the expertise to set up appropriate classroom environments for *comprehensive individualized instruction*.

Learning is a function of the "goodness of fit" of learn units received by each student. By "goodness of fit" we mean that the learn unit includes material that the student is ready to receive and that all of the components of the learn unit are present. The teacher provides the best antecedent, the best response time, and the best consequences for each student. Because no student is at the same level as another student at the same moment, the group approach is a nonproductive way to teach. As the teacher and the classroom environment meet the requirements for comprehensive individualized instruction, each student will receive many individualized one-on-one learn units as they rotate in and out of individual play and one-on-one instructional settings. As the teacher becomes more facile, she can alternate learn units between pairs of students and gradually increase the number of learn units received by each individual child. Simultaneously, the students are taught to play or work independently when they are not receiving teacher-initiated learn units. Students at this stage are best served when they receive 700 or more learn units in a typical school day (Kelly, 1994; Kelly & Greer, 1996; Martinez & Greer, 1997). We also suggest that the parents supplement the number of learn units received by the child by doing 300 or more learn units at home when the parents are taught to present learn units by the parent educator.

The teacher's facility at doing this "juggling act" constitutes the contingency-shaped repertoires of an expert teacher of students who are at the basic communication stage. Teachers and their teacher assistants who are expert in the contingency-shaped component of teaching can present 700 learn units per child in a standard school day in a classroom of six who are at the prelistener and early speaker stages or 15 to 20 students with reliable speaker/listener repertoires who are learning the first steps of reading. Those students who are not receiving individualized instruction at any given moment are engaged in activities as a function of the reinforcement of the stimuli associated with the activities, because of the instructional history that the students received in which the activities were continuously paired with prosthetic reinforcers. Novice teachers learn these repertoires best under the close supervision of a supervisor. Alternately, they practice following the instructions outlined in this and other chapters in this book until their performance is accurate and fluent.

We teach students to be engaged independently by using group activities of listening to stories, engaging in artwork, and related tasks, but individualizing the instruction in the groups. After the students are set up for the activity, the teaching assistant observes each student for 5-s intervals rotating around the group until a preset time period has elapsed (e.g., 5 min). (See "interval recording" in the Glossary, if you are not familiar with this rotating whole interval procedure for collecting data.) The number of intervals that the student is engaged in is a measure of his or her individual engagement in group activities (e.g., a curricular objective is established for the child and a scripted program, and graphic display is put in place).

This interval observation procedure sets the stage not only for measuring the passage of time in which the student is appropriately engaged, but also for teaching professionals to teach the student to acquire increased *reinforcement value* for the tasks in which they are engaged. The teacher as group leader (alternately, the teacher assistant may be the group reader while the teacher collects the data) reinforces students continuously for engagement. As the student appears to come under the control of the activities, the teacher and teacher assistants gradually fade their reinforcement. More extended intervals of engagement mean that the reinforcement of "doing the task" becomes stronger. The purpose of this "individualized group instruction" is to teach activities and the associated stimuli such that the tasks acquire increased reinforcement value. A second purpose is that once the activities have greater reinforcement value, the student can engage in independent activities while individualized instruction is occurring for other students. Of course, instruction that leads to independent play or work is an early step in the self-management curriculum.

All of the instruction in the classroom is individualized with the group instruction setting seen as a means to teaching independence and not as an end to the process. Even the group instruction is individualized. This type of group instruction also prepares the student for future classrooms that are not comprehensively individualized. Students who acquire these repertoires and the enlarged community of reinforcers that control individual engagement will be able to function in classrooms that are not individualized later in their education.

There are two types of engagement in group settings—(1) the engagement individually in activities or self-instruction *that is different than the activities that the other students' are engaged* (i.e., one student works on spelling while another works on matching to sample) and (2) the engagement in the same activities as all of the other students (i.e., all of the students sing the same song). The optimum condition occurs in classrooms in which each student is engaged in the right task for them. Unfortunately, the student will encounter situations in which the classroom is arranged such that each student is not receiving the same material at the same time. In either case, the group instruction procedures described above provide the instructional history students need to derive the most from what is available. It is probably useful to design instructional programs to teach each type of individualized responding in the two types of group settings.

We have set the stage for a smooth functioning classroom for our target students and, at the same time, described the range of contingency-shaped teaching repertoires needed by the expert teacher to manage individualized instruction for students who are at this level of the verbal behavior hierarchy.

Over a decade of research and applications in classrooms has shown this approach to be feasible. Moreover, the approach has resulted in significant increases in learning by both teachers and students compared to group approaches (Selinske, Greer, & Lodhi, 1991).

INDIVIDUALIZED INTERACTIONS BETWEEN TEACHER AND STUDENT

The more effective teacher will present learn units to the student with very few teacher errors (i.e., teacher presentations meet the requirements of the learn unit in all or most cases). For the experienced teacher, the flawless presentations of learn units occurs automatically. The teacher will perform them and at the same time be in touch with everything going in the classroom. The following are directions for the process described above.

Present each learn unit such that the student is attending to the teacher or the target stimulus that is to function in the S^d role. The antecedents or S^ds are unambiguous for the student, the student has an appropriate and usually consistent opportunity to respond (i.e., 5 s or another scripted or predetermined time period), and the student's response or lack of response is consequated immediately and accurately (e.g., the correction is appropriate or the reinforcing operation acts to reinforce). Of course, the learn unit and instructional objective must fit the student's learning stage and the objective must have a place in the long as well as short term goals of the student. These goals must also have immediate or eventual utility for the student.

Learn units are typically presented in massed form (e.g., one right after the other), such that the time or inter-learn unit (ILT) between learn units is short. Short inter-learn unit times function to keep the student engaged and they result in greater learn unit productivity and fewer student errors. Teachers also want to *capture* learn units with students whenever the natural events present themselves (i.e., in the cafeteria, on the bus, or embedded in other tutorial programs). These latter opportunities provide generalization opportunities and increase the number of learn units received. The expert teacher will take advantage of every opportunity and record student responses to captured learn units with the same diligence that they record massed learn unit responses.

Different instructional program call for different antecedent presentations, just as different students require individually tailored presentations. This is why the availability of written or scripted programs is so crucial for these students. Modifications should also be written on each pupil's graphic displays in order to make the student's history of instruction available later.

The range of teacher behaviors we have described is best taught in settings in which they are needed. That is, the supervisor, consulting teacher, or professor uses the teacher performance rate/accuracy observation and training procedure (TPRA) to provide the teacher with the necessary *in situ* instruction (Ingham, 1984). Alternately, the teacher may videotape his or her own presentations and subsequently observe him- or herself using the TPRA procedure described in the chapter on teacher training and supervision.

While the accurate presentation of learn units needs to be automatic per standard practice for students in general or a particular student, the teacher must not teach in such a way that his or her responses are insensitive to the responses of the individual student. The instructional repertoire needs to be automatic in order that the teacher can perform the more complex skills of contingency analysis as she teaches. Just as the student who is performing a complex operation in mathematics must not stop to think about the component math skills (i.e., addition, subtraction, or multiplication) in order to concentrate on what operation will solve the next step of the problem, so must the teacher perform the basic teaching operations automatically while assessing how the student is doing and whether or not changes in pedagogy are needed. The contingency-shaped repertoire frees the teacher to use the analytical repertoire needed to solve recalcitrant instructional problems as they are encountered.

TACTING THE EVENTS IN THE CLASSROOM AS A SCIENTIST: TOWARD ANALYTIC EXPERTISE

While the teacher is performing individualized instruction and keeping everyone engaged simultaneously (contingency-shaped repertoires), he or she must begin to tact the events occurring in the classroom using the vocabulary or the science. That is, he or she applies the vocabulary of the science by thinking (speaking to herself) the terms as she does the operations or observes what is occurring in the classroom.

The following are some examples of these scientific tacts. The occurrence of a ''spontaneous'' request is recognized as a true mand. The contingencies leading to inappropriate behaviors are identified in terms of the antecedents and the consequences. The results of planned ignoring are identified as negative punishment effects. In a second case, the occurrence of an inappropriate behavior by one student acts to evoke reinforcement of students who are engaged because the teacher tacts that the student's misbehaviors reinforced by attention. The teacher then uses approval of other children as an occasion for making her approval and hence her ignoring more powerful to eliminate the misbehavior. The teacher tacts that an establishing operation has occurred for the misbehaving child such that the reinforcement effects of attention are enhanced. In another case, he or she tacts that limiting the amount of a reinforcer or saving a particular preferred item for difficult instruction is seen as an establishing operation. Reinforcing the accomplishment of a nonpreferred activity by brief exposure to a preferred activity is seen as an incident of the application of the Premack principle. Incorrect responses are seen as problems in the presentation of the learn unit (e.g., ''oops, I presented the wrong antecedent'').

The continuous characterization of what is occurring in the classroom using the verbal labels of the science sets the stage for thinking analytically. The occurrence of the scientific tacts is necessary to the use of *verbally mediated* solutions to learning problems as they are encountered. Unless the vocabulary is accurate, extensive, and fluent (e.g., the terms come automatically), the teacher does not have the necessary component expertise for analyzing the existing contingencies in order to identify what tactics are needed for this student at this particular time. The degree of analytic skill of the teacher is tied necessarily to his or her grasp of the vocabulary.

Not only must the teacher see or characterize the behavior and the environment of the student using the scientific vocabulary, but also he or she must do the same with her own actions and environment and those of her teaching assistant. Was the teacher's action a contingent or noncontingent response to the student? Was the intraresponse time accurate? Did a feature in the environment reinforce the assaultive response? Was that an incident of negative punishment? Did that disapproval error function to reinforce off task responding? Was that an appropriate antecedent for a student with prelistener behavior?

A critical part of the scientific vocabulary also incorporates that special terminology of the *behavioral data language*. As is the case with any science or technology, measurement is the key to effectiveness in practice and in subsequent analyses of the effectiveness or ineffectiveness of the current operations in effect. The teacher and other professionals must maintain continuous measurement of the correct/incorrect responses of the student. The data are necessarily transformed and interpreted via the visual displays or the graphs of each student's responding for each learn unit of each instructional program. The instructional tactics and strategies have meaning only in terms of the data. The behavioral data language is the key to making use of the graphs and a necessary component for analyzing data and making decisions. In the chapter devoted to the analytic repertoire of the teacher, we presented the decision protocol that the teacher can use to analyze the data and make decisions. The following reviews some of those components.

The terms that characterize and summarize the student's progress include trend, criterion, stability, variability, level, phase change, ordinate, abscissa, rate or number in time, number, blocking, trends within sessions, trends across sessions, and characteristic patterns (i.e., extinction pattern, cyclical patterns, schedule patterns, stability), to name a few key terms. As the data language becomes part of the scientific tacting repertoire of the expert teacher, verbal–nonverbal correspondence evolves, and the tools of decision making fall in to place. By verbal–nonverbal correspondence, we mean that the words or scientific tacts conform to what actions occur or are to occur. They initially reflect action, and later direct action, such that the teacher him- or herself, or another with the same vocabulary, knows precisely what is occurring.

Is that an ascending trend? Did the student meet criterion? Do the data show a new tactic is needed? Is the reinforcement schedule fixed or varied, intermittent, or FR1? Does the speed of responding constitute fluent? Is the variability of the data an indication of problems for the teacher or student in the learn unit and the learn unit setting? Do sessions need to be blocked to detect effects that are cyclical but noncritical? Do these three sessions constitute a trend? Should the current tactics continue? Should tactics change? Is the problem strategic (i.e., is there a new question that needs to be asked, not simply a need for a change in tactics associated with the former question for which a solution was sought with the tactic in use)? These are only a few uses of the terms in the behavioral data language vis-à-vis its role in analysis. Once the teacher is fluent in characterizing the data and the actions of students she is fluent with the vocabulary of the science. In turn, a fluent vocabulary sets the stage for the development of the scientific analytic strategies that differentiates a teacher who is an excellent data based teacher (but one not analytically proficient) from a teacher who is a strategic scientist of instruction.

You will need to be familiar with the list of tactics that follows (Table 2). The list includes tested tactics from the published research literature and tested tactics from our research programs that have not been published at the time this book was written. You will want to match tactics with particular problems that students encounter. In many cases particular tactic may work with one of several problem areas. The process of matching tactics with particular problems is currently a matter of logic rather than science. Brief descriptions of these tactics are found in the Glossary or introductory texts in the applied behavior analysis. However, the student will want to be familiar with the research literature that provided the tactics.

As you learn to use these tactics for individual problems encountered by you students, your use of them will shift from verbally mediated (i.e., you follow the verbal instructions for what to do) to contingency shaped teaching responses (i.e., you use the tactics fluently because classroom events evoke them automatically). However the process involves decisions on your part—decisions that draw on all of your teacher repertoires.

EXAMPLES OF TACTIC SELECTION OPERATIONS

The Table 3 lists some common problems encountered in the instruction of students who are the target population of this chapter. We also list one possible source of the problem and a corresponding tactic from the research literature that may rectify the problem. The chapter on the analysis of instruction describes the process involved and the chapter on supervision describes how we have been successful in teaching these repertoires to teachers.

TABLE 2
List of Tactics for Students with Listener through Early Reader Repertoires

- Learn units
- Response prompts
- Response delay
- Modeling
- Peer reinforcement conditioning
- Visual feedback ("Jack climbs the beanstalk")
- Generalized reinforcers
- Tokens
- Preferred activities as reinforcement (Premack principle)
- Activity deprivation as establishing operation
- Time delay (for response)
- Extrinsic stimulus prompt
- Intrinsic stimulus prompt
- Contingent reinforcement
- Planned ignoring
- Contingent observation
- Brief time-out from reinforcement
- Time-out toy/item
- Good behavior ribbon
- Interspersal of known items
- Matching-to-sample as perquisite repertoire training
- Topography of response changed for same function
- Interrupted chain as establishing operation for mands
- Brief motivation procedure as establishing operation for mands
- Incidental "trials" as establishing operation for mands
- Mand deprivation with listener response momentum procedure to establish vocal verbal behavior (Ross & Greer, in press)
- CABAS

- Self-monitoring
- Game instruction for social skills
- Simultaneous stimulus prompt (fixed and progressive)
- Rule-governed responding
- Teacher performance rate accuracy observation
- Supervisor rate
- Data-based teacher decision analysis
- Peer mediation
- The "bad" behavior functional analysis protocol
- DRO, DRA, DRH, DRL
- Overcorrection
- General case shaping
- Self management of time-out
- Matching law and learn units for treating bad behavior
- Chaining (forward, backward)
- Textual prompts to eliminate echolalia
- Behavioral momentum
- Distributed programs or learn units
- Massed learn units
- Contingent escape
- Noncontingent reinforcement for bad behavior
- Training mands for socialization for conversation units
- Teaching conversational units as socialization
- Anthropomorphic toys to induce speaker as own listener
- Reinforcement sampling
- Thinning reinforcement
- Rate of responding criterion for maintenance
- Mastery learning
- Personalized System of Instruction (Keller plan)

(continues)

TABLE 2 (*continued*)

- Graduated response requirements
- Five procedures for inducing or increasing food consumption: (a) respondent eliciting stimulus on tongue, (b) reinforcement with preferred food, (c) peer reinforcement conditioning, (d) modeling, and (e) forced feeding (a pre-data-based procedure)
- Satiation (for operant rumination/vomiting), calorie related
- Mand and tact training procedures
- Tolerance for delay in reinforcement; tolerance for noncontingent punishment
- Picture prompts
- Signing or pictures as topographies for verbal behavior
- Verbal immersion
- Beginning writer immersion
- Function reinforcement versus topography reinforcement
- Multiple control (verbal behavior term)
- Generative rate criterion for complex tasks
- Approval, ignoring, rules as tactics
- Vibrotactile stimulus as reinforcement
- Novel stimulus reinforcement
- Opportunity to respond
- Stimulus fading
- Good behavior game
- Lottery for cash awards for staff
- Backup reinforcement rentals
- Tandem reinforcement
- Multiple schedules (e.g., praise FR 1, opportunity to mand VR 3)
- Teacher nonvocal approval
- Errorless learning through fading and chaining
- Pyramid training
- Programming across environments
- Escape extinction
- Peer-mediated establishing operation
- Flash card instruction rate

- Peer incidental teaching (alternating learn units between peers)
- Video feedback
- Vicarious reinforcement
- Self-recording to decrease/increase behaviors
- Social praise to condition reinforcement value
- Conditioning play to decrease stereotypy
- Increase learn units to decrease SIB and stereotypy
- Contingent music listening as reinforcement
- Prior reinforcement conditioning to decrease learn units to criterion
- Timing to increase performance rate
- Presentation rate of learn units and achievement to decrease off task
- Rapid toilet training procedure
- Independence of selection, production
- Independence of tacts/mands
- Independence of tacts; multiple control and selection
- Side effects of DRO
- Extinction bursts
- Negative-punishment-induced assault
- Stimulus satiation
- Response satiation
- Satiation to shift activity reinforcement using low-preference activities
- Peer motivation groups
- Learn units versus differential reinforcement
- Verbal–nonverbal correspondence training
- Student reinforcement dispensation to increase socialization
- Self-prompting rule-governed responding
- Rate training for contingency shaping behaviors
- Social skills training procedure

(*continues*)

TABLE 2 (*continued*)

- Children prompting reinforcement from teachers
- Response cost
- Contingent free play
- Reinforcement of novel responding
- Graduated physical guidance
- Contingent instruction
- Teaching alternative textual control of verbal behavior to decrease echolalia
- Textual control to develop appropriate intraverbals
- Situation training for weight control
- Reinforcement of appropriate eating for weight control
- Reinforcing exercise for weight control
- Contingent edibles
- Removal of sensory consequence for SIB
- Bell-signaled nighttime training for bed wetting
- Variable interval schedules for academic performance
- Prompted S^d in learn unit
- Shaping components of vocal topographies
- Picture prompts for conversation
- Rapid tacting procedure

- Response blocking for SIB
- Fading restraints
- Food-induced SIB
- Scripts for peer interaction
- Wrist weights, gloves, sensory screening for SIB
- Masking for SIB
- Response cards
- Learn unit distribution for delayed reinforcement
- Access to high-probability responses to increase social interactions
- Generalized imitation training (behavior momentum)
- Task analysis and chaining
- Establishing operations to teach yes/no, location, quantity, quality autoclitics
- Inducing conversational units via removal of adult interaction opportunities when student is in peer setting
- Task demand and escape as reinforcement
- Pure tact versus impure tact training
- Contingent exercise
- Positive practice
- Operations to program "generalization"

SUMMARY

This chapter described pedagogical operations that are particularly important for teaching early forms of verbal behavior. In order to describe the pedagogical operations, we frequently related the pedagogy to curricula. Whenever possible you will want to use scripted curricula and other tested curricula. You will need to design learn unit assessments and graphic displays for programs that do not include those instructions. The Fred S. Keller Scripted Curriculum for Verbal Behavior, Teaching Operations for Verbal Behavior, Edmark, *Reading Mastery*, and the Morningside generative curricula are some of the scripted curricula that are tested. Using them will simplify data collection and will provide tested sequential material. Using tested material that is scripted will allow the teacher to devote more time to pedagogy and less time to designing curricula. Alternately, you can script out the instructional procedures yourself.

TABLE 3

The Repertoires of Analysis: Forty-Nine Learning Problems and Exemplar Solutions for Prelistener through Early Readers

Detection of problem (by viewing the data and the student and scientific tacts)	Possible source of problem and behavioral principle[a]	Tactics from research (some of many possibilities)
(1) The graphs of the data show three data paths with descending, or flat, trend and the student responds slowly.	Consequence does not reinforce correct responses periodically. **Reinforcement**	Vary reinforcers; limit the opportunities to work for the reinforcer used with the program; shift to variable reinforcement schedule.
(2) Descending, stable, variable, or slow ascending trend with variability; experiment or perform probe to isolate component of learn unit containing the likely problem.	Consequence not reinforcing periodically. **Reinforcement**	Identify preferred activity, toy, or edible that functions as reinforcer; reinforcement assessment.
(3) Stable correct responding (few or no incorrect responses) but at slow rate; maintenance probes show the student does not maintain.	Performance is mastered but does not meet a rate-in-time objective; difficulty with maintenance is a function of lack of fluency; consequence not contingent on correct and a rate requirement. **Reinforcement schedule**	Reinforcement made contingent on correct with a number in time performance requirement.
(4) Problem not in learn unit presentation: analysis suggests that prerequisites are mastered but each time the complex task (i.e., matching printed words to pictures) is encountered, the student has a slow rate of correct and high incorrect responding.	Component skill performance was mastered but does not meet the normative-number-in-time criterion. **Instructional history**	Reinforcement made contingent on correct responses with a number per minute rate requirement.
(5) Student requests items when asked "what do you want?" but does not request items independently; the data show that the student has met criterion on requesting but does not generalize.	The student has learned an intraverbal response but has not learned pure mands; *a different antecedent, setting event, and consequence needs to be taught with the existing response.* **Wrong verbal operant taught**	Eliminate verbal antecedents; present item when deprivation has likely occurred (later fade presence of item); present the requested item only when requests occur (a few echoic teaching seasons may need to occur); use no praise with presentation of item requested (e.g., do not say "nice talking").
(6) The data show that the student makes good progress with a single item used as reinforcer, but after the student becomes satiated no progress occurs.	Limited conditioned or unconditioned reinforcers. **Inadequate history of reinforcement conditioning experiences**	Pair other stimuli with presentation of the existing reinforcer; shift to variable schedule but present other item alternately with the item that does reinforce; pair approval with both; if feasible use tokens with the functioning reinforcer as the backup reinforcer for the token
(7) Student learns only with food reinforcers.	The student appears to have few if any conditioned reinforcers in his repertoire. **Inadequate history of reinforcement conditioning experiences**	Use minimeals; pair all food presentations with praise; shift to variable schedule; thin schedule, but maintain FR1 praise; pair food with toy play or other activities that are potential conditioned reinforcers.

(8) Student engages in self-injury, stereotypy, or assaultive behavior; the number of learn units received daily is low (i.e., less than 400 per day for a student acquiring listener and speaker repertoires); the well-being of the student and classmates can be assured by monitoring and "redirecting the student."	Response/reinforcer relationship for the assaultive behavior produces frequent response reinforcer occurrences (the consequences for the bad behavior are easily attained, few conditioned reinforcers). Matching law. **Inadequate opportunities for reinforcement for instructional responding**	Increase learn units two to three times over baseline rates; spread instruction across many curricular programs; rotate teachers; avoid reinforcing assaultive behavior by stopping teaching; work to have the response/reinforcer relationship of learn unit responding replace the assaultive behavior consequence relationship.
(9) Student engages in self-injury, stereotypy, or assaultive behavior; the number of learn units received daily is low (i.e., less than 400 a day for a student acquiring listener and speaker repertoires); the well-being of the student and classmates can be assured by monitoring and "redirecting the student."	Response/reinforcer relationship for the assaultive behavior produces frequent response reinforcer occurrences (the consequences for the "bad" behavior are easily attained, few conditioned reinforcers). Matching law. **Insufficient reinforcement for alternative responding or escape from responding to academic tasks unwittingly reinforced**	Use differential reinforcement tactics such as DRO, DRA, DRL, and DRI; probably best used in conjunction with doubling or tripling of learn units.
(10) Student echoes teacher instruction when he should not; the data show frequent inappropriate echoic responding (e.g., teacher instructs "sit still" student sits still but also says "sit still"); Probe shows that the student can mand and tact using the words, but uses the "words" echoically when there is a vocal antecedent.	The student is under the wrong or multiple antecedent controls, shaped perhaps by reinforcement of the correct response that included the unwanted echoic appendage. **History of faulty intraverbal control: Incorrect antecedent stimulus control**	Use written word or other nonvocal antecedent, until the student responds to the new antecedent and then fade in vocal command with the printed or visual antecedent; if no echoic behavior occurs fade printed stimulus and keep vocal; teach numerous pure tacts and mands where the antecedent is nonverbal; if possible, teach reading repertoire for silently reading written questions.
(11) Student ruminates or "voluntarily" spits up food during instruction (vomiting is operant not respondent) and free time and in various settings.	Student not sated on food, still food deprived. **Deprivation of food: Setting Event**	Increase eating time and consumption until the student stops eating, even if the other children are finished and return to class.
(12) Student emits assaultive or self-injurious responses when given a command; functional analysis shows that the response does not occur when the student is left alone.	Response to verbal antecedents reinforced by escape from demands. **Response avoidance is being reinforced: Instructional history**	Increase learn units for instruction that is at the student's level while not allowing the student to escape; use brief moments of free time as reinforcer, while conditioning other reinforcers in the learn unit presentations
(13) Student has erratic responses with a new teacher; functional analyses comparing teachers spots difference in rate of learn unit presentations.	Problem located in the teacher component of the learn unit. **The teacher operant component of the learn unit is flawed**	Teach teacher to present faster and accurate rate of learn units.
(14) Student emits correct responses to a particular instructional program that is run by one staff member but incorrect response occur with other staff members; TPRA observations show inconsistencies in antecedent presentations (i.e., touch red, as opposed to point to red).	Inaccurate antecedent presentation by teacher; problem located in the teacher component of the antecedent. **The teacher operant component of the learn unit is flawed**	Provide consistent antecedents; be sure that programs are scripted and that the teacher adheres to script; once mastery occurs systematically introduce antecedents you wish to have equivalent effects (i.e., show me red, give me red, where is the red one).

(continues)

TABLE 3 (continued)

Detection of problem (by viewing the data and the student and scientific tacts)	Possible source of problem and behavioral principle[a]	Tactics from research (some of many possibilities)
(15) Student has difficulty acquiring correct responses to concepts (discrimination within and between stimuli); probes show that reinforcement is adequate, teachers are accurate, and student has the prerequisites.	Problem located in the curricular design of learn unit presentations; student does not receive range of positive and negative exemplars adequate for shaping stimulus classification. **Instructional history; specifically, the prior antecedent control is not established**	Use general case programming where irrelevant features of the positive exemplar are varied while the range of negative exemplars are presented (e.g., when teaching color, vary size, shape and texture of target color presentations and present a different color for the "not target color" each time the negative exemplar is presented; teach student yes and no responses to positive versus negative exemplars.
(16) Student has difficulty acquiring correct responses to concepts (e.g., shapes, colors); probes show that reinforcement is adequate, teachers are accurate, and student has the prerequisite; general case programming procedures used correctly; probes for asking the student to "put same with same" show that the student does not have same with same prerequisite.	Problem associated with the student's instructional history—that is, the student does not have perquisite "same with same" repertoire. **Instructional history; specifically, the prior antecedent control is not established**	Teach sameness through match to sample instruction (e.g., "same with same") repertoire across the range of stimulus objects and then return to general case programming once mastery is achieved and if necessary teach until the student performs at a rate objective in addition to mastery.
(17) Student uses speaker behavior at school, but does not do so at home; probes at home show that there are few opportunities to use the school repertoire.	The home environment does not provide available antecedent listener and speaker opportunities are rare. **Wrong setting events in the home (i.e., deprivation is responded to with non vocal responding)**	Parent educator teaches parents to provide speaker opportunities and longer wait time for antecedents and how to reinforce speaker behavior; Parent avoids reinforcing nonspeaker behaviors (i.e., whining, crying, assaultive or self-injurious behavior gets reinforced).
(18) Student assaults peers when in group setting for free play; analysis shows that the behavior occurs when student attempts to take toy from another student.	Lack of response that is appropriate for obtaining toy or for waiting turn. **Inadequate mand repertoire**	Design and implement program to teach appropriate mands to peers and also to wait for turn.
(19) Student does not work independently while not receiving individualized instruction; likely source of the problem is lack of conditioned reinforcement control of the task.	Problem of lack of conditioned reinforcement control by the stimuli and activities of independent play or work. **Instructional history with reinforcement pairing**	Increase teacher assistant reinforcement of student behavior during independent activities; develop learn unit programs that target the child playing for longer and longer periods before teacher reinforcement; while differential reinforcement of nonengagement is occurring you are really seeking to shift reinforcement control from the teacher to the activity stimuli.

(20) The graph shows that the student has performed on two successive sessions at criterion levels, yet more sessions were done and response starts to drop.	Criterion achieved and adequate reinforcement occurs even when the student is not attending to and responding to the antecedents. **Teacher component of the learn unit is inadequate: Teacher not under the discriminative control of the visual display** (alternately, the teacher is not graphing immediately after the program was run)	Move the student to the next level in the curriculum; the program in question is done again only for maintenance purposes or in generalized settings.
(21) The graph shows three descending data paths.	The current operations or curricula are not working. **Probe for prerequisite repertoires using the verbal behavior stages as a guide**	Draw on the strategic analysis repertoires that are listed in the prior chapter to locate the source of the problem.
(22) The level of ascent is slow and variable.	The operations of pedagogy are not optimum. **Teacher component of the learn unit is inadequate: Teacher not under the discriminative control of the visual display**	Draw on the strategic analysis repertoires to look for the source of the problem, but in this case it may be simply a minor problem in the presentation or in the student's listener repertoire.
(23) The rate of correct responses is high but so is the rate of incorrect responses.	A change of operations is called for. **Probe the range of strategic questions relative to reinforcement schedules**	Draw on the strategies of the science to look for a possible source.
(24) The student acquires an echoic form but cannot acquire the mand function.	The motivational conditions for acquiring the mand function are not present. **Manipulate setting events**	Use an establishing operation that incorporates a well-known chain, such as completing a puzzle; withhold the last piece until the student mands it.
(25) The student acquires an echoic form but cannot acquire the mand function.	The motivational conditions for acquiring the mand function are not present. **Manipulate setting events**	Catch the student in a natural setting where the mand item is wanted but made inaccessible and then use a delay tactic as an establishing operation to evoke the mand function.
(26) The data show that the student does not work independently or play with a toy independently.	The object present for independent activity has little or no conditioned reinforcement value. **Inadequate conditioned reinforcement history**	Run a program wherein the student plays or works with activity/toy for 5 s while pay is continuously reinforced by the teacher; allow a 5-s period for free choice and record choice intervals with the target object; continue to alternate until the student chooses the target activity for three sessions of 10 consecutive choices of the target object; expand the time frame as needed until the conditioned stimulus control occurs.
(27) Student runs around the room and does not sit while other students do.	Reinforcement value of sitting is preempted by competing reinforcers. **Setting events not appropriate**	Remove the chair, and do not allow the child to sit until he mands; Continue the same procedure for each occurrence of out of seat.

(continues)

TABLE 3 (continued)

Detection of problem (by viewing the data and the student and scientific tacts)	Possible source of problem and behavioral principle[a]	Tactics from research (some of many possibilities)
(28) Data show that the students are accurate but slow in teacher-presented learn units.	Teacher stimulus control interferes with speed. **Teacher operant in the learn unit is flawed**	Set up learn units in a learner-controlled format; reinforce for successive approximation of a rate criterion (e.g., 1-min timings or even 10 s timings).
(29) Data show that the students is accurate but slow in teacher-presented learn unit.	Teacher stimulus control interferes with speed. **Teacher operant in the learn unit is flawed**	Increase teacher rate of presentations using TPRA observations by the supervisor or videotaped observations done by the teacher of her own instructional sessions in which the teacher does her own TPRA.
(30) Student performs optimally only with one teacher—when others teach the student the performance is poor; Others work with the student infrequently.	The student is not under generalized stimulus control with others. **Generalized stimulus control not programmed across individuals: Rotate components of the setting events**	Arrange instruction such that teaching is rotated across all available instructors. Provide limited access to reinforcers outside of the individualized settings with the rotated instructors.
(31) Student has met criterion on several colors but when a third or fourth color is added the data plateau.	The number of objects to discriminate are too many in number; the antecedent is flawed or the student is not fluent with the prior discriminations. **Antecedent is flawed, given the student's instructional history**	Decrease the number of stimulus objects or introduce the new object with fewer but rotated foils.
(32) Student has met criterion on several colors but when a third or fourth color is added the data plateau.	The number of objects to discriminate are too many in number; the antecedent is flawed or the student is not fluent with the prior discriminations. **Instructional history**	Teach prior objects to an adult rate per minute criterion.
(33) The student meets all subobjectives in two sessions with no or few errors.	The instruction may not be appropriate. **Teacher error: The repertoire is already in the student's repertoire**	Probe across the range of subobjectives; do a long-term objective test; do a complete behavioral inventory, and then determine the true deficits and teach those objectives.
(34) The student has reinforcers available for independent play but makes little or no progress.	The instructional programs are too advanced. **Instructional history with Response or antecedent**	Probe across the range of behaviors; begin programs at prerequisite levels where the student succeeds before moving on to new objectives.
(35) The student is echolalic and has acquired some textual stimulus control.	The student has faulty intraverbal control. **Instructional history in which echoing is reinforcement or the intraverbal control was never established**	Write appropriate intraverbal responses on an index card; Provide a vocal antecedent (hi, bob), but before the student has a chance to echo hold up the index card for the correct intraverbal response ("hi, teacher"); when the student responds correctly reinforce; if not, correct with the printed stimuli; gradually fade the printed stimuli and fade the use of the index card across various intraverbal settings until mastery and the elimination of echolalia.

(continues)

110

(36) The student reads words in a story context slowly.	The student has not mastered and is not fluent in textual responding to individual words. **Inadequate instructional history with speed of responding**	Teach fluent responding to phonetic sounds and words, and also teach fluent responding to Dolch words; When mastered at the specified rate criterion, reinsert words in story format.
(37) The student reads beginning reading material slowly but accurately. Responses to comprehension questions are unreliable.	The student is not fluent in responding to textual stimuli in story format. **Inadequate Instructional History for the Textual Components of Reading**	Teach fluent responding to words in story format; do not ask comprehension questions until the student can read the story at normative rates of speed; once she does so, ask comprehension questions.
(38) The student misreads letters (i.e., she reads *b* for *d*)	Student has not learned to discriminate between letter shapes. **Incorrect stimulus control from instructional history**	Use intrastimulus prompts to prompt discriminations in massed learn units; repeat until a rate criterion is achieved.
(39) Student avoids reading.	Print stimuli have not acquired reinforcement properties. **Conditioned reinforcement history**	(a) Read stories to the child, reinforcing pointing to pictures. (b) Pair reinforcers with looking at books until the student, alternate 5 s of looking with the child and sec. free operant looking by the child over 5-minute sessions; Criterion occurs when he or she looks at books for 90% of the free operant intervals. (c) When he or she selects books in free time and meets the above criterion, return to reading instruction. (d) Use a reading program such as Edmark that carefully programs print stimulus control.
(40) Student does not show ascending trend for early print discriminations in *Reading Mastery* or Edmark.	The student does not match print stimuli reliably; He is under instructional control, receives a known reinforcer, and there are no teacher errors in the presentation; however, there are no data records of a history of instruction with matching objects or pictures. **Probe for prerequisites stimulus control**	Probe for picture matching; if the repertoire is deficit, train to criterion on common objects, and then abstract pictures; after criterion is achieved on these prerequisites, return to the reading curricula.
(41) First attempts at textual reading (e.g., "decoding") result in correct responses at very low levels.	Student has no experience with reading, but likes looking at books; however, this is his first attempt at decoding. **Intervening repertoire needed**	Develop or use a scripted program to teach the student to discriminate words from pictures or nonwords as well as absence or presence of words (i.e., "point to the card with the word on it"); Alternately, use the Edmark reading program or follow the above programs with the Edmark curriculum.
(42) The student reads the textual words from the "whole word" texts, but cannot read words that are not in the text.	The student may not be under the control of the phonetic properties of the words; probe for phonetic components of words. **Inadequate instructional history**	Teach the phonetic alphabet to fluency; then, train phonetic word reading. Use the Direct Instruction materials that have been tested in the research literature.

(continues)

111

TABLE 3 *(continued)*

Detection of problem (by viewing the data and the student and scientific tacts)	Possible source of problem and behavioral principle[a]	Tactics from research (some of many possibilities)
(43) The student reads fluently, but performs poorly on comprehension questions.	Probes suggest that the student responds to pictures in the book rather than the text itself, possibly due to a history of making up stories while looking at books with adults. **Wrong stimulus control for comprehension**	Use the Edmark comprehension exercises or any text without pictures that requires the student to respond to print rather than illustrations (e.g., the Sullivan reader)
(44) The student reads fluently, but performs poorly on comprehension questions.	Probes suggest that the student does not have correspondence between words and activities specified by the words. **Correspondence between reading and doing**	Give instructions for obtaining reinforcing activities or items using written text; the student must read the text and follow instructions to receive a preferred event or item; conduct written compliance exercises much like the early listener skill exercises.
(45) The student reads first words fluently, but performs poorly on comprehension questions.	Probes show that the student cannot match words with pictures. **Inadequate word and picture correspondence**	Student receives instruction on matching words and pictures, but no success; probes of matching pictures to pictures shows this repertoire is missing; picture matching taught to criterion, followed by a return to word–picture matching.
(46) The student matches written words to pictures, but cannot say or textually respond to the written words reliably.	Probes suggest that the student does not equivocate pictures to vocal word sounds. **Inadequate picture to word correspondence**	Student taught to label pictures, and saying of words emerges.
(47) The student moves adequately through learning to sound the text but his pronunciation is faulty. This results in delays in his progress.	Probes show that comprehension is at criterion levels and that the textual responding is fast.	Do not delay moving the student through the reading lessons because the pronunciation is not clear, as long as it is understandable to the teacher. Work on pronunciation as a separate or parallel program for pronunciation using consonant vowel blend combination, first to percentage mastery and then to a fluency rate criterion.
(48) The student reads adequately when asked to do so, but does not read in free time.	Systematic observations of the student in free time shows that he does not read. **Inadequate conditioned reinforcement**	Pair praise, tokens, preferred activities, and even edibles with his reading in class; use tokens and other prosthetic reinforcers whenever the student reads in free time; use a token economy to reinforce reading books independently and answering comprehension questions at criterion levels.
(49) The student reads and comprehends at grade level, but writes poorly.	Probes show that the student can copy text satisfactorily, but cannot compose. **Writer as own reader deficit**	Set aside periods in the day when all requests for token exchanges and preferred activities require written mands; respond in written form, requiring corrections until the composed mands are effective; begin exercises that require the student to explain to a peer how to do something or describe something such that the peer can do the task described or until the peer can identify the object described.

[a] Bold type indicates one probable component of the learn unit context associated with the problem.

However, you will need to adapt learn unit data collection as well as determine curricular sequence for much of your instruction, because there are only a few tested curricula. The chapters on curriculum will assist you in arranging learn units for all curricular materials. In addition, mastery of the next chapter on teaching students with self-editing repertoires will provide a sense of direction for the next steps for teaching students who are the subject matter of this chapter. You will want to prepare your students to have the necessary prerequisites for advanced forms of verbal behavior. The next chapter is devoted to the specific repertoires that are required of teachers who instruct students who have or who are acquiring advanced verbal functions.

References

Becker, W. (1992). Direct Instruction: A twenty year review. In R. West & L. Hammerlynck (Eds.), *Design for educational excellence: The legacy of B. F. Skinner*. Longmont CO: Sopris West.
Becker, W. C. (1977). Teaching reading and language to the disadvantaged. *Harvard Educational Review, 47*, 518–543.
Becker, W. C., & Carnine, D. W. (1981). Direct Instruction: A behavior theory model for comprehensive educational intervention with the disadvantaged In I. S. W. B. R. Ruiz (Ed.), *Behavior modification: Contributions to education* (pp. 145–207). Hillsdale, NJ: Erlbaum.
Chu, H. C. (1998). *Functional relations between verbal behavior or social skills training, and aberrant behaviors of young autistic children*. Unpublished Ph.D. dissertation, Columbia University, New York.
Donley, C. R., & Greer, R. D. (1993). Setting events controlling social verbal exchanges between students with developmental delays. *Journal of Behavioral Education, 3*(4), 387–401.
Edmark Corp. (1992). Redmond, WA: Pro-Ed. www.proedinc.com.
Greer, R. D. (1986). *Teaching operations for verbal behavior*. Yonkers, NY: CABAS and the Fred S. Keller School.
Greer, R. D. (1994). The measure of a teacher. In I. R. Gardner et al. (Eds.), *Behavior analysis in education: Focus on measurably superior instruction*. Pacific Groves, CA: Brooks/Cole.
Hart, B., & Risley, T. (1996). *Meaningful differences in the everyday life of America's children*. New York: Paul Brookes.
Ingham, R. J. (1984). *Stuttering and behavior therapy*. San Diego: College Hill Press.
Johnson, K. R., & Layng, T. V. (1994). The Morningside Model of Generative Instruction. In I. R. Gardner et al. (Ed.), *Behavior analysis in education: Focus on measurably superior instruction* (pp. 283–320). Pacific Groves, CA: Brooks/Cole.
Kelly, T. M. (1994). *Functional relations between numbers of learn unit presentations and emissions of self-injurious and assaultive behavior*. Unpublished dissertation, Columbia University, New York.
Kelly, T. M., & Greer, R. D. (1996). *Functional relationships between learn units and maldaptive behavior*. Manuscript submitted for publication.
Keohane, D. (1997). *A functional relationship between teachers use of scientific rule governed strategies and student learning*. Unpublished Ph.D. dissertation, Columbia University, New York.
Lodhi, S., & Greer, R. D. (1989). The speaker as listener. *Journal of the Experimental Analysis of Behavior, 51*, 353–359.
Martinez, R., (1997). *Reducing aberrant behaviors of autistic students through efficient instruction: A case of matching in the single alternative environment*. Unpublished doctoral dissertation, Columbia University, New York.
Ross, D. E., & Greer, R. D. (in press). Generalized imitation and the mand: Inducing first instances of functional speech in nonvocal children with autism. *Journal of Research in Developmental Disabilities*.

Selinske, J., Greer, R. D., & Lodhi, S. (1991). A functional analysis of the Comprehensive Application of Behavior Analysis to Schooling. *Journal of Applied Behavior Analysis, 13*, 645–654.

Skinner, B. F. (1957). *Verbal behavior.* Cambridge, MA: B. F. Skinner Foundation.

Skinner, B. F. (1968). *The technology of teaching.* New York: Appleton-Century-Croft.

Teaching Practices for Students with Advanced Repertoires of Verbal Behavior (Reader to Editor of Own Written Work)

TERMS AND CONSTRUCTS TO MASTER

- Self-management repertoires
- Personalized System of Instruction (PSI)
- Reader/writer repertoires
- Writer as own reader
- Determining ratio of teachers from verbal behavior repertoires of students
- Fluent reading and writing repertoires
- Self-editing
- Recycling
- Mathematical repertoires and self-editing
- Mathematical fluency
- Aduction
- Academic literacy
- Problem solving as verbal mediation
- Methods of history, logic, and science (as verbal communities)
- Self-monitoring
- Setting goals for self
- Self-reinforcement for goals
- Frequent learn units and "bad behavior"
- Comprehensive point system
- Counts versus tokens/points
- Renting backup reinforcers
- Token auctions
- Sequence for teaching learner independence
- Verbally governed and event governed behavior
- Evolution of self-editing repertoires

TOPICS

Repertoires for 2000

Design and Teaching Operations for Academic Literacy

Design and Teaching Operations for Discipline-Based Problem Solving

Learner-Controlled Instruction and Time Management

Design and Teaching Operations for Expanding the Student's Community of Reinforcers

Summary

Appendix: Tactics for Teaching Advanced Verbal Repertoires

REPERTOIRES FOR 2000

We propose that individuals who will flourish and contribute to society in the next millennium will have the following repertoires. They will have:

(a) *Fluent reader repertoires* and a wide range of reading interests (literature, history, science, leisure subjects);

(b) Verbally and event-governed *repertoires for self-management and self-instruction*;

(c) Repertoires of speaking, *writing, and of self-editing that produce desired effects on the writer's target audience (*e.g., a relevant audience can do the written instructions, fix the problem based on the writers solution, or in other ways be affected as the writer intends);

(d) *Fluent* computational and conceptual *math repertoires*;

(e) Repertoires for analyzing and solving problems by choosing the relevant tools for the problem from historical methods, logic (i.e., mathematics), and the scientific method; and

(f) An expanded community of reinforcers to maintain and expand the above repertoires (interests in literature, the arts, science, history, and leisure subjects and activities).

The capable high school graduate will have the skills to set goals and manage the events in their lives such that these goals are achieved. They will eschew short-term benefits when working for long-term gains. They will skill-fully apply sophisticated and discipline-based inquiry to determine solutions to complex problems.

This chapter describes the pedagogical repertoires needed to teach students the instructional content consistent with the outcomes described above. The pedagogical repertoires presume that the teacher's students have some, most, or all of the repertoires outlined in the prior chapter (although teachers will continuously find missing prerequisite repertoires and should be prepared to teach those prerequisites to their students).

While the *principles* of behavior described in applied behavior analysis texts and the *strategies* for searching the learn unit context are the same for the students described in both this chapter and chapter 5, the tactics from the research literature and the contents of the learn unit and its context will differ. They differ because of the existing verbal capabilities of the students. The verbal capabilities of the student change both the *objectives* of instruction and the *manner* in which they are taught. Each increase in the sophistication of students' verbal behavior status increases their independence. The increases in verbal capabilities correspond with increases in independence and allow for corresponding increases in the number of students per teacher. When increases in class size occur as a *result* of learner independence, there will be no lapse in instructional effectiveness, provided that the classroom is managed according to the operations we describe in this chapter. We described a similar graduation in student/teacher ratios for those students learning the listener, speaker, and beginning reader/writer repertoires in the prior chapter. Table 1 extends this principle to the students who are the subject of this chapter.

Table 1 outlines the broad functional the repertoires for students. The curriculum associated with these repertoires is described in greater detail in the chapters that are devoted to the analysis and design of functional curricula and its delivery. Examples are included here to provide readers with a context in which they can place the pedagogical operations that are described in this chapter in proper context. These pedagogical operations include the teaching and classroom design operations to expand:

- *Fluent reading repertoires;*
- *Effective writing repertoires;*
- *Self-editing repertoires;*
- *Self-instruction and self-management repertoires;*
- *Problem-solving repertoires;* and
- *Each student's community of reinforcers.*

The curriculum connects the evolving verbal repertoires with the desired outcomes presented at the beginning of this chapter. The curriculum falls into four broad categories. They are widely recognized by numerous behavior analysts and educational philosophers. They include:

1. *Academic literacy* as the tool or component skills needed to achieve competency with more complex repertoires;
2. *Self-management* for learner independence;
3. *Problem-solving operations;* and
4. *An expanded community of conditioned reinforcers* and related behaviors that benefits the individual and society, as well as the stimulus discriminatives that are *selected out* by a multitude of conditioned reinforcers.

TABLE 1
Evolution of Self-Editing Repertoires and Their Effects

Verbal repertoires	Effects in the evolved verbal world
Prerequisite repertoires: fluent listener, speaker, and speaker as own listener; basic reader repertoires; and basic computational repertoires	
1. Reader status (including mathematical behavior as verbal behavior)	Individuals who have reading repertoires can locate useful, entertaining, and critical responses to environmental contingencies from print or electronic text. The verbal material may be derived by the reader without the time constraints that control the speaker–listener relationship. The reader can exert effective control over the writer just as the listener can for the speaker—the behavior of a good writer or speaker is controlled by the effect of his or her writing on the target audience. Similarly, individuals with basic math literacy both can follow and produce mathematical communications as forms of verbally governed behavior.
2. Writer status (including language and computational repertoires)	A competent writer may control environmental contingencies (e.g., the behavior of the reader) through the mediation of a reader who reads the written material in real time or across centuries. A writing repertoire changes the potential impact of the writer on his environment. Writing is inextricably tied to the successful efforts of persons in business, science, technology, the arts and humanities, and politics. In an information society, the inability to write effectively will function as a severe handicap for the individual.
3. Writer as own reader: The self-editing status	As writers increase their ability to read their own writing (or analyze their own ideas) and computational products from the perspective of the *eventual or target audience*, writers grow increasingly independent of frequent reliance on other instructional audiences (e.g., teachers, supervisors, colleagues). A more finished and more effective behavior-evoking repertoire provides the writer with wide-ranging control over environmental contingencies such that time and distance can be bridged.

(continues)

TABLE 1 (*continued*)

4. Verbal and mathematical mediation for solving problems: An expansion of the self-editing repertoire	A sophisticated self-editor has the verbal expertise associated with formal approaches to problem solving (e.g., methods of science, logic/math, history): He or she can solve complex and new problems independently and can characterize a situation using precise verbal descriptions (including mathematical descriptors). The verbal descriptions occasion other verbal behavior that can, in turn, direct the action of the writer as her own reader to solve the particular problem. The verbal community that constitutes disciplines (including a community of mathematics) is based on verbal expertise that is tied to other verbal and nonverbal environments (e.g., the physical world or complex ideas).

DESIGN AND TEACHING OPERATIONS FOR ACADEMIC LITERACY

There are three principal tools of literacy that drive all of the more complex repertoires. They are fluent reading, writing, and mathematical performance. Each step in the evolution of these tools or *pillars of literacy* builds on the fluency of the prior component skills. The pillars determine the students' eventual competence in science, social, or global studies and the arts and humanities. Emphasizing the pillars should not detract from the students valuing or gaining complex problem-solving abilities in science, mathematics, social studies, literature, and the humanities. The component skills are not ''inert forms of knowledge'' (Whitehead, 1929) if they are taught expertly *and if they are applied to problem solving or verbally mediated repertoires*. In fact, fluency or automaticity in the building blocks is necessary to the performance of complex and creative behaviors, just as technical facility is required for making music (see Johnson & Layng, 1994, for a thorough treatment of this issue). The pillars can be taught as part of the other subjects but only when the component skills are contingency shaped. In this context ''contingency-shaped'' responding is the automatic use of component skills, wherein the skills are done without out recourse to verbal rules, as when one performs a set of mathematical operations effortlessly without following a written formula. When the components are fluent all of the broad categories of the curriculum are taught in a spiral fashion. That is, as soon as fluency is acquired, the ''tool'' repertoire is applied to problem solving, self-management, and an expanded community of conditioned reinforcers. However, since their is no standard scientific definition of fluent as yet, we shall define fluent as the performance of component steps in problem-solving or creative behavior such that the task has complete stimulus control—there are no pauses in responding, there are no observable usages of verbally governed strategies (e.g., using number lines or referring to written directions), and the responses are quick and accurate.

Teaching Operations and Curriculum Design for Fluent Reading

The previous chapter described the teaching operations and design for establishing beginning reading skills as fluent repertoires. They included "whole word" instruction (e.g., Edmark) and phonetic instruction done simultaneously. We prescribed operations to teach fluent "decoding" repertoires (i.e., textual responding or saying the printed word), including exercises to teach students to read not simply at their grade level, but at the rate of adult readers, and operations to teach comprehension. We also outlined the sequence of curricula to lead students to fluent reading.

Once the student is reading grade level material at the speed of an adult reader (or at a speed that results in maintenance and comprehension for that student) and has mastered the *SRA Reading Mastery* and Edmark curriculum, we emphasize expanding the student's vocabulary and application of that vocabulary to other components described in the material that follows. Students at this stage are capable of and should engage in reading and following written instructions. They can function in classes of 15 to 30 students, *if* they have the independent reader and writer repertoires described above. The contingencies in the classroom are arranged to increase the reader and writer repertoires, while simultaneously increasing the numbers of learner-controlled learn unit presentations possible for individual students in classes with increasingly larger numbers of students. Teachers can begin early applications of the Personalized System of Instruction or PSI.

Personalized System of Instruction

The PSI approach was developed by Fred S. Keller for university classes (Keller, 1968) and extended to elementary and secondary students (Greer, 1980). The approach includes features of:

- Individualized pacing;
- Communication between students and teachers in written form;
- The teacher as designer of instruction rather than lecturer (the student is at the center);
- Tutoring or proctoring to assist in individualization;
- Arrangement of the subject matter in units or long- and short-term instructional goals; and
- Mastery of each short-term objective before proceeding to the next objective in the hierarchy of the curriculum.

In the PSI approach, students receive folders, electronic or paper folders, that prescribe their individually appropriate tasks and the learning materials to perform those tasks. Instruction for *new concepts* or objectives of the curriculum and the related vocabularies in science, literature, or social studies are presented in written form followed by written responses from the student. Once the student reviews the written consequences of the teacher (e.g., confirmation of correctness, teacher-prompted corrections by the student of her errors) to the relevant responses of the student, a learn unit is completed. Correct responses, marked with check marks or other written or graphic commendations, serve as reinforcement for the student's written response *when he or she reviews them*. Incorrect responses do not become learn units *until the student performs the corrections with or without tutorial assistance*. Some students can make the corrections without assistance; others will ask for assistance, initially by raising their hands and eventually by communicating the questions to the teacher in written form (see the section on teaching effective writing).

Communication between the student and the teacher is increasingly in *written* form. This arrangement sets the stage for teaching *learner independence* as well as the conditions for teaching all of the curriculum. Communication may occur on traditional paper, but the medium will become increasingly electronically based. Running the classroom instruction in PSI format *prepares our learners for distance learning* and learner-controlled self-instruction. It also increases the rate of a student's correct responses and allows the student to go at his or her own enthusiastic pace by providing the needed reinforcers.

PSI is used as a method to arrange the use of all of the other tactics from the science as individual students need them. It is a basic vehicle that allows the teaching of all of the four broad curricular areas—academic literacy, self-management, problem solving, and an enlarged community of reinforcers. The content of literacy, self-management, and problem solving consists of verbal communities consisting of functional vocabularies.

We shall use the term *vocabulary* to include all of the component skills needed for literacy that lead to more complex repertoires for solving problems using the needed disciplines of inquiry. Vocabulary encompasses the vocabularies of mathematics, literature, scientific subjects, history, composition, and the arts (i.e., music, art, theater, dance). Vocabulary learn units can be increased significantly by establishing *vocabulary building instruction* as a major component of reader/consumer instructions as well as writer/producer instruction (e.g., vocabularies of social studies, math, scientific findings). Vocabularies expand *the verbal communities for the student*. Thus, in this sense vocabularies are to lead to verbal repertoires or verbal operants, not just the traditional treatment of vocabulary as structural responding. That is, rather the verbal behavior is to lead to effective behaving not a simple litany of linguistic meanings. This is accomplished by arranging classroom instruction such that individual students master vocabulary at their own speed, moving on to the next list of words at their own rate of mastery and using the vocabulary as a *functional* repertoire. In earlier chapters we described the vocabulary of the teacher as scientist and how that vocabulary became functional in the analytic repertoire. The vocabulary is initially an intraverbal repertoire and a tact repertoire and becomes functional when the vocabulary functions in a verbally governed capacity in both the reader and the writer modes. This will require the teacher to design the vocabulary for large blocks of work in advance. As the students master the component skills of disciplines, they are introduced to more complex verbally governed repertoires that build on fluent intraverbal and tact repertoires that constituted the vocabulary. Two basic tactics are used to increase the rate of vocabulary learn units mastered by the students—peer tutoring and self-instruction. Of course, other tactics from the science will be used on an individual basis as needed.

Tutoring

First, *classroom-wide tutoring* increases the learn units received (Greenwood, Delquardri, & Hall, 1989; Heward, Heron, Hill, & Trapp-Porter, 1984). Teaching to present learn units to each other (i.e., reinforcement with tokens, approval, and full corrections requiring the tutee to respond correctly to the word while viewing the word) provides incremental increases in learn units as well. Teach and monitor the tutors' delivery and *recording* of correct and incorrect responses, as well as start times and stop times. Tutees will plot their own rate correct/incorrect. (The monitoring component is part of the students' self-management and self-instruction curriculum, as well as part of the mathematical and scientific problem-solving repertoires to be described later.)

Next, as the students gain independence on the self-monitoring objectives, they receive vocabulary lists that they study (using self-study guides prepared in advance by the teacher and learner-generated learn units) until they are ready for mastery/fluency exams with the teacher or tutor. The students will time and graph their study exercises themselves. In this second tactic, the students move to PSI vocabulary self-instruction. Responding should be written rather than vocal in most cases, since that is the form of communication that the student will need. That is, vocabulary definitions and spelling responses are written rather than spoken.

Combining Curriculum with PSI, Tutoring, and Self-Management

In summary, first the students engage in vocabulary instruction via teacher-supervised tutoring. Next, when the students have mastered related self-instructional and tutoring skills, they receive vocabulary assignments through the PSI arrangement and monitor their own progress with increasing independence while exchanging tutor and tutee roles. In the latter procedure, they receive their daily folders, with assigned words to master, along with other instructional objectives for a single period or subject area and eventually in a progressive fashion all of the instructional objectives for the day. Each student moves through their own units of instruction individually. The students tutor themselves using the self-monitoring scheme described.

The teacher role is:

1. *To design* the sequence of instruction and continuously modify the sequence for individual students as needed.
2. *To monitor and reinforce tutoring accuracy* and tutor data collection accuracy and, eventually, self-tutoring accuracy.
3. To identify students who are having difficulty and *provide strategic analyses and tactics from the research literature to eliminate the learning obstacles* when they are encountered.
4. To ensure that *the POINT SYSTEM (or token economy) for the classroom is reliably run at all times*. This latter function is critical since the token economy will provide prosthetic reinforcement that can be delivered by the teacher, the tutor when so assigned, and eventually the students themselves as they acquire learner independence (e.g., expert self-monitoring skills, goal setting repertoires, and self-reinforcement capabilities). The design and management of point systems is described in the section in this chapter devoted to self-management and is still another basic component added to the use of tutoring and PSI. It is applied simultaneously with the use of PSI, tutoring, and self-management instruction.
5. To ensure that the vocabulary becomes part of the student's community of reinforcers by seeing that positive reinforcement is present and that the student makes progress with as few errors as possible.

Thus, the design and teaching procedures for teaching academic literacy combine the method of instructional transmission for literacy and problem solving and part of the second component of the curriculum—the learner independence *and self-instruction curriculum itself*. The function of these procedures, like the ones described in the prior chapters, is to place the student rather than the teacher at the center of the stage. The teacher occasions effective and fluent performance by designing and implementing the classroom contingencies.

Teaching Operations for Fluent Mathematical Computation and the Component Skills of Thinking

The *introduction* of new mathematical operations and increasingly complex computations is presented to the student in *written* format, not *lecture* format. This approach promotes response opportunities that can than be converted into learn unit instruction with less delay than is required by the typical lecture format. The student is learning to respond to written instructions, graphics, and other textual means of introducing new concepts and subject matter. We are teaching our students to flourish as their own teachers for a lifestyle of continuous education via distance learning. The use of *written instructions* as the verbal control for learning new material provides the learner with the repertoires that will allow her to teach herself to be *progressively* independent of the presence of a teacher. When students do not yet have the necessary learner fluency to proceed independently, peer tutoring or teacher-intensive interventions are used. The level of independence is determined by the frequency and types of learn units that the student requires. The direct teacher intervention is to be faded as soon as possible and replaced by indirect teacher interventions and the evolving independent repertoires of the students. This is done by following the instructions for identifying learn units described in the chapter devoted to the learn unit (chapter 2).

There are only a few complete sequentially related and empirically tested curricula. For mathematics some of them are the Morningside Generative Model of Instruction curricula, the Saxon curriculum, and the Direct Instruction math curriculum (Johnson & Layng, 1992, 1994). The Saxon materials provide the sequence and the Morningside curriculum provides component skill exercises and fluency criteria that are used to develop fluency in the Saxon or Direct Instruction sequence. For each objective of the curriculum, the mastery level is set for 100% correct responding, followed by a rate (number per minute) or fluency objective per the Morningside curricula. Again the tactics combine trained tutoring (Greer & Polirstok, 1982) and eventually self-instructional methods using the token economy/point system together with the PSI mode of instructional transmission that we describe later.

The Teaching Operations and Classroom Design Operations to Promote Effective Writing Repertoires

The writing repertoire of the student should function to direct a reader to perform verbal and nonverbal operations or to identify the operations that they will need to locate from other written sources. Instruction is designed to teach the writer *how to influence the behavior of a reader*. Some types of writing have *technical* outcomes others have *esthetic* outcomes (Skinner, 1957). Students need to learn both.

The technical outcomes range from preparing a shopping list to describing the steps involved in a complex scientific chain of events (e.g., nuclear fission, evolution). They include descriptions of phenomena as well as steps to follow in assembling electronic devices or mechanical devices or the programming operations for developing software. The writing will involve the use of mathematical representations and scientific methods/tactics, as well as linguistic forms. An example of a scientific statement is:

Efficiency is some measure of benefit divided by cost. (Dawkins, 1996, p. 7)

Still other types of writing are designed to affect emotional responses (i.e., esthetic outcomes). Such writing is different in purpose and requires written responses that affect the readers' emotions (e.g., poetry, fiction forms, advertising, and inspirational writing). The following sentence is an example of verbal behavior designed to affect the emotional response of the reader:

> The world was so recent that many things lacked names, and in order to indicate them it was necessary to point. (Marquez, 1970, p. 1)

Both scientific and esthetic writing require instruction in not only the structural *form* of the writing (the component found in most writing curricula), but also in the *function* of the writing. The function of writing is ignored in most educational curricula. However, instruction that is based on scientific analyses of the *writing function* will make the function the centerpiece of the curricula. In order to teach function, curricula must include the particular antecedent events that evoke writing, including the setting events and motivational variables(i.e., the "need to write"), the writing topography and form, and the effective consequence. Thus, writing is taught as operants. The teaching of writing operants characterizes a functional as opposed to a structural curriculum (see the curriculum chapters for a more extensive treatment of functional curriculum) (Johnson & Layng, 1992). The student is learning repertoires, not inert subject matter (Whitehead, 1929). Just as the students described in chapter 5 learned the functions of communication at the level of acquiring mand or tact repertoires, the student learning to write technical or literary forms must learn the functions of technical and literary writing as repertoires of behavior as they master the component tools of literacy.

Function is the preeminent objective, and *form is learned to follow that function*. Thus, at this point our approach is consistent with the pragmatic function of education advocated by Dewey (1910) and Whitehead (1929). There is, however, *a difference*. The difference is that a science of behavior provides the scientific basis for how the pragmatic repertoires can be taught.

Even though function is preeminent, the student must be fluent with the forms used for function and the component functions needed for advanced functions. The forms or structure of the behavior includes the topography of forming letters or typing and the topography of linguistic structure. As a component of the writing function the teacher provides adequate instruction in fluent form-building exercises. However, as the student acquires fluent component skills for each function, the function is inserted immediately. The forming of letters and words is the component tool for using writing to affect the learning of function that is to affect the behavior of a reader.

In order to learn the *function of writing*, writing instruction is designed to require the reader to affect environmental changes by changing a reader's behavior (i.e., the engineer directs the builder, the comedy writer evokes laughter from the reader). Thus, writing instruction is arranged to engender each of these functions. Of course, the writer must be fluent in the component skills (i.e., penmanship clarity and speed, typing speed and accuracy). A writer who is attempting to affect his target audience must not be disadvantaged by slow component skills. Thus, the writer must write legibly and at a speed commensurate with that of a competent adult or at whatever rate frees the student to functionally affect the reader. Some early writing exercises call for the student to copy text as quickly as she will need to write to compose. Because vocal verbal behavior is always faster than writing, it is particularly important that the writing mode, typing or handwriting, be fast and *automatically accurate*. Each time that the speed of forming the text interferes with the writer's own *writer as own reader repertoire the "train of thought" (really intraverbal chains) is impeded*. Thus, from the outset the student needs to learn both accuracy of form (i.e., the writing forms must be legible) and speed. First, achieve objectives of accuracy, and then achieve objectives of rate of responding, similar to the hierarchy described for mathematics. Once the writing forms are accurate provide exercises to increase the production of the forms of writing or typing such that they are automatic. Of course the particular subject matter of the writing needs to incorporate the verbal communities or vocabularies of each of the disciplines including history, science, and literary writing. Function is the selection by the consequences of the writing that is the effect on the reader's behavior. That is, historical writing affects a reader of history, as scientific writing affects a reader of science, and fictional writing affects a reader of fiction. Each type of writing is for an audience that shares the same verbal community of the writer, and the verbal community of each audience is the discipline. Thus, disciplines are taught as the function of writing.

When the student has the writing topography, instruction in function begins. One of the first steps for teaching function can be undertaken by using written communication as the *only* form of communication for certain periods in the day. To learn this function a strategy is used in which all requests for backup reinforcers used in the point exchange system or classroom materials or privileges is done in *written form only*. Next, the students are required to ask for help from the teacher or tutors using written forms of communication, rather than raising their hands. The initial objectives of writing function require one or two word written mands (e.g., free time, please, help me with subtraction). Initially, a few errors in spelling or syntax are allowed, followed by increasingly stringent requirements for accurate spelling and syntax. As the student's writing evolves, more extensive autoclitics (use of more sophisticated specificity in the writing function) and related structural forms are required to obtain desired items or activities, including correct syntax (e.g., ''I wish to watch the videotape on lions. I will purchase the free time to do so for 100 points.''). Note that the motivational conditions for ''wanting the videotape'' provided the basis for the functional effect the writer seeks—to implore the reader to exchange tokens for the videotape. Correct renditions at the *appropriate response level for the particular student* should result in obtaining the reinforcement from the teacher who *functions as the reader*. Each correct rendition is a correct response to a learn unit; each rendition requiring a rewrite/correction is an incorrect response to a writing learn unit. When the response has incorrect components, the student recycles the response making correction as the teacher's comments specify. Each corrected response is a learn unit. The teacher then requires more elaborate communications commensurate with the student's skill. Teacher *corrections* and prompts to the student occur in *written form*; thus the student learns to respond to verbal antecedents in order to change the teacher's behavior as a reader (Hogin & Greer, 1994). Some vocal prompts will be needed at the outset if the reader is not yet a fluent reader, but progressively, as the student's reading becomes more fluent the teacher's correction consequences to the student writer take written forms. Jadlowski (2000) found that having the target student serve as editor of other students' writing affects the writing of the target student more than having the student's receive the editing process from other pupils or from the teacher.

Other writing exercises may be used in which the student must write directions to a fellow student. The tutor/reader returns the students manuscript to the reader *until* the reader can perform what is asked (Madho, 1997). Alternately, the student may be asked to describe something (i.e., an orange) without using the word *orange*. He rewrites the description until the reader identifies the object. Alternately, the student describes a mathematical concept, scientific construct, or historical construct. The latter is an exercise for learning technical functions. In still other cases the student writes a joke and rewrites the joke until the reader laughs. In the latter case the function is to affect the emotional behavior of the reader. Poetry and short stories also function to affect the emotional response of a reader. Fellow students and the teacher are the audience that the student must influence. Some research suggests that when the student's audience is other students, the instruction is more effective (Jadlowski, 2000).

As the student becomes more effective in writing functions, all assignments associated with all subject areas can require communications that affect the teacher as a reader. This means that the teacher or student acting as editorial tutor *requires rewrites until* the desired effect occurs. The process of rewriting is referred to as *recycling*. Writing must involve rewrites to lead the writer to effective self-editing. If the student writes a report for science social studies, or literature, the report is rewritten until the effect on the reader is as it was intended to be. This process leads to the most critical outcome of writing, which is for the writer to progressively edit his or her responses such that the reader is affected with fewer and fewer rewrites. The student comes to read his or her own writing as the target audience would read it *before* the target audience comes in contact with the text. The process is referred to throughout as *self-editing*. It is both effective writing and effective thinking, where thinking is defined as effective self-editing toward minimizing the number of rewrites required to obtain the desired effects on a teacher/reader. Intervening strategies incorporate other students as tutor/readers, the PSI system, and the token economy or point exchange. In addition specific tactics from the science are applied to individual students as they are needed. These tactics are taken from the applied literature introduced in basic texts in applied behavior analysis and include those listed at the end of this chapter.

The Teaching Repertoires and Classroom Design Operations to Promote Self-Editing Repertoires.

The process of rewriting is a *necessary* component of teaching self-editing. One of the biggest deficits in writing instruction in normative classrooms is that there is little effort devoted to editing or rewriting. The classroom needs to be designed around the frequent use of rewriting exercises. This can be done in conjunction with the token economy *both*:

1. For the student to gain *access to backup reinforcers* and
2. As a *behavior which is reinforced* by tokens.

Initially any rewrite *attempt* receives reinforcement, but gradually reinforcement is obtained only for correct rewrites. Thus, the student's writing behavior is increasingly reinforced for editing and rewriting before submitting the written request. The number of learn units relative to the number of operants decreases as described in the chapter on learn units (Jadlowski, 2000; Madho, 1997; Marsico, 1998). Assignments in various subject areas are used not only to teach accuracy in content but also to teach the desired effects of writing, a process that teaches effective use of content, form, and function, by the effects of the writing on the reader. Effective writing instruction incorporates scientific writing and mathematical computation as well as literary communication. The same effects on the audience are taught in teaching musical composition, writing fiction, or creating visual art products; however, in the latter case the effects are of a different nature than those sought in technical and scientific writing.

DESIGN AND TEACHING OPERATIONS FOR DISCIPLINE-BASED PROBLEM SOLVING

The process of solving problems is described at length in the chapter on analyzing instructional problems (chapter 4, ''The Strategic Analysis of Instruction and Learning''). In that chapter we characterized the repertoires as verbally mediated repertoires.

When we teach students fluent performance of the component skills in any subject area, we are preparing the student for problem solving and ease in acquiring more complex skills that require the tool skills. If the component skills of any given problem area are fluent we increase the probability that the learner will respond to the immediate contingencies. This is analogous to the contingency-shaped repertoires needed by teachers to use behavior analytic teaching skills as automatic responses to student behavior. When teachers are well trained in this repertoire, they solve many problems as a result of their immediate responses to problems on a moment-to-moment basis. However, students like teacher/scientists or other discipline-based professionals will need discipline-based inquiry repertoires (e.g., the analytic repertoires of the teacher described in chapter 4). We call these verbally mediated repertoires. Students who have self-editing repertoires or who are beginning to learn them will need to learn discipline-based verbally mediated repertoires also. The students learn the vocabulary of the discipline, the contingency-shaped repertoires of the discipline (fluency in component skills), and the verbally mediated repertoire of the discipline.

Students who are learning to solve problems mathematically will need to have mathematically mediated repertoires. Here we treat math as a form of verbal communication, but a special form that has utility different from that of linguistic-based verbal behavior. They will need to tact events mathematically (e.g., number–object correspondence, operations). Next, those mathematical tacts will suggest formulas that can be tested by the student to determine their viability. This process needs occur only a few times for each student. For example, we do not want the students to continue to count on their fingers or use number lines after they have learned number and object correspondence as a verbally mediated repertoire. Once this particular problem is solved via math mediation similar *events in the future* will need to evoke contingency-shaped responding. Thus, in order to teach mathematically mediated problem solving, the teacher as instructional designer will need to arrange a series of problems that require math-mediated responding. As the student develops fluency in the component math skills, he or she should be introduced to the next and more complex math-mediated responding. The phenomenon described as adduction results. That is, a student who is fluent in the component skills will master the more complex repertoires with fewer learn units from the teacher (Johnson & Layng, 1992).

In mathematics, word problems are often the primary form used in basal texts to teach problem solving. The texts are not designed to teach problem solving as mathematical forms of verbally mediated responding. The teacher as instructional designer will need to do this. The first requirement is that the students be fluent in all of the component skills. They must have:

1. Fluent intraverbal math repertoires (e.g., fast and accurate basic math facts);
2. A large vocabulary of math tacts (e.g., object/number correspondence);
3. A large vocabulary of math mands (e.g., formula producing repertoires to affect the behavior of fellow mathematicians);
4. Fluent number/formula writing skills (e.g., write numbers/formulas fast, accurate, legible);

5. Fluent language text reading skills and fluent correspondence between language and mathematical operations (e.g., conversion of verbal language textual stimuli to mathematical operations);
6. Fluent skills in identifying the steps for solving the problem; and
7. Frequent learn units that evoke the repertoires of verbal math-mediated behavior.

The process of solving problems mathematically calls for the problem solver first to identify the problem as one that can be solved via mathematical operations. Next the problem must be characterized in the "vocabulary of math." The math "tacts," in turn, evoke math formulas or operations that can potentially solve the problem. The possible operations or formulas can then be "proved" and eventually empirically tested. When proofs or empirical tests show that the problem is not yet solved, other operations are evoked and subsequently tested.

We have described a task analysis of the repertoires of solving problems mathematically. *The process of teaching these repertoires* is similar to that of teaching all of the repertoires described in this chapter. That is, the token PSI point system, combined with progressive independence training based on progressive math fluency, are used to provide learn units that result in response opportunities to problems that can be solved mathematically. The instructional designer will need to do analyses of basal texts and arrange the material in the formats that will provide the component fluency learn units and the problem-solving learn units. Existing strategies and tactics from the science of individualized instruction can be applied to various learning problems encountered by the individual student.

Mathematical problem solving like other modes of problem solving can be categorized in a more generic manner. In the chapters devoted to the behavior analysis and design of curriculum (chapters 7 and 8), we suggest three broad categories of problem solving extrapolated from the writing of Pierce (1935). The categories are true disciplines in that they require that the student be a part of a verbal community. Verbal communities are associated with operations for solving particular types of problems under the verbally mediated repertoires of the very broad disciplines. Pierces' category/disciplines are

1. Historical methods;
2. Logical methods; and
3. Scientific methods.

Mathematics is one of the *logic-based methods* of solving problems. Other logic-based inquiry includes philosophy, scientific interpretations of data, and the related inductive and deductive methods. Issues concerned with the law, historical precedence, politics, artistic analysis, literary criticism, analyses of cultures, esthetic inquiry, and ethics inquiry, to name a few, call for applications of the *historical method*. Scientific methods can be generically characterized as applications of the methods of agreement, concomitant variance, the method of differences, the joint method, and the method of residues. Each subject area of science (e.g., sociology, chemistry, biology, basic and applied behavior analysis, the study of the behavior or the distribution of behavior in populations, physics, psychological constructs) requires different vocabularies but uses the same basic methods of science (i.e., posing verbally mediated questions based on tacts from the relevant verbal communities).

Teaching problem solving as discipline-based inquiry requires the instructional designer to arrange the contingencies that will produce the necessary learn units in a logically or empirically arranged curricular sequence or a sequence that leads to the relevant verbal mediation within the relevant scientific verbal community. The specific verbal community being introduced must be well understood by the instructional designer. Outlining that specific expertise is beyond the scope of this or any other single text. However, the behavior analysis expertise and the pedagogical system described in this text can be used by the subject matter specialist to provide the instructional design and related strategies and tactics to teach students fluent problem-solving repertoires in any specialized verbal community. This level of instructional design is complex. Perhaps this is why the teaching of true discipline-based inquiry is so rarely described in the literature. Typically, this inquiry is developed in apprenticeships with those who have mastered the inquiry and those who are actively engaged in the pursuit of research. Much of this text is devoted to the instructional design and implementation of discipline-based inquiry to solve individualized instructional problems and system-wide instructional problems.

However, teachers who are designing instruction for the development of early repertoires of problem solving can use the operations described in this text for task analyses of discipline-based inquiry. Teachers can teach basic operations associated with basic historical methods, basic logical methods using mathematics or other forms of logic, and basic applications of the scientific method.

The chapters devoted to the design and analysis of curriculum from a behavior selection perspective provide additional treatment of the teaching of problem-solving repertoires. It remains for those who are expert in specific disciplines to draw on the suggestions we have outlined to design, implement, test, and revise curricula to teach specialized and advanced forms of problem solving in their respective verbal communities.

LEARNER-CONTROLLED INSTRUCTION AND TIME MANAGEMENT

One of the major goals of education is to teach students the repertoires that will allow them to instruct themselves in an efficient and effective manner. The achievement of advanced self-editing repertoires is a key step in learning to learn independently of the teacher. Simultaneously students need to learn:

- To *monitor* their own behavior objectively and reliably (measurement);
- To *set* relevant and achievable *long-term goals* and intermediate goals; and
- To determine what and when they should administer reinforcement (events, activities, and leisure that the learner controls) for their own short-term and long-term achievements.

Of course, all three of these goals are initially under the control of verbal rules (i.e., verbally governed responding)—first from others and then under the control of the student's own verbal "rules" (Catania, Matthews, & Shimoff, 1982). As the same problem is encountered the solution responses move from verbally governed to event-governed or contingency-shaped behavior.

A key collection of tactics used to teach these repertoires is a complex token economy. The token system is *a point system for teaching learner independence*. The token economy was developed early on in applied behavior analysis (Ayllon, Layman, & Burke, 1972). It was used as a means of reinforcing behaviors via generalized reinforcers. Later the notion of the token economy was extended to a point system where points where substituted for tokens. For individuals with more advanced verbal repertoires, the point system can function as a system of generalized reinforcers, too. But we shall describe how we can also use the point system to teach the three repertoires of self-management (Greer, 1980).

In the 1970s and 1980s, we (Greer, 1980) combined point systems with PSI for teaching music repertoires to elementary and secondary students. The point system was used in ways that allowed the student to progressively take charge of the measurement of their own learning and to determine what those points were to be exchanged for and when to exchange them. This research and that of others (Mithaug, 1993) has shown that when students set their own goals, determine their reinforcers, and measure their own progress, they progress more quickly. However, we suspect that the latter benefit occurs only as the student gains the necessary verbally governed repertoires to reap the benefits of self-management. Studies and wide-scale applications demonstrated and replicated the viability of the procedures for teaching learner independence and the repertoires students need to teach their own target goals to themselves (see Greer, 1980, for a summary of that research and a list of the references). Subsequent applications of this research showed that the opportunity to receive instruction or to teach oneself could become a back up reinforcer for learning another task. We called the procedure "A Comprehensive System of Personalized Instruction for Teaching Comprehensive Musicianship." It became one of the predecessors of CABAS (a behavioral model of schooling called the Comprehensive Application of Behavior Analysis to Schooling). For the purposes of this book. that research and related applications provided the tested operations for teaching self-management and learner independence. The first key behavioral strategy was the token economy; this was then extended to the point system, and we then combined the point system with PSI. Finally, we used the new combination of behavior procedures to teach the steps to learner independence.

A Comprehensive Token Economy/Point System

The comprehensive nature of the token economy/point system is critical for all instruction. We present it here not only for reading, writing, and mathematical literacy but for the teaching of problem solving and learner independence. *The point system is used not only for the particular subject matter and problem-solving activities, but it is a critical component of teaching self-instruction and self-management repertoires.* That is, the point system is not only a means for delivering generalized reinforcers but it is the basis for teaching learner independence.

For younger students or students who are entering the self-editing stages of instruction, actual tokens will be delivered. These can be any item that can be accumulated in a container on the student's desk (but never items that can be counterfeited). The container reminds the student of how well he or she is doing from moment to moment, a kind of early form of teaching him to self-monitor. As students develop more self-editing independence (i.e., verbally governed behavior), the tokens shift to a tally of points. The points can be recorded in each student's log:

- First by the teacher using a special marker;
- Then by the tutor and teacher; and
- Finally by the students themselves and the teacher using separate marking devices for the teacher and the students themselves.

Tokens and points are distributed for following classroom rules and appropriate social behavior as well as each correct response. However, tokens *cannot be substituted for counts of accurate student performance or inaccurate student performance* (these are best calculated and graphed as number per unit of time, usually number per minute). The teacher will want to provide reinforcement for students for moment-to-moment learning outcomes (i.e., correct/incorrect rates) and decrease reinforcement for social behavior as the students come increasingly under the control of their accomplishments. The teacher for each student will determine this evolution individually.

Tokens and points are exchanged for backup reinforcers. There must be an extensive array of backup reinforcers. What these are will depend on the individual composition of the class. The research shows that they can be classroom activities, free time in the reinforcement area (a good idea), preferred activities, events, or items, and other instructional activities. Items may be *rented rather than sold.* The token economy may need to be continually enhanced for students to spend their tokens, because the work itself will require reinforcement effects. So, in order to maintain the use of the token economy for self-management purposes, you may want to have auctions as an establishing operation, where the student can bid against their peers for special item rentals or events. The important principle is that students earn points for *their performance at their particular level*. For example, a more advanced student will earn points for doing all of the vocabulary at adult speeds, while a less advanced student will earn points for each correct word in excess of his prior performance. Students who are performing at different achievement levels will be working side by side in the same classroom. Both the core curricula and the token economy are adjusted for each student's ability level (i.e., principle of successive approximation). The only *competition that you are teaching the student is self-competition, because self-competition is a necessary component of learner independence*.

The three components of learner independence involve self-monitoring, goal setting, and self-administration of reinforcement or its analog (Malott & Heward, 1995). That is, by the time the student is performing many operants before receiving a learn unit, the responses of the operants are automatically reinforced by the outcomes of the responses themselves. The points in a token economy act as possible reinforcers when the automatic consequences are not reinforcing (i.e., at the point where the student requires a learn unit). When the learner does not need the points or tokens to respond he or she still receives them, but in the latter case the points are used to teach learner independence and self-management.

A key component of self-management is one's own skills to reliably monitor one's own skills and to measure one's own behavior. Students are taught this important repertoire first by teaching students to monitor the behavior of others, and as they become proficient at monitoring others, we then teach them to monitor their own behaviors reliably. The following outline provides an example of how to arrange a series of instructional objectives that lead to competence in self-instruction.

An Exemplar Sequence for Teaching Self-Management

Basic Self-Monitoring

A. Obtain agreement between teacher measurement and the student's self-monitoring/measurement.
   1. Obtain agreement with *teacher-dictated answers* (e.g., correct answers to a spelling test scored by the target student).
      a. Observe all student marks for the student's own correct/incorrect responses in a single subject (e.g., spelling). The first short-term objective (STO 1) calls for 100% agreement across three consecutive sessions. There should be no *instances of unsupervised erasures or mark-throughs*.
      b. Obtain 100% agreement across two additional subjects (e.g., mathematics, global studies) for two consecutive sessions for each subject.

    c. Arrange periodic checks on the student's "reliability" at monitoring her own performance. Each probe observation by the teacher that is undetected by the student (e.g., the student is unaware of the teacher monitoring the student's responses) that results in 100% agreement constitutes achievement of maintenance STO. Sessions with less than 100% agreement result in a return to instruction at STO 1 and STO 2 levels (e.g., recycle).

    d. Continue until three consecutive maintenance probes are achieved that do not result in a recycle (e.g., 100% agreement on maintenance probes following the achievement of STO. In addition, the student should be self-monitoring all of his or her relevant academic responding by this time.

Self-Monitoring Using Written/Textual Forms

A. Check concurrently the student's answer correction behaviors (STO 1). When the student agrees 100% with the staff members while the student is directly observed to be marking her/his own papers on two consecutive sessions, he or she may be recorded as achieving the first STO. As in the case of all instruction devoted to monitoring objectives, the student is given no feedback concerning the agreement *until the session is completed*. The teacher may and should monitor training using prompts. The student is then graduated from prompts by fading the prompts until the conditions of STO 1 are met.

B. (STO 2). Monitor student papers completely marked by the student. Mark-throughs or erasures by the student are treated as errors. When the student's own paper corrections are errorless for a preset number of occasions the STO 2 is achieved (e.g., 100% agreement with the staff on two consecutive sessions). Permission for making changes must be initialed by a staff member. At more advanced levels students may provide footnotes that describe the rationale for erasures or mark-throughs.

C. Monitor the student for more extended evidence that she or he is reliable at checking his or her own work with an answer form. The terminal product agreement objective should demonstrate that the student is a reliable recorder of his or her correct and incorrect responses for checking extended performance products against an answer form from the teacher. For example the student might provide five consecutive paper corrections at 100% accuracy for probe sessions meeting the criterion specified in STO 2.

Learning to Monitor the Responses of Other Students Reliably

A. The student is taught to collect reliable data while the teacher is tutoring and collecting data on another student. The target student independently collects data simultaneously with the teacher. (The STO 1 might be 100% point-by-point agreement on five consecutive sessions.)

B. The student then functions as the tutor and data collector while the teacher functions as the reliability observer. (The STO 2 might be the achievement of 100% agreement on five consecutive sessions with checks done at the end of the sessions.)

Teaching Students to Graph and Evaluate Their Observations and Combining Self-Measurement with Literacy Instruction

The following are possible steps for teaching students to graph their own performance on self-monitoring tasks.

A. First, it is best to teach the student to calculate and plot rate of correct and incorrect responding as part of mathematics instruction. The student learns to accurately and neatly plot his/her data on the relevant graph under the supervision of the teacher for the consecutive learn units associated with the first objective listed under the first self-monitoring objective. Neatly means that the student used a straight edge to connect the data points, a pencil, and a closed circle at the precise data point. Note that when the student plots the data accurately and accurately monitors his or her own behavior the result is a correct response to a learn unit in self-monitoring and a learn unit in mathematics. Thus, achieving a preset number of correct responses to these learn units *results in a count of two objectives*—one under self-management and one under mathematics. Note also that the student has engaged in repertoires that will later allow him to solve scientific and mathematical problems associated with solving problems. Thus, we have a self-management objective, a literacy objective, and a problem-solving objective. Additionally, when the instruction is controlled by positive reinforcement we act to expand the student's community of reinforcers.

B. The student learns to graph the performance of his or her tutee. The tutor graphs the performance of the tutee for five consecutive sessions on the tutee's graph under the supervision of the teacher, but does so independently, according to the criterion set in the self-graphing objective (e.g., neatness and accuracy).

C. The student then learns to graph his or her own performance in math, reading, social studies, and self-management daily for 2 consecutive days at 100% agreement (STO 1) according to independent checks by the teacher. STO 2 would incorporate the student graphing her daily self-monitoring for 5 consecutive days; STO 3 might include the student graphing her own response accurately for 10 consecutive days at 100% agreement. STO 3 would call for the student to accurately graph his or her behavior for 1 month. Again, two objective counts may be recorded, one for math and one for self-management.

D. After the student has met the STO 3, he or she is placed on maintenance. Thereafter, each time the student is accurate on teacher probe checks of self-maintenance a maintenance objective is counted. Student errors result in recycling to an earlier STO.

Recording Learn Units for Self-Management or Learner-Controlled Self-Instruction

The section above provides a brief outline of the sequence that may be used to teach self-monitoring. Concomitantly, the students are taught to set their own goals and subobjectives associated with that repertoire, to determine the number of points for each accomplishment, and to determine the backup reinforcers for which they will spend their points. Setting goals and reinforcing one's own accomplishments is broken down into subobjectives also. In this manner, the teacher arranges a systematic curriculum for teaching the setting of goals and self-reinforcement.

By this point in the instruction, it is likely that the results of self-measurement and goal setting will consist of operants that are reinforcing in and of themselves. The delivery of backup reinforcers will probably not be the functional reinforcers. But teaching the student to exchange points for items or activities is done to teach time management for work and leisure arrangements. The goal is for the student to be individually engaged in learning activities dispersed across periods of time with newly conditioned reinforcers. In this manner we provide instructional contingencies to monitor and expand the community of constructive reinforcers for the student.

DESIGN AND TEACHING OPERATIONS FOR EXPANDING THE STUDENTS' COMMUNITY OF REINFORCERS

Prior sections described the design and implementation of instruction for teaching academic literacy and fluency, self-management and learner-controlled instruction, and discipline-based problem-solving repertoires. We provided the design for arranging the contingencies for expanding the student's community of reinforcers. If the repertoires of self-management are to be maintained and used in new settings that the student has not encountered, all of these repertoires need to be taught in the appropriate motivational settings (i.e., situations the student needs to know in order to be successful). In addition, if the students acquire the components of the repertoires as true operants, they will build new reinforcers. The *treatment of conditioned reinforcers and the process of the conditioning new reinforcers* has not been extensive in mainstream behavior analysis. However, there is a rather large body of research on conditioned reinforcement as an applied instructional objective in the fields of the psychology of music from a behavior analytic perspective (Greer, 1980; Madsen, Greer, & Madsen, 1975). That research provides a reasonable scientific basis for the tactics and strategies presented in this chapter.

The use of positive reinforcement operations as part of frequent learn unit opportunities implicitly provides operations to condition the effects of behavior as reinforcement. In our early research on expanding music preferences (i.e., new conditioned reinforcement), we found that if we paired approval and unconditioned reinforcers with listening to initially nonpreferred music with individuals from 2 years of age through college age, they would choose to listen to that music as a function of the pairing procedure. Moreover, when the students learned to prefer music as a function of the conditioning operations *before* being taught discriminations about the music, the students required half as many learn units to learn specific discriminations in the music than those students who were taught the discriminations before the reinforcement value was taught. As we developed the CABAS preschools (a comprehensive behavioral model of schooling that we will describe later in detail), we applied the same procedures to looking at books and listening to stories. The students typically increased their free time choice of looking at books after having reinforcers paired with looking at books and listening to stories. In the previous chapter we described the teaching and measurement operations for doing this.

The process that we introduced systematically in the CABAS schools naturally occurs or is supplemented in many homes where parents read to their children in settings that function to condition the texts and print stimuli as reinforcement. The actual behaviors that are reinforced are the looking and listening behaviors (i.e., observing behaviors) and the stimuli of the print and other stimuli in books. The procedures for educating parents that we describe in subsequent chapters describe how to teach parents to do this using systematic behavioral teaching operations. Reading research demonstrates that the frequency with which parents read to their children and the number of books found in the home correlate positively and significantly with children's reading success in school. Other research found a strong relationship between numbers of language interactions in the home and children's subsequent vocabulary (Hart & Risley, 1996). Although it is not clearly specified in their text, it is not unreasonable to postulate that the interactions are very close to learn units. Thus, more "language" learn units in the home leads to better vocabularies. One of their other findings concerned the importance of the positive nature of the interactions. That is, children in homes with more language interactions also had interactions that where characterized as positive. This meant that there were more approvals and physical reinforcements in homes with high numbers of language interactions. It is likely that language is being simultaneously conditioned as automatic reinforcers (Skinner, 1957).

Thus, the lesson that we obtain from these studies and numerous others is that the *teaching process should be characterized by the occurrence of frequent positive reinforcement and the avoidance of coercion*. The instructional operations that we have described throughout the text are built on the use of positive reinforcement and learn units as positive interactions. If learn units are properly designed and implemented, most of the aversive contingencies associated with schooling can be avoided. In CABAS schools the students are eager to come to school. From time to time we encounter difficulties with children when it is time to go home—they want to stay at the school. School is a happy place and the children are happy learners. We measure and document the approach and avoidance behavior of children in our schools with systematic observations of that behavior in behavioral inventories (i.e., portfolios of the reliable attainment of instructional objectives according to the four components of our curriculum). That is, we observe the emotional responses and approach behaviors of children to school in the morning as well as free time engagement with newly conditioned educational reinforcers.

Data on emotional responses to arriving at school, truancy, absenteeism, and free time behavior can provide the relevant information on whether or not the school setting and activities reinforce approach and observing behaviors. If there are problems, the instructional designer should pay close attention to the numbers of positive reinforcement operations received by an individual and the number of coercive interactions. If need be, eliminate the aversive interactions and increase the positive ones.

In a series of studies (Chu, 1998; Kelly, 1994; Kelly & Greer, 1996; Martinez & Greer, 1997), we found that providing frequent learn units or conversational units was effective in reducing or eliminating self-injurious and assaultive behavior with young autistic children and with adolescents with autism who had high incidences of aberrant behaviors. DRO operations were also successful, but were much less effective than learn units, probably because we could generate more reinforcement with learn unit presentations than can be done with DRO operations. Martinez (1997) applied the matching law formula to data on the maladaptive behavior, learn units presentations and responses, DRO, and measures of the identified reinforcers for the maladaptive behaviors. All of the behaviors and reinforcements matched very closely. Chu (1998) also fit the matching law to the aberrant behaviors of 4-year-old autistic children and found that teaching the children to mand in social interactions lowered the maladaptive behavior and its reinforcement effects. Teaching a social skill procedure was also effective, but not as effective as teaching the mand in social contexts. More reinforcement for social behavior accrued as a function of social manding. The manding evoked more reinforcement than the maladaptive behavior and served to displace the maladaptive behaviors.

The above findings are consistent with those of Sulzer-Azaroff and Mayer (1986) who found that schools with positive reinforcement had lower incidences of truancy and vandalism. In addition, school that produced better achievement had lower incidences of truancy and vandalism. Coercive procedures are prevalent in nonbehavioral schools, and coercive behaviors beget other coercive behaviors (Polirstok & Greer, 1977; Patterson, 1982). Positive reinforcement operations beget nonviolent but effective means of interaction.

In summary, the accrual of positive reinforcement pairings and the *development of reinforcement obtaining responses through effective and efficient learn unit presentations* are critical to the habilitation of students and to the expansion of communities of reinforcement. The reinforcers from moment to moment "select out" the behaviors of the student. The greater the number of reinforcers and the related S^ds that are in the student's repertoire, the greater the variety of behaviors. When students' are attracted to literature, art, science, music, athletics, games, and puzzles, to name a few behavior reinforcer relations, the competition for maladaptive behavior/reinforcement relationships is intense. When the number of behavior/reinforcer relationships is narrow in range, there is likely to be a higher incidence of maladaptive behavior. While the old cliché "empty hands are the devil's work" seems to express the above statement in a straightforward manner, it is not quite accurate. It is the reinforcement community of the individual, taken together with the existing setting events in the environment, which set the stage for the behaviors that are selected out by the environment. It is not the community of reinforcer/behavior relationships alone, however, that determines aberrant behavior. The setting events, or motivational conditions, act on those reinforcer/behavior relationships, also. A hungry, impoverished, or socially ridiculed student is placed under conditions that immediately enhance behaviors that are not in the best interests of others, but are simultaneously in the best interests of the individual student, who at the moment seeks escape from an aversive setting. However, we as teachers can expand the reinforcer/behavior repertoires and attain school settings that ameliorate the effects of poor social conditions. Perhaps we can even join forces with other professionals to redesign social contingencies to alleviate the kinds of setting events that evoke maladaptive behaviors.

SUMMARY

In this chapter we outlined some of the instructional designs and related operations that are needed to arrange optimum learning conditions for students who are acquiring advanced repertoires of verbal behavior and the related repertoire of managing their time and their own learner-controlled instruction. These teaching operations constitute systematic steps toward teaching *students to learn and to teach themselves with progressive independence from direct teacher interaction*. This process teaches learner independence and simultaneously develops procedures to maintain efficient individualized instruction in classrooms with ever increasing ratios between the number of students and the number of instructional staff. Efficient is defined as low instructional costs (e.g., teacher energy and costs) relative to educational outcomes (e.g., fewer learn units to criterion). Thus, as verbal behaviors are mastered and become fluent, the student becomes increasingly independent of specialized teacher intensive interventions. A student who has mastered these repertoires can function effectively in most school settings that do not have teachers with advanced repertoires in behavior analysis. They can learn effectively in school settings and in postschool settings where there is a need for the individual to arrange her own instruction and study. In time they will learn even when the teaching is not expert.

APPENDIX

Tactics for Teaching Advanced Verbal Repertoires

There are numerous tactics from the applied research literature that can be used as standard operations for teaching students who are at the advanced verbal stage. In addition, there are tactics that can be used to solve learning or instructional problems that occur for students who are the subject of this chapter. Many of the tactics listed in chapter 5 can be used also; however, the following list identifies tactics that are particularly relevant to the students who are the subject of this chapter.

Reader/writer learn units	Brief activity deprivation	Time-out toy/item
Response prompts	Time delay (for response)	Token removal
Response delay		Good behavior watch or jewelry
Modeling	Extrinsic stimulus prompt	
Peer reinforcement conditioning	Intrinsic stimulus prompt	Hero group contingency
Visual feedback		Total group contingency
Generalized reinforcers	Contingent reinforcement	Trained behavioral tutoring (peer, cross age, whole class)
Tokens	Planned ignoring	
Token economy	Contingent observation	Interspersal of known items
Preferred activities reinforcers	Brief time-out from reinforcement	Response replacement

CABAS

Self-monitoring

Game instruction for social skills

Trained rule-governed responding

Teacher performance rate accuracy observation

Supervisor rate

Data-based teacher decision analysis

DRO, DRA, DRH, DRL

Overcorrection

General case and multiple exemplar

Shaping

Self-management of reinforcers/punishers

Matching law and learn units

Chaining

Textual prompts for intraverbal vocalizations

Reinforcement sampling

Thinning reinforcement using changes in backup reinforcer costs

Rate-of-responding criterion

Mastery learning (accuracy with a rate criterion or without)

Personalized System of Instruction (Keller plan)

Graduated response requirements

Mand tact training procedure

Tolerance for delay in reinforcement

Tolerance for noncontingent punishment

Written prompts for self-management

Writer immersion (all interaction in class occurs in written form)

Generative rate criterion for complex tasks

Approval and ignoring as tactics to teach following rules

Opportunity to respond (as part of learn unit)

Stimulus fading or stimulus shaping

Good behavior game

Lottery to staff for awards

Backup reinforcement rentals

Backup reinforcement auction

Teacher nonvocal approval

Programmed Instruction

Pyramid training of staff

Contingency contracting

Programming across environments

Escape extinction

Home-to-school and school-to-home token economies

Tutoring training to increase achievement of tutors

Peer conditioning procedure to condition tokens/points

Flash card instruction rate training

Video feedback

Vicarious reinforcement

Self-recording to decrease/increase behaviors

Social praise to condition reinforcement value

Increase learn units to decrease inappropriate classroom behavior

Continent music listening as reinforcement

Contingent music instruction to increase reading achievement

Prior reinforcement conditioning to

decrease learn units to criterion

Timing to increase performance rate

Independence of selection and production responses (multiple choice versus production)

Contingent access to tutoring opportunities

Extinction bursts

Stimulus satiation

Response satiation

Satiation to shift activity reinforcement using low preference activities

Verbal and nonverbal correspondence training

Student reinforcement

Student reinforcement dispensation

Self-prompting rule-governed responding

Contingency-shaped versus rule-governed responding

Time-out ribbon

Self-time-out

Desensitization– relaxation training

Social skills training procedure

Students recruiting reinforcement from teachers or peers

Student shaping teacher behavior

Doomsday contingencies in PSI

Peer tutoring in group contingencies

Response cost

Contingent free time

Reinforcement of novel responding to teach creativity

Contingent instruction

Teaching alternative verbal behavior to decrease

inappropriate verbal behavior (inaccurate tacts or assaultive verbal behavior)

Textural control to develop appropriate intraverbals

Contingent edibles

Variable interval schedules for academic performance

Role of S^d in learn unit

Scripts for peer interaction

Response cards as prompts

Learn units for delayed reinforcement in self-management curriculum

High probability response to increase social interactions

Task demand and escape

Operations to program generalizations

References

Ayllon, T., Layman, D., & Burke, S. (1972). Disruptive behavior and reinforcement of academic performance. *Psychological Record, 22*, 315–323.
Catania, A. C., Matthews, B. A., & Shimoff, E. (1982). Instructed versus shaped human verbal behavior: Interactions with nonverbal responding. *Journal of the Experimental Analysis of Behavior, 38*, 233–248.

Chu, H. (1998). *Functional relations between verbal behavior or social skills training, and aberrant behaviors of young autistic children*. Unpublished Ph.D. dissertation, Columbia University, New York.

Dawkins, R. (1996). *Climbing mount improbable*. London: Penguin Books.

Dewey, J. (1910). *How we think*. New York: Heath.

Greenwood, C. R., Delquardri, J., & Hall, R. V. (1989). Longitudinal effects of classwide peer tutoring. *Journal of Educational Psychology, 81*, 371–383.

Greer, R. D. (1980). *Design for music learning*. New York: Teachers College Press.

Greer, R. D., & Polirstok, S. R. (1982). Collateral gains and short term maintenance in reading and on-task responses by inner city adolescents as a function of their use of social reinforcement while tutoring. *Journal of Applied Behavior Analysis, 15*, 123–139.

Hart, B., & Risley, T. (1996). *Meaningful differences in the everyday life of America's children*. New York: Paul Brookes.

Heward, W. L., Heron, T. E., Hill, D. S., & Trapp-Porter, J. (Eds.). (1984). *Focus on behavior analysis in education*. Columbus, OH: Merrill.

Hogin, S., & Greer, R. D. (1994, March). *CABAS for students with early self-editing repertoires*. Paper presented at the annual convention of The International Behaviorology Association, Guanajuato, Mexico.

Jadlowski, S. M. (2000). *A functional analysis of the effects of peer editing and teacher editing on acquisition of self-editing skills*. Unpublished doctoral dissertation, Columbia University, New York.

Johnson, K. R., & Layng, T. V. J. (1992). Breaking the structuralist barrier: Literacy and numeracy with fluency. *American Psychologist, 47*(11), 1475–1490.

Johnson, K. R., & Layng, T. V. (1994). The Morningside Model of Generative Instruction. In I. R. Gardner et al. (Ed.), *Behavior analysis in education: Focus on measurably superior instruction* (pp. 283–320). Pacific Groves, CA: Brooks/Cole.

Keller, F. S., & Schoenfeld, W. N. (1950). *Principles of psychology*. New York: Appleton-Century-Crofts.

Keller, F. S. (1968). Goodbye teacher . . . *Journal of Applied Behavior Analysis, 1*, 79–90.

Kelly, T. M. (1994). *Functional relations between numbers of learn unit presentations and emissions of self-injurious and assaultive behavior*. Unpublished dissertation, Columbia University, New York.

Kelly, T. M., & Greer, R. D. (1996). *Functional relationships between learn units and maladaptive behavior*. Manuscript submitted for publication.

Madho, V. (1997). *The effects of the response of a reader on the writing effectiveness of children with development delays*. Unpublished Ph.D. dissertation, Columbia University, New York.

Madsen, C. K., Greer, R. D., & Madsen, C. H. (Eds.). (1975). *Research in music behavior*. New York: Teachers College Press.

Mallott, R., & Heward, W. (1995). Saving the world by teaching behavior analysis. *The Behavior Analyst, 18*, 342–354.

Marquez, G. G. (1970). *One hundred years of solitude*. New York: Harper.

Marsico, M. J. (1998). *Textual stimulus control of independent math performance and generalization to reading* [on-line]&l;. Doctoral dissertation, Columbia University, New York. Abstract from: UMI Proquest Digital Dissertations. Dissertations Abstracts Item: AAT 9822227.

Martinez, R., (1997). *Reducing aberrant behaviors of autistic students through efficient instruction: A case of matching in the single alternative environment*. Unpublished doctoral dissertation, Columbia University, New York.

Mithaug, D. E. (1993). *Self-regulation theory: How optimal adjustment maximizes gain*. Westport, CT: Praeger.

Patterson, G. R. (1982). *Coercive processes*. Eugene, OR: Castela.

Pierce, C. S. (1935). Collected papers of Charles Saunders Pierce. In C. Hartshorn and P. Weiss (Eds.), *Pragmatics and pragmatism* (Vol. V). Cambridge, MA: Harvard University Press.

Polirstok, S. R., & Greer, R. D. (1977). Remediation of mutually aversive interactions between a problem student and four teachers by teaching the student to reinforce her teachers. *Journal of Applied Behavior Analysis, 19*, 321–344.

Skinner, B. F. (1957). *Verbal behavior.* Cambridge, MA: B. F. Skinner Foundation.

Sulzer-Azaroff, B., & Mayer, R. G. (1986). *Achieving educational excellence.* New York: Holt, Rinehart & Winston.

Whitehead, A. E. (1929). *The aims of education.* New York: Philosophy Library.z

Functional Repertoires: Curricula from the Perspective of Behavior Selection and Verbal Behavior

Behavioral Selection and the Content of Curriculum

TERMS AND CONSTRUCTS TO MASTER

- Behavioral selection and curriculum
- Programmed Instruction
- Direct Instruction and direct instruction (Direct Instruction, the curriculum; and direct instruction as a generic procedure)
- Mill's canons of the scientific method
- Verbal behavior as treated by B. F. Skinner
- Linguistic analysis of language
- Functional curricula
- Structural curricula
- Context, setting events, and establishing operations as they relate to curriculum
- Antecedent (as it relates to the S^d in curriculum)
- Consequence (as it relates to curriculum)
- Learn units (as they relate to curriculum)
- Structural approaches to curriculum
- Multiple antecedent control

- Exemplar identification
- Exemplar production/construction
- Pedagogical operations (adequate and inadequate)
- Describe function
- Identification function
- Design of curriculum
- Thinking from a behavioral selection perspective
- Solving problems from a behavioral selection perspective
- Self-motivation from a behavioral selection perspective
- Creative solutions from a behavioral selection perspective
- Independence from a behavioral selection perspective
- Responsibility from a behavioral selection perspective
- Prognosis
- Complex human repertoires
- Pierce's ways of fixing belief

- Method of a priori
- Method of authority as a functional curriculum goal
- Method of science as a functional curriculum goal
- Method of tenacity as a functional curriculum goal
- Method of difference as a functional curriculum goal
- Method of agreement as a functional curriculum goal
- Method of concomitant variables as a functional curriculum goal
- Joint method
- Method of residuals
- Speaker behavior as a curricular goal
- Listener behavior as a curricular goal
- Writer behavior as a curricular goal
- Reader behavior as a curricular goal
- Speaker as listener as a curricular goal
- Writer as reader as a curricular goal
- "Spontaneous" verbal behavior from a behavioral selection perspective
- Nonverbal antecedents and their role in so-called "spontaneous" speech
- Deprivation as a component in functional curricula

- Mands as curricular goals
- Tacts as curricular goals
- Multiple control
- Multiple choice responding as a curricular goal
- Editing functions as curricular goals
- Contingency-shaped behavior as a curricular goal
- Verbally mediated behavior as a curricular goal
- Natural contingencies as a curricular goal
- Intraverbal behavior as a curricular goal
- Autoclitics as a curricular goal
- Social conversational units as a curricular goal
- Academic conversational units as a curricular goal
- Individualization of instruction as a method of teaching
- Independent repertoires as objectives of instruction
- Self-discipline as a functional curriculum objective

TOPICS

Behavioral Selection and Curriculum Analysis

Function versus Structure

The Functions of Academic Responses

Curriculum Design for Complex Human Behavior

Contributions of Verbal Behavior to Curriculum Design

Individual versus Group Instruction: A Functional Perspective

Natural Fractures in the Educational Curriculum

Summary and Conclusions

BEHAVIOR SELECTION AND
CURRICULUM ANALYSIS

The content of instruction or curriculum can be addressed from the unique and powerful perspective of the philosophy of the science of behavior called *behavior selection* (Donahoe, Burgos, & Palmer, 1993; Donahoe & Palmer, 1994; Greer, 1996a, 1996b). Behavior selection and the science of the behavior of individuals have a history and literature that is closely tied to pedagogy; however, the potential contributions of both the science and the philosophy of the science to curriculum have been overlooked. One indication that the design of curricula can benefit form the perspective of the science comes from the successful effects of the authors of Direct Instruction materials who wedded curriculum analyses to behavioral pedagogy. A large research effort undertaken by the federal government showed that when behavior analysis was applied to the curricula used by teachers students did better then nonbehavioral approaches. However, when carefully analyzed and empirically based curricula were wedded to the use of applied behavior analysis for pedagogy, students did even better (Engelmann & Carnine, 1982; Greer, 1989). Effective pedagogical interventions very quickly lead to the identification of weak curricular design. In this chapter we shall explain why and how the philosophy of behavior selection and the research in verbal behavior provide the means to improve curricula. The behavioral selection perspective is particularly important because the locus of the difficulty experienced by many students is associated with their curricular histories and curricular presentations.

Engelmann, as the principal designer of the curriculum component of Direct Instruction drew on the logic of the canons of science proposed by the English philosopher John Stuart Mill to the analysis of curriculum. The Mill's canons of science summarize the *logic of the experimental method.* (Mill, 1950). Engelmann showed the utility of this scientific logic for curricular design. Still another logic comes from behavior selection and its functionalism as an extension of *pragmatism*. This perspective was clearly presented by B. F. Skinner who addressed communication as a form of operant behavior in his landmark book on verbal behavior (Skinner, 1957).

Skinner introduced behavior selection and functional perspectives to verbal or communicative behavior (Skinner, 1957). His conception differed so radically from the structural or linguistic analysis of language that it was dismissed out of hand by some (Chomsky, 1959), because it was either not read or not understood (MacCorquodale, 1969) (see MacCorquodale, for a treatment of the misunderstanding). Research began to test the validity of Skinner's view of language as verbal behavior in the latter part of the 20th century (Vargas, 1988). Research suggested that certain components of his conception replaced vague and nonoperational terms like "thinking" with operations such as the speaker as his own listener (Lodhi and Greer, 1989) or "spontaneous" speech with verbal operants such as mands, tacts, autoclitics, and intraverbals (Williams & Greer, 1993). While much more needs to be done with an analysis of verbal behavior in terms of the pragmatics of language (e.g., the use of language to communicate), the conceptions found in verbal behavior proved useful to the analysis of curricular functions in a generic way, growing out of the research that resulted from tests of the verbal operants Skinner proposed (Williams & Greer, 1993).

Jointly, the perspectives of verbal behavior, the Mill's canons of science and Pierce's categories of "ways of knowing," provide new and useful approaches to the analysis and design of curriculum. The purpose of this chapter is to describe and elaborate on the behavioral selection perspective and its potential for redesigning the content of what is taught. Many of the points argued in this chapter are tied to findings from the science (e.g., the research in verbal behavior) while some of the content is a theoretical extension of the logic of behavioral selection. This extension, however, is much less speculative than many currently held views of curriculum that are based on a structural rather than a functional philosophy. There are at least four reasons why the behavior selection perspective is more useful than the structuralist view.

First, a behavioral selection approach allows the content of curriculum to include complex forms of human behavior as functions—functions made accessible to instruction. These complex behaviors and their stimulus control have been ignored because they have *not been accessible to treatment as curriculum* (e.g., thinking). This has been the case because these complex behaviors have been treated as inert capacities *rather than presented as operations that incorporate what we know about the verbal and nonverbal operant and the setting events for these operants*. Thus, complaints that our schools have not taught independence, thinking, creative behavior, and self-management are accurate. They are accurate because there have been no effective curricular designs for these important educational goals.

Second, a behavioral selection analysis of curriculum allows one to view the traditional content of curriculum in terms of its operant functions. Our treatment of *function* incorporates *the effects of the behavior*, not simply the topography of what is taught. Moreover, the function includes the context, or setting events, and the antecedents of what is learned as well as the effects of the behavior (i.e., the operant and its setting). A behavioral selection view of content suggests new classes of responses and new sources for improving curriculum.

Third, the components of operations or constructs and the controlling variables of stimuli organize the contents of what is taught in new ways. Thus, curriculum must be considered in light of the types of antecedent control (e.g., verbal, nonverbal, multiple antecedent control), the behavior categories (speaker/listener, writer/reader), and the types of consequences (nonverbal, verbal vocal, verbal scripted).

Finally, the behavior selection perspective provides a basis of individual instruction for large or small groups of learners. The perspective takes into account the importance of teaching *the settings* that will require the *learned contingencies*.

In summary, the behavioral selection perspective on instruction provides:

- A *functional approach* to curricula for complex behavior;
- The consequence as part of *function;*
- The role of the antecedent and setting events or context as an integral part of curriculum; and
- The role of individualized and group instruction as curricula.

FUNCTIONS VERSUS STRUCTURE

The teaching, supervisory, and organizational operations described in this book are the basis of a science of pedagogy. They have been applied successfully to teaching traditionally arranged curricula. Curriculum designers have used a *structural* approach (Johnson & Layng, 1992). The science of pedagogy *is equally applicable* to either structurally based curricula or functionally based curricula. Indeed, the structural analyses and the functional analyses may prove to be complementary, not unlike the complementary perspectives on language afforded by the linguistic view (e.g., structuralism) and the selection perspective of verbal behavior (e.g. functionalism) (Catania, 1998). It is, however, possible that the inductive tests of each approach will identify the true and separate repertoires that result separately from each treatment. This occurred in one experimental comparison of the two approaches to language training (Williams & Greer, 1993). In this experiment, behavioral pedagogy was held constant but curricula based on the two views of languages were rotated. In each view the presumption was that "functional speech" would result from instruction. The result was the structuralist approach to "spontaneous speech" was found to be one kind of verbal behavior, and the responses taught as *verbal operants* resulted in the functions of spontaneous speech. Increasingly, research that draws on perspectives from verbal behavior has shed light on the actual controlling variables for a range of responses (Carr & Durand, 1985; Lamarre & Holland 1985). We believe that this is but the beginning of an expansion of the selection perspective to the complete range of curricula.

THE FUNCTIONS OF ACADEMIC RESPONSES

Students respond to curricular materials by engaging in exemplar identification, exemplar production, or construction. The particular responses for the classes of behavior include matching-to-sample (i.e., put red with red), pointing to (point to red with distracters), circling (i.e., circle red), or checking or underlining (underline red), for exemplar identification; or the production of exemplars by drawing (color in red), writing (write the word for the color), or speaking (say the name for the color). Some have called the exemplar identification "selection-based responding" and the exemplar production "topographically based" responding. As an example of the dilemma raised by the lack of a distinction between these response classes considers the fact that multiple choice responding in exams is a pervasive activity in American schools. Yet in applications of learning to life we need to produce or construct responses. Much instruction is devoted to exercises requiring pointing to, drawing a line between, circling, matching exemplars, or filling in blanks. These latter responses are more like the responses of listening or reading. Such responses are quite different from those that require a writer or speaker to affect the behaviors of a listener or reader by producing behavior that affects other listeners or readers. Writer or speaker requirements are associated more typically with short answer and essay responding. Both responses are needed, but the writer/speaker behaviors receive less instruction and yet they are often the responses most needed to be successful.

All of the above responses can be viewed as functions of verbal behavior—both the listener/reader and the speaker/reader functions. Verbal behavior is *behavior mediated by another;* it is also behavior that has both verbal and nonverbal components. Verbal behavior affects the environment for the speaker/writer through the behavior of the listener/reader. To complicate matters even more, the effects or *results* of the behavior and the antecedents may themselves be verbal or nonverbal behavior. As we shall see later on, the concepts of verbal behavior have far-reaching potential for the analysis of curriculum. For now we shall analyze the current types of curricular responding from the functional perspective of verbal behavior, thus highlighting the differences in the structural and functional views.

Take the case of exemplar identification. If the student is identifying classes of microorganisms, leaf types, or anatomical features of mammals by inspection of the natural environment, the antecedents are likely to be *both* verbal and nonverbal. An initial verbal direction (textural or vocal) sets the stage for a series of nonverbal antecedents. We shall refer to this as multiple control—short for multiple antecedent control. That is, the student reads "locate the mammals" and then must come under the stimulus control of pictures of animals that are mammals, followed by the response of saying or circling. The consequences for the student range from receiving corrections (vocal or textual/verbal behavior usually) as consequences to incorrect responses or praise (vocal or written check marks) from the instructor for correct responding. The target objective is for the student to come under both the nonverbal antecedent control of the object to be categorized and the natural reinforcement of seeing the environment in a new way. The instructor has arranged prosthetic antecedents and consequences to achieve the desired objective. Such prosthetic contingencies include prompts, frequent praise, and corrections that should function as prompts. Still another important component of the scenario is the setting events that function to enhance or detract from the reinforcing consequences (prosthetic or natural) associated with the instruction. What are taught are the setting events and the three-term contingency.

However, *regardless of how effective the instruction* is, if the wrong contingencies are taught, the student will not have learned a useful repertoire.

The value of our behavioral selection perspective in viewing the above instructional scenario is that we can identify what actually is being learned. Given the antecedents and consequences provided and the provision of pedagogically effective instruction, the students will learn and what they learn is exactly what was taught. Thus, if instruction described above occurs, the designer of the curriculum and the instructor must be certain that this result is the target repertoire. Given a particular setting, specific verbal and nonverbal antecedents, and the consequences that reinforce the student, the student will emit the responses consistent with his instructional history. Traditional analysis might presume that something else entirely is being taught. That is, we might have presumed that the structure of the subject matter is learned independently of the functions. That is, after learning to identify a specific stimulus, the student is presumed to have learned how to name it or produce the taught stimulus when it is needed. This error in presuming that the student has learned the "subject" is why some critics have referred to this type of learning as inert knowledge (Whitehead, 1929, p. 342). The reason that it is inert is that the construction function has not been taught under the antecedent and consequent conditions needed to produce the response for the antecedent and consequent conditions called for.

The *behavior selection perspective* assists us to pinpoint the *actual repertoire that will be learned*. This will of course raise the question of whether or not this is indeed the repertoire sought. If the instructor assesses the repertoire by asking students to name prepared slides (e.g., point to, circle the letter associated with the slide), the contingencies should be the same as the instruction. It is important to know that what is taught is no more or less than the designed instruction effectively presented and consequated. *If exemplar identification is a critical repertoire for functioning effectively with the subject matter then that is what is to be taught*. However, if one uses exemplar identification to represent or sample the student's repertoire in describing the concepts (e.g., stimulus categories) taught, then *exemplar identification is not* the repertoire that should be taught. Learning exemplar production does not result necessarily in an exemplar production repertoire. The structuralist perspective holds that the student has learned the *concept when the multiple response is correct*, while the functional perspective tells us that the student should not be expected to have the exemplar production repertoire. Thoroughgoing instruction will teach *all of the antecedent consequent and setting event conditions* under which the student will need to respond to be successful.

If the object of a unit of instruction in geography is to "picture" the shape and boundary nations of a particular country, the learn units and responses should involve constructing a drawing of the country and labeling the boundary countries. Instead, if one asks multiple-choice questions to represent an assessment of the instruction, the assessment repertoire differs from the instructional goal. In the traditional view the student is at fault, but in the selection view the instruction has not taught the desired goal. The first goal is one concerned with production. The second is an intraverbal function (e.g., name countries that border India). The student is not at fault, nor is the instruction if it is done using sound pedagogy, rather the repertoire that was needed was not part of the student's curriculum.

If the objective of instruction is for the student to arrive inductively at, or construct, exemplar categories (e.g., one of science's objectives), teaching repertoires of exemplar identification are not likely to result in the repertoire desired. The selection view provides the designer of instruction with a new view of the components of curriculum: one that allows the redesign of instruction tied to functional effects. When students learn functional repertoires they learn when and how to use the behaviors for specific functions. They have learned dynamic repertoires rather than inert ideas.

If the objective is for the student to *describe* theories and the pros and cons of those theories, then that is what should be taught. Identifying the various theories from an array of alternatives does not teach the describe function and vice versa. The responses that should be taught are written textural responses in which the students performs the *describe* function in a way that effects the reader in the desired way. If, however, *identification* is the long-term goal (e.g., finding the absence or presence of certain cells), then the identification stimulus control will probably not be forthcoming from teaching only the describe function.

Teaching a child to *point to* the color red is a *different* response than teaching a child to *name the color* of the object. Thus, not only do we need teach the qualities of "redness" but also the particular antecedents and consequences. To presume that the student *knows about colors* because he points to or matches colors accurately does not mean that the student will *name the colors* when presented with the color. Both may need to be taught. The false assumption that one antecedent response consequence function ensures that the student *knows* colors, leaves the student who has not been directly taught the naming response with some presumed intellectual deficit. The danger resides in the structural view that that the student *understands* the concept based on teaching and assessment of *only one set of contingencies*. Similarly, the acquisition of an exemplar identification repertoire does not necessarily result in repertoires that produce the exemplar. What is learned is not the subject rather what is learned are the contingencies. When a pilot learns to identify from an array of multiple choices what she is to do in the case of wind shear is not equivalent to responding in the cockpit to wind shear conditions. What is learned are the contingencies that are taught.

Teaching a student certain mathematical constructs should not presume accuracy in mathematical calculations or vice versa. Students who are taught the concepts involved in the multiplication of 4×2 such that they learn to add by summing across *sets* of two is an important *verbally governed mathematical* repertoire. This does not mean that the student has implicitly learned reliable accurate addition repertoires. Both repertories and their interrelationship must be taught. The designer of curriculum should include all of the necessary contingencies associated with the mathematical repertoires that the student will need. Of course, this expanded view calls for teaching the desired antecedents, consequences, and setting events *as well as the response* to each repertoire. By doing so, we teach the function such that the learner is prepared for applications and extensions.

The *settings or context* in which the antecedent and consequence will evoke the response is critical. If the *need to know* (e.g., the motivational conditions of deprivation) is not present or taught along with the three-term contingency, the probability of the response being emitted when the student *needs* the response is not certain. We suggest that this view is a scientific explanation of what Dewey meant when he introduced his views on curriculum (Dewey, 1910, p. 343). Such motivational conditions determine the student's use of the learned repertoire in the context of the "real world." Thus, that aspect of the traditional so called "discovery approach" that has to do with the "need to know" is valid. Fortunately, we need not wait until the actual real-world experience to learn the repertoire; the need can be arranged in an instructionally effective presentation by drawing on tactics that function as establishing operations. Dewey's pragmatic experiential curriculum recognized the "need to know." The science of the behavior of the individual provides the means to incorporate experiences into learn units, now that we know that their needs to be certain establishing operations incorporated in the teaching of the three-term contingency.

In summary, reflections on the typical repertoires taught across various subject matter from the behavioral selection perspective suggest that functions and contingencies must be taken into account in designing curriculum and that much of these contingencies involve arrangements of verbal and nonverbal contingencies. Presumptions about *what has been taught* must be reconsidered. To presume that something has been taught beyond the context of learn unit presentations is a faulty assumption. Rather, an accurate analysis will pinpoint all of the components of the learn unit as the actual result of instruction. This will in turn result in a revision of the curricular content and pedagogy to reflect the objectives actually sought. Some of these considerations are listed in Table 1.

The design of curriculum should take into consideration all of these components for the potential student's operant *as function regardless of the particular topic or logical structure of subject matter*. Useful information can be obtained about a particular student or the design of curricula by grouping student responses by *function* in addition to the logical structure of the subject matter. Some students will perform well across exemplar identification, but perform poorly in *describe* functions. In this case more curricular and learn unit presentations should be devoted to describe functions rather than more attention to the structure of the subject. Most importantly, new sources of errors in curricular design are made available as a result of the selection perspective.

TABLE 1
The Contingencies for the Student Implicit in Typical Classroom
Instruction for Intact Learn Units

Antecedent	Response class	Consequences
• Verbal (vocal/spoken, gestural, written)	• Exemplar identification	• Prosthetic reinforcement (praise, generalized reinforcers, activities/events)
• Nonverbal	• Production	
• Verbal and nonverbal	• Describe	• Corrections
• Multiple antecedents (written, spoken, picture, object, musical).	• Construct	• Natural effects (unconditioned and conditioned)
	• Complete	
	• Determine operations	• reinforcements, punishments as natural effects
• Teacher presented	• Verbally mediated operations	
• Nonteacher textual or other stimulus presentation	• Contingency-shaped operations	
• *All of above in specific settings (e.g., motivational conditions)*	• Say, gesture, write, type, draw	
	• Respond musically (voice, instrument)	
	• Echo (as in the vocal/verbal echoic)	
	• Imitate (nonvocal duplication)	
	• Copy	
	• Respond to verbal (vocal, signed, or written) question	
	• Transcribe	
	• Draw line between	
	• Circle	
	• Check	
	• Underline	
	• Fill in box	
	• Respond fluently to all	

CURRICULUM DESIGN FOR COMPLEX
HUMAN BEHAVIOR

Some of the most important criticisms of those who call for school reform have included the following. Critics maintain that schools *do not to teach students* to

(a) Think for themselves;
(b) Solve problems;

(c) Be self-motivated;

(d) Arrive at creative solutions;

(e) Perform independently of continuous supervision; and

(f) Take responsibility for their products.

These are important concerns, but they are *not operationally explicit*. The use of nonoperational constructs such as short-term or long-term memory deficits, decoding deficits, attention deficits, and expressive speech deficits, to name a few, have a long history in psychology and in the vocabulary of the layperson. While these constructs may be useful to certain research questions, they are not teachable without the specificity that is needed to teach. The use of tautological constructs shifts the locus of a learning problem for the student from instruction to a categorical deficit. The result is that no new instructional solutions will be found nor sought.

Yet, the need for students to achieve the kinds of repertoires described as "the ability to think independently" is critical to the long-term prognosis for our culture and our species. In addition to problems such as illiteracy, the basis for the crisis in schooling is that our students lack complex human repertoires in a world that demands them.

When we replace goals based on nonoperational constructs with goals based on learning the contingencies of the behavior, we provide useful ways to teach. These are alternatives that can lead to curriculum design based on teachable constructs.

What constitutes some behavioral replacements for the notion of thinking? Earlier we described forms of thinking as particular verbal repertoires (i.e., speaker as own listener, writer as own reader). In the chapter on analysis of instruction we showed that within the verbal community of behavior analysis the analytic repertoires of instructional analysis is verbally governed behavior. But there are several verbal communities that are used to solve different kinds of problems. These *various communities of verbal behavior* are the accumulated disciplines of western culture. We shall use one of what are probably several ways of identifying these repertoires into particular classifications.

One potential source for an operational approach to thinking as a particular community of verbal behavior comes from the American philosopher Charles Pierce (1935). According to Pierce, when one goes about solving problems, one goes through what may be described as operations that determine the accuracy and outcomes of propositions (i.e., other verbal behavior). Pierce described this as the methods one goes through in *fixing belief regarding outcomes*. He identified the four methods of fixing belief as:

(a) The method of a priori;

(b) The method of authority;

(c) The method of tenacity (logic); and

(d) The method of science.

Three of these can be described as methods of problem solving, but more importantly they can be made operational as *ways of behaving* using specific operations given certain conditions and a selection perspective.

The method of *a priori* concerns operations that people perform as a result of *a priori* beliefs. That is, the source for the behavior is belief not based on *authority*, not based on *tenacity* (logic), and not based on the *scientific method*. Most of us engage in behavior associated with this method to some extent no matter how much we might wish to deny it. Often these are simply the subtle controls of our particular culture. The process of education is one that replaces the number of behaviors controlled by a priori assumptions with operations associated with the three other forms of fixing belief (e.g., estimating the probability of future events in order to make decisions). Thus, our discussion will be devoted to methods of authority, tenacity, and science since these are the types of "thinking" or verbal communities that our students need. All three of these methods evolved as tools and evolved into disciplines (i.e., verbal communities) for making decisions when one is confronted with certain problems. These verbal communities are the generic disciplines of curricula associated with thinking as forms of verbal behavior. They guide our "belief" in which particular sources of action will work. That is, they are the ways in which we can use other repertoires that we have mastered to solve problems. While the categories are not from the epistemology of behavior selection per se they are from one of the basic foundations of behavior analysis and behavior selection—the philosophy of pragmatism.

We can "objectify" the method of authority as the verbally mediated operations performed by individuals who determine courses of action based on the most useful sources of verbal authority. The appropriate authority is defined for each type of problem and the related "discipline" or verbal community that is to provide the relevant verbally governed operations (e.g., history, law, theology). Authority operations are ones that result in the selection of particular authoritative sources for actions. Finding the solution to particular problem and the *determination of which problems call for the method of authority* is done under the particular events or verbal mediations associated with the authority driving the search for a solution. These are important educational repertoires and serve as the basis for much of what is taught as content in schools (e.g., legal training, history as discipline). Examples include the use of legal precedence in law or the fixing of primary sources in history or looking for the relevant scientific research for the solution to a medical problem. Students learn to use this verbal community to guide their verbal and nonverbal behavior in projects designed to build this repertoire to solve particular problems by drawing on the accumulated authorities of particular subdisciplines.

However, the operations of determining, comparing, and questioning authorities (method of tenacity) are equally as important as learning the content of one or more authorities. Curricula based on task analyses of these operations and related contingencies can be designed to teach these operations as repertoires of function controlled by specific antecedents and consequences. The need for doing so is not a new notion in American education. Dewey and others have made this point but they did so uninformed by a selection perspective and the science of the behavior of the individual (Dewey, 1910). What was lacking was a perspective, which took into account the *contingencies of instruction* that could teach students how to engage in these repertoires given a particular problem. In other words behavior analysis can provide the means to develop the relevant verbal controls needed by a student to solve a particular problem that calls for tenacity, authority, and their combined usage.

The contingencies include the behavioral analysis of curriculum design in this chapter and the pedagogy described in the chapter on teaching. Without this perspective there are no systematic means to teach individuals directly to function effectively given complex problems. We suggest that the deficiency of schools in teaching these repertoires is the absence of an application of the behaviorally based contingencies of instruction as well as the perspective of verbal behavior that Skinner provided and that we extend to the curriculum of problem solving.

The method of tenacity or logic involves the operations of weighing the pros and cons of information and using the results of weighing those pros and cons to arrive at a proposition. Again these are operations that can be designed as functional curricula across various subject matters. While much of curriculum content is devoted to teaching students particular theories, students are expected to "think for themselves" in weighing those theories. If, however, they are to do so, the particular operations and their related contingencies (setting events, antecedents, responses, and consequences) must be taught not as thinking but as task analyzed operations that are programmed into the curriculum across various domains of human behavior (e.g., literature, chemistry, history) as complete contingencies. Much of this repertoire involves the kind of technical writing that an author uses to convince a relevant audience. The verbal community that the writer draws on to convince the audience is the particular vocabulary built up in the subdiscipline that the student is learning.

Finally, the methods of science represent a major grouping of operations for predicting the probability of certain outcomes or in describing the world precisely. In fact, much of this book is devoted to the verbal community of the science of the behavior of the individual applied to pedagogy. Much of the content of curricula studied by students of our science is devoted currently to the identification and description of the result of scientific researches. So much so that the layperson erroneously believes that science consists of the findings of science. Actually, what are being taught are the appropriate authority sources for the current findings of the science not the method of science. But if students are to perform or "think" (e.g., solve problems using scientific methods) as scientists, the task-analyzed contingencies surrounding those operations must be taught. The Mill's canons of science (Mill, 1950) provide a description of the process as:

(a) Method of agreement (if one locates a phenomenon found in common in all instances of the event under investigation, this variable has a higher probability of a possible relationship to the event);

(b) Method of concomitant variance (if one locates a correlation or indices of agreement between occurrences of events and degree of the occurrence of other variables, an even higher probability of a relationship is located);

(c) Method of differences (if one demonstrates through experimentally adding and withdrawing a variable as it relates to the event and finds when all other conditions are constant a *functional relationship* of necessity or sufficiency is demonstrated and this results in the strongest probability of an existing relationship between the variable and the existing event);

(d) The joint method (if one shows that the variable and the event are related at the levels of methods of agreement, concomitant variance and difference, affirmation of the functional relationship are enhanced); and

(e) The method of residues (those remaining sources of variability not associated with the variable tested remain a source for further testing to determine still other relationships).

Each of these may be treated as *functional repertoires* (e.g., repertoires of contingencies) for students to learn and they can be task analyzed or broken down into learn units and subobjectives. As such, the objective and learn units can be taught across subject areas once the desired antecedent, context, and consequences are analyzed and presented sequentially until the repertoire is present for the student. Given a particular problem, the need for solving it, and certain antecedents, the student is taught to perform the specified scientific operations to arrive at a solution. Contemporary behaviorism provides the curricular analysis tools and the pedagogical procedures to analyze and subsequently teach such scientific repertoires. This must be done, however, as is the case with all of these complex repertoires, across particular subject matters or subdisciplines of science (e.g., chemistry, physics, behavior analysis). While the principles of the scientific method remain the same, the practices or tactics vary between the natural sciences, social sciences, and behavior analysis, for examples. The chapter on the analysis of instructional problems (chapter 4) provides numerous examples of the scientific and logical methods of finding solutions in the science of behavior. Similar operations and decision trees may be invoked for each of the scientific disciplines.

The student must determine which of these complex repertoires is needed, from moment to moment, based on what is known and what is accessible at a given time and the time limit in which action must be taken. It is even debatable, although experimentally testable, whether the particular epistemic analyses based on Pierce's work are the best. We can test their validity through good behavioral instruction with its use of research design, and if they are wanting, we can draw on other analyses.

Thus, rather than treating complex human processes as a function of nonoperational constructs such as thinking, we can act on the processes as contingencies and teach them through thoroughgoing curriculum design coupled with adequate pedagogy. We can only do so however if we treat the curriculum content to the kinds of analyses described above.

CONTRIBUTIONS OF VERBAL BEHAVIOR TO CURRICULUM DESIGN

The conceptual scheme found in Skinner's book on verbal behavior (Skinner, 1957) provides still another unique perspective on the complex human behavior of language that was previously treated only from a structural linguistic view. Skinner's functional categories include distinctions between the contingencies surrounding speaker versus listener, writer versus reader, nonverbal control versus verbal control of verbal responses, intraverbal behavior (verbal behavior directly controlled by other verbal behavior), and verbally mediated versus contingency shaped behaviors. His conceptual framework makes still other complex forms of human behavior accessible as operations, rather than leaving their origins to prescientific constructs. Once they are accessible as operations under the control of reinforcement contingencies, they can be incorporated into the curriculum as learn units. The objectives of a verbal behavior curriculum can complement those of traditional treatments of the grammar, syntax, and topography of language. This often involves converting the structural analysis into repertoires of behavior.

Verbal behavior is behavior that involves two or more individuals—a speaker and a listener or a reader and a writer. It is behavior *mediated* by another person. That is, at its basic level verbal behavior *functions for the speaker* as a way for the speaker to act on the nonverbal world and the verbal world indirectly through the behavior of the listener. The evolution of repertoires of verbal behavior allows the speaker or writer to obtain reinforcement without directly acting on the nonverbal environment. One can call, write, or "click" for food and other goods rather than produce them directly. One can make verbal contact with nonverbal behavior of others or one's self and one can manipulate the nonverbal and verbal world. Verbal behavior may be vocal, gestural, or textual (e.g., written). Educational goals often overlook the actual functions of verbal behavior. In retrospect, it is not surprising that the results of current approaches to curriculum are found wanting.

Spontaneous Verbal Behavior

One category of verbal behavior concerns verbal behavior that is under the control of nonverbal antecedents. Pure mands, for example (Skinner, 1957), are verbal operants that are under the control of deprivation (brief or extensive) and the presence or potential presence of the phenomenon wanted (e.g., that which is deprived). The mand *form, or topography*, may involve numerous words or a single word, gesture, or symbol. The mand specifies its reinforcer. Someone under deprivation of water (e.g., after a long run, after a portion of salty food) asks for water: "May I have a glass of water please" or simply points to a pitcher of water. A listener consequates the person manding water with the nonverbal reinforcer—water. The speaker avoids getting up to obtain the water since the person as mediator obtains the water. The listener mediates between the verbal behavior of the speaker and the nonverbal reinforcer. Commands are still other forms of mands as in the case of the teacher who mands, "Do the addition problems on pages 16–18 in your text." The students (e.g., listeners) reinforce the teacher nonverbally by following the mand.

The nonoperational construct of "spontaneous speech" can be replaced by the *functional description of the mand* that specifies the source of the behavior in such a way that the source is accessible. Spontaneous is a term that says the behavior of speaking has an unknown origin (sometimes thought to be a mental function). However, arranging instruction for developmentally disabled students who have little or no spontaneous speech in a manner that allows the nonverbal antecedents to evoke the behavior and consequating the behavior with that specified by the verbal behavior results in repertoires of "spontaneous speaking" (Williams & Greer 1993).

Tacts are still other forms of behavior (e.g., spontaneous speech) that are under the control of nonverbal antecedents. This function of verbal behavior is one in which the speaker makes contact with the nonverbal world. This verbal contact results in a generalized reinforcer from a listener. The speaker says, "There is an indigo bunting!" The listener says, "Oh, yes, I see it, and I have seen it before; but I didn't know the name for that kind of bird." The reinforcer for the tact response is a generalized reinforcer such as attention, praise, or confirmation. By teaching students to be controlled by nonverbal antecedents and the particular consequences associated with mands and tacts, the repertoire sought was actually taught. If the instructional antecedent taught is the verbal query, "what do you want?" the response will not likely come under the control of nonverbal antecedents.

Speaker versus Listener Behavior

The distinction between whether an objective of instruction is to teach a student speaker versus listener behavior is an important one. On the one hand, the purpose of the instruction is for the student to respond to the behavior of a speaker; one set of contingencies is in effect. If, on the other hand, the objective of the instruction is for the student to function as a speaker, then other contingencies are in effect. Responses to verbal commands (e.g., antecedents) such as "look at me" or a series of multistep commands have as their objective for the student as listener to come under the instructional control of the commands. If the objective is for the student to produce certain environmental effects on a listener, however, very different contingencies must be used and taught.

Teaching students to come under the instructional control of the behavior of the teacher as speaker is an important listener repertoire. This listener behavior is often a prerequisite to teaching students some speaker repertoires. Listener repertoires embrace following instructions such as "point to," "find the——," "underline all of the——," or "write your name on your paper," to name a few responses controlled by the teacher as speaker. Much of what is referred to as classroom management or compliance concerns listener behavior. If the students are not under good instructional control, few learn units will be emitted and few objectives obtained. The teaching operations for instituting good listener behavior are described at length in excellent texts that introduce teaching as applied behavior analysis (Sulzer-Azaroff & Mayer, 1986; Cooper, Heron, & Heward, 1987) and in chapter 5 of this text. Scripted curricula in some cases are used to teach these programs to individuals with developmental disabilities and young children. These particular programs include step-by-step objectives and teacher operations for teaching, "sit still," "look at me" (Lovaas, 1977), or "do this," to name a few intentional curricula for teaching listener compliance to verbal commands.

Speaker behavior is, however, verbal behavior in which the speaker uses the verbal medium to manipulate verbal and nonverbal environment through the medium of a listener. It is shaped by the effects that the speaker's behavior has on the listener. In some cases, the listener is the teacher; in other cases, the listener is another student or students. If the speaker behavior is to develop and expand, it must affect listener behavior. If newly developed speaker behavior is not reinforced by the listener, it falls out of the speaker's repertoire particularly in the cases of young children and those with developmental disabilities. If students learn vocal forms of mands and tacts and those mands and tacts are not reinforced by listener responses from teachers, parents, or peers, the repertoire will decline. If teachers do not reinforce students speaking out in class, speaking out will stop. Reinforcing tantrums, noncompliant, or assault responses will teach those responses as forms of mands and they will be used in lieu of standard forms of verbal behavior.

Writer versus Reader Behaviors

Written or textual repertoires have similar distinctions as do speaker and listener repertoires. Reader behavior reinforces the behavior of the writer and *must do so immediately at early stages.* Thus, a curriculum that is designed to teach writing must arrange contingencies such that the function and form of writing is shaped by the natural contingencies of effects on another reader initially. Vargas (1978) has described contingencies that are suited to teaching writing responses. Arranging contingencies of writing that require the writer to describe operations that are to be undertaken by a reader should be followed by having a reader immediately respond to the writer so that the writer's behavior can be shaped by the listeners responses. Unsuccessful written instructions are revised until the reader performs the operations that the writer seeks (see also the research testing J. Vargas' idea by Madho, 1997).

Portions of the school day are arranged such that all communications between teachers and peers require written responses. Doing so increases the probability that writing will come under the controls of appropriate and natural consequences. Similarly, the reader will need to be fluent. Jokes can be written in order to obtain laughter from the reader; instructions in how to play games or how to construct or draw a piece of art work can be arranged. Students can be required to write communications with the teacher concerning questions that they have about instructional material or requests for material or permission to engage in activities. The syntax, spelling, and vocabulary required of the student are adjusted consistent with the student's ability level. More advanced responses will require greater specificity (e.g., what Skinner terms autoclitics), and more accurate spelling and syntax. Of course, the teaching of these repertoires successfully must incorporate the pedagogical operations that we described earlier.

Multiple choice responding has more of the characteristics of listener behavior, while short answer responses or, more specifically, essay responses have more of the characteristics of writer behavior (i.e., an extension of speaker behavior). Thus, to teach good writing (e.g., the lack of this repertoire is a frequent criticism of school critics), the curriculum must provide learn units devoted to writing as a functional repertoire designed to affect the behavior of a reader. One of the problems with standardized tests involving multiple choice responding is that they do not predict the writer repertoire and all that it entails. The writing repertoire is needed for the complex behaviors that are required, if individuals are to function effectively as scientists, engineers, businesspersons, teachers, and private citizens. Substituting multiple-choice responding, because it is less time-consuming for the scorer, will not teach students to construct and affect their environment through writer forms of behavior. Listening repertoires are more characteristically multiple choice in nature, while writer behavior calls for the writer to affect the responding of a reader.

Editing Functions

One is said to edit one's own speaking or writing behavior when one functions as her own reader or listener (Skinner, 1957). Lodhi and Greer (1989) found that young children functioned as their own listener when certain contingencies were arranged. It is likely that the repertoires of self-editing are extensions of this capacity (Jadlowski, 2000; Madho, 1997; Marsico, 1997). More advanced writing repertoires require that the writer function for long periods as his or her own reader. Similarly, the presentation of lectures or speeches requires that an advanced speaker function as her own listener. This function can be intentionally trained by progressively lengthening the writing or speaking requirement before another reader or listener responds. This is, of course, what happens as students are asked to write longer and more complex reports or papers and to read them aloud or have others read them aloud. The notion of training good self-editing skills by design, however, suggests that the length and complexity of responding required to receive learn unit consequation should be based on the careful mastery of shorter units of communication before longer units are undertaken involving someone else as the who is a more expert reader or listener and then the writer as his or her own expert reader. Attention to these functional components of writing and speaking will ensure more adequate command of the repertoires by students. Thus, eventually the student functions more effectively as her own reader/listener as a result of an intentional shaping process, not simply by chance. We described these steps at length in the chapters devoted to teaching students with advanced forms of verbal behavior and in the chapter on instructional analysis. Unless, or until, the speaker and writer behavior controls the behavior of readers or listeners (self or others), the repertoire will not be sufficient for the students' eventual needs for long periods of time in which he or she must function as his or her own listener/reader.

Verbally Mediated versus Contingency-Shaped Behavior

As described earlier in the chapters on teaching, the distinction between contingency-shaped and verbally mediated repertoires is critical. Reading about the operations necessary to accomplish a task results in a different facility with the natural contingencies in the task than does *coming under the control of the nonverbal contingencies*. It is important in some types of instruction for the student to be allowed to come under the control of the natural contingencies. Laboratory courses and practicum courses have that purpose. But, the practicum courses and settings can be designed more effectively to evoke the contingency control desired. Repeatedly emitting responses that do not affect the natural consequences in desired ways will punish the student or teach inappropriate contingency-based responses. For example, students who practice musical instruments do not learn the effective behavior unless the right "musical" consequences control the behavior (Zurcher, 1975). Thus, it is important to arrange or design the instructional environment for achieving contingency-shaped behavior. In still another example, learning to ride horseback is made less difficult for the riding student when the student is assigned the right horse for the existing achievement level of the student. The teaching of contingency-shaped repertoires is not left entirely to chance.

In many cases, it is important to allow the natural contingencies to shape behavior as in the case of someone who is learning to drive an automobile by coming under the control of the contingencies of the road rather than the verbal mediation applied by the instructor. There are important responses associated with almost every endeavor from cooking to brain surgery that should be taught to allow the control of the natural and nonverbal contingencies. Contingency-shaped repertoires must be learned such that they come under the control of natural contingencies. However, this does not mean that learn units are any less important. Airlines have developed flight simulators that not only simulate the natural contingencies but also are capable of programming functions that present learn units. The training procedures for teaching teachers contingency-shaped behavior, described in the chapter on supervision, is still another means of designing programs of instruction for the teaching of contingency-shaped behavior. The teacher's use of verbal behavior such as prompts and social praise are instances of verbal mediation and should be used sparingly. Ideally, programs for shaping contingency-shaped behavior would manipulate the responses of students such that the student responses would shape teacher behavior without recourse to verbal mediation.

Verbally Mediated Behavior

Skinner (1957) used the term *rule-governed behavior*, but it was a term with which he was not entirely satisfied (Skinner, 1984). Earnest Vargas (1988) proposed that the term *verbally governed* is a more appropriate to the effect Skinner was describing and we have used the term verbally mediated together with Vargas term because both terms characterize the controlling stimuli for the class of responses that Skinner described. Verbally governed behavior is behavior that is under the control of verbal contingencies rather than the direct control of natural contingencies. Examples include constructing an assembly according to written or vocal instructions, performing a series of operations controlled by verbal directions (e.g., using methods of authority, methods of tenacity, methods of science, designing curriculum based on verbal contingencies), following a course of ethical behavior in certain settings, and responding to data in a given setting rather than the natural contingencies.

The fact that humans can perform complex operations for the first time by coming under the control of written or verbal behavior allows us to avoid reinventing the wheel, so to speak, every time a wheel is needed. The admonition that "the student is always right" for example can serve as a verbal contingency mediating between the teacher and the student. Another set of verbal statements could also control the teacher's behavior such as "it's the responsibility of the student to learn and the student's learning difficulties are the result of the student's cognitive deficit." Here the particular verbal mediation used by the teacher results in very different courses of action as we described in the chapter on the analysis of teaching and learning difficulties. Verbal mediations govern important repertoires in all professional and person endeavors. Of course, if the verbal mediation (e.g., verbal rule or prescription) does not control responding, it is not truly functioning as a verbal governor. The verbal admonition to take advantage of the art or science programs on television may not exert strong enough control over the contingencies evoked by viewing alternative channels on the television. Setting events or contexts serve to enhance or detract from the verbal and nonverbal contingencies.

Thus, it is important that repertories intended to function as "*verbal governors*" do so. If one simply repeats or duplicates a formula to themselves, a set of operations (e.g., principles of research design, structures of grammar, rules of spelling), the behavior actually being learned is an intraverbal; it is not yet a verbal governor of other behavior. To be verbally governed the verbal behavior must actually control nonverbal behavior (or other forms of verbal behavior) such as prescribing the actual performance of scientific operations, using spelling rules, and rules of grammar for functionally effective behavior consequences. Simply teaching the verbal behavior *about* doing a task (e.g., intraverbal behavior) in response to a question will not alone provide the repertoire. Nor will asking the student to choose a component of or a representative sample of the consequences of following the prescription teach the necessary repertoire. Curricula that teach verbally governed repertoires need to program nonverbal response control resulting from verbal contingencies in a systematic manner.

The use of the term *verbal mediation* in this book to describe one of the major repertoires of teachers who function as strategic scientists is an elaborate extension of the term verbally governed behavior. The teacher views student responses within a given instructional or natural environment. The observation evokes a scientific description (e.g., that behavior should be a scientific tact) rather than the less precise lay term. The verbal description in turn suggests that the difficulty encountered by the student is due to the fact that the response desired is not occurring because the wrong contingencies are in place (e.g., the antecedents are verbal rather than nonverbal or the consequence is changing the tact function to a mand function). This verbal scientific description suggests new teacher operations such as making the nonverbal antecedent salient and avoiding a verbal antecedent, while reinforcing the correct response with a generalized reinforcer. Thus, the teacher's behavior is verbally mediated by verbal behavior about the science. The verbal behavior mediates between the student and teacher resulting in operations that the teacher would not normally do if he were strictly under the control of the behavior of the students.

Examples of verbally mediated responses for students include incidences when students uses rules (e.g., verbal governors) to control their spelling responses (e.g., *i* before *e* except after *c*) or cases in which a student follow sets of operations (e.g., applying particular operations associated with using the method of difference) to the question of whether functional relationships exist between a measure of an environmental event and a set of variables in the environment. In still other situations, the student confronts the question concerning whether a set of behaviors performed in a particular setting is legal or illegal, and the student argues or acts on the basis of authoritative verbal mediators for certain behavior and environmental interaction.

Effective instruction of verbally mediated repertoires will result in verbal behavior which functions as a *tool* for individuals to manipulate the environment in ways that are not possible without the verbal mediation. Much of the present discussion has itself involved verbal mediation in which the writer (e.g., person needing a solution) functions as his or her own listener or reader. This repertoire resulted from someone else initially functioning in the listener role until this writer developed the editing repertoire (Vargas, 1978). But a teacher, as listener or reader, must teach the control of the verbal behavior over the student's speaker/writer behavior until that student's own listener/reader behavior controls the student's speaking or writing. Moreover, a particular context and set of contingencies must evoke the verbal behavior followed by the person performing that verbal behavior. One way to do that is to have the student state or write the verbal behavior in the appropriate settings and then follow that verbal behavior by actions consistent with the verbal behavior. The overt statement of the verbal directions can be faded eventually. The correspondence between saying and doing needs to be taught directly (Baer, 1987).

Those who describe themselves as cognitive behaviorists refer to the learning of the above repertoires described as instances of cognitive learning. But no special cognitive function has been evoked necessarily, rather verbal behavior is controlling the nonverbal behavior of the student in what Skinner described as rule-governed behavior as early as 1957. If the student does not emit the nonverbal behavior, invoking cognition as an explanatory construct, it leaves one with the belief that something is wrong with the cognitive function. No one knows how to affect cognitive functions, but with a verbal behavior perspective we can make the process operational and apply behavioral pedagogy to teach the repertoire. Viewing the same lack of control from a behavioral selection perspective leaves one with a solvable instructional problem; that is, *verbal behavior is not controlling the nonverbal behavior of the student.* The behavior analyst as teacher can go about finding the correct prerequisites and contexts that lead to the student coming under verbal control. The behavior analyst can look to the pedagogical history of the student, the lack of ambiguity (e.g., Engelmann & Carnine, 1982) of the presentation of the curriculum, and the functionality of the presentation as operants. Indeed, the purposes of the behavioral selection and the cognitive psychologists are different; the former seeks instructional control, and the latter, cognition. In one case, behavior and its instructional control are the focus; in the other, behavior and instructional control are simply means to finding a physiological mechanism associated with the environment behavior relationship. The latter search may have important outcomes for neuropsychology and can lead to useful findings that can be converted into behavioral operations for instructional purposes too, but is it wrongheaded if one wishes to locate instructional controls for teaching or research on teaching by referring to cognitive constructs alone.

Intraverbal Behavior

Verbal behavior controlled by other verbal behavior is involved in the description above. *Verbal behavior about subject matter* falls within this category. Its eventual usefulness is, however, tied to the degree with which it is incorporated within verbally mediated repertoires. Autoclitics (a form of intraverbal behavior) specify, locate, affirm, negate, or quantify other forms of verbal behavior. Their eventual utility and maintenance in the student's repertoire will come to rest on how they affect the consequence of verbal behavior.

Conversational Units of Verbal Behavior

The importance of teaching social repertoires to individuals becomes critical when these repertoires are deficient. Emitting verbal behavior for the reinforcement of the verbal behavior of another is an important component of one's lifestyle. Students who are withdrawn or are reluctant to engage in verbal conversational units are not under the reinforcing control of the verbal behavior of others. This functional control can be taught or programmed given the appropriate establishing operations and context (Donley & Greer, 1993). Thus, not only can verbal behavior provide a unique and powerful perspective on academic curriculum design, it can also provide a set of functional repertoires to train social skills that are often essentially verbal. In fact, the best measure of social skills may be conversational units between individuals (Chu, 1998). While many tests of the application of the concepts and operations specified in Skinner's conception of verbal behavior remain to be accomplished with curriculum design, the possibilities for its utility in educating children and adults in the performance of complex human behavior are vastly superior at present to that of invoking the currently used tautological constructs. Verbal behavior may yet prove to be one of the most important innovations in curriculum design in the history of education.

INDIVIDUAL VERSUS GROUP INSTRUCTION: A FUNCTIONAL PERSPECTIVE

The individualization of instruction has played a prominent role in behavioral models of instruction and curriculum (e.g., Programmed Instruction, Precision

Teaching, PSI, Ecobehavioral Analysis, and Applied Behavior Analysis, and CABAS, and The Morningside Generative Model of Instruction) (Greer, 1989). Direct Instruction is also built on the behavior of individuals but provides group settings for programming learn units.

The importance of the individualization of instruction is almost a cliché in the educational literature. Yet, one of the major criticisms against teaching as behavior analysis is the criticism that individualized instruction is just not feasible in public schools because of the size of classrooms. It is just not cost-effective given the current design of American schools (Brophy 1983).

If that were the case, and we do not think it is, then schools should be redesigned such that individualization is cost-effective and feasible. Several examples of the feasibility of comprehensive individualization are given in the studies and summaries of applications of the CABAS (Greer, 1996a, 1996b). There is also extensive research showing the viability of the Personalized System of Instruction to individualize instruction in classrooms for students with self-editing repertoires (Buskist, Cush, & de Grandpre, 1991). There are numerous tactics from applied behavior analysis that can be used to make individualization cost-effective as described throughout this text. There is still another argument for individualization beyond the fact that it is the most effective form of instruction. That is, behaving as an individual is a critical objective of education.

Adults go about doing the business of life as individuals. Scientists, businesspersons, medical doctors, lawyers, architects, teachers, law enforcement officers, word processors, and other professions and endeavors require that much, if not most, of the work is accomplished as individuals. It is only in schools that we engage in work in a subject matter all at the same time. In schools, students wait to begin en masse, they "get ready to get ready" to go to class, go home, go to lunch, go to the buses, and other lockstep movements as groups. In the postschool world, people begin and stop work individually, initiate changes in activity, arrange meetings individually, and in general accomplish tasks individually. It is true that we engage in group sports, play in music ensembles (e.g., individual parts), and socialize in groups. But most of the repertoires that should be taught by schools are ones that are intended to be done by the individual.

It is no wonder, then, that people often encounter difficulties at the individual level with complex or even simple tasks. Moreover, the kind of "self-discipline" (a form of stimulus control relative to ones learned repertoire concerning the correspondence between saying and doing or reading and doing) needed to work individually, as well as locate assistance or sources of information, is not taught in schools geared to teach groups. If schooling is to begin to prepare individuals adequately to support their own habilitation (Hawkins, 1986, p. 343) and culture, schools must be designed to teach through individualization and *to teach independent individual performance as a goal*. Moreover, a critical component of individualization is the individual measurement of performance guided by preset criteria. Teaching continues until measurement shows that the student is fluent.

Students can learn to act independently and to work at their own optimum pace as a natural outgrowth of teaching through individualized instruction in the manner we will describe. Part of this instructional process and curriculum involves shifting part of the responsibility for the measurement of learning from the teacher to the student. These procedures will result in self-evaluation and self-motivation where the notion of "self" is removed from a tautological or circular reasoning and placed in the contingencies of the environment.

Learning to Function in Group Problem Solving

People do need to function as contributing members of groups. When this is a goal for education one should not confuse the goal with the method of instruction. There is an educational literature that has touted the use of group problem-solving assignments as a pedagogical tactic. The tactics that are used are drawn from the applied behavior analysis research on group contingencies. These tactics include hero group contingencies, total group contingencies, and target student within group contingencies, among other variants in which the total group receives reinforcement contingent on the performance of certain or all members of the group. When the group contingencies are effective, they are usually reducible to tutoring interactions between students. The tutoring literature contains tactics that are more effective and ensure that the interactions between peers are positive and arranged to accommodate the performance levels of both tutor and tutee (Axelrod & Greer, 1994).

When the goal of instruction is to teach individuals to function as effective contributors to the solution of group problems, the same individualized instructional operations that are described throughout this text are necessary. Without careful monitoring and the clear delineation of each students learning objectives, there is no way to determine whether or not one or two of the members of the group did most or all of the steps leading to the outcomes of the group endeavor.

NATURAL FRACTURES IN THE EDUCATIONAL CURRICULUM

The unique function of schools is to teach certain repertoires *by design*. In particular, the repertoires are ones that the individual or collective members of the *culture do not acquire without specialized instruction*. Interestingly, the types of repertoires that need to be taught are closely aligned with categories of verbal behavior. The long-term goal of schooling is for the student to function effectively in the world independent of school. There is also a strong relationship between verbal behavior repertoires and self-regulatory and self-management repertoires. Interestingly, the four modes of curriculum (e.g., academic literacy, self-management, problem solving, and an expanded community of reinforcers) are related to the student's advancement in verbal sophistication. Note that the verbal behavior scheme is not interchangeable with linguistic conceptions. That is, it is the way in which verbal behavior functions on or for individuals that ties together the four modes of curriculum.

The stages of verbal behavior suggested in chapters 5 and 6 are useful categories or even *possibly natural fractures* for describing the student's progress along the continuum from dependence to independence. Verbal behavior repertoires are tools that when properly taught result in the mastery of the content of the three modes of the curriculum. The advancement along the verbal behavior continuum suggests certain instructional or pedagogical strategies as well as content. To wit, the lack of listener behavior dictates total dependence on the parent or teacher. Acquisition of listener behavior decreases dependence (e.g., the students can be told to do something for themselves as opposed to being led by the hand).

Speaker behavior allows the student to manipulate the environment in cases of deprivation or to expand other environmental controls that were formerly lacking. When the speaker can function as his or her own listener, alternatives to having to directly experience contingencies are available. The child who speaks to him- or herself as a listener can say "The light is red; wait for the green light." The teacher need not always be present to protect the student from danger through nonverbal experience shaping alone. At the level of writer, the flexibility of manipulating the environment is no longer limited to the movement of verbal behavior with the listener present. The antecedent responses and consequences can be delivered/received across time.

The acquisition of the reader repertoire multiplies the student's options over that of simple speaker or listener behavior. Speaker or listener behavior is anchored to fixed time periods requiring the presence of both listener and speaker. A reader can come under the control of written contingencies any time that the print is present. The writer repertoire provides the individual with the means to affect the behavior of readers. The writer can "speak" to the reader across time or immediately.

Once the writer functions as his or her own reader (e.g., self-editing), the repertoire of speaker as own listener is also multiplied. As the writer, who is his or her own expert reader, draws on the problem-solving repertoires (e.g., authority, logic, science), the effects of those repertories on solving problems become more independent of the presence of all of the contingencies at any given time.

Each time the student adds a repertoire of verbal behavior, or increases his or her sophistication within a repertoire, the student gains a wider voice in the control of his/her life and his/her environment. Each step in the evolution of verbal repertoires prescribes both the goals to teach and the prerequisites for teaching each goal. Progressive advancement in the repertoires requires proportionately less teacher-directed instruction. If the design of curriculum, pedagogy, and the pedagogical support is arranged to address these functions of verbal behavior, the response to learn units will increasingly be more indirect between the student and teacher, and the *number of teacher-directed learn units* will become less frequent. The responses become consequences in the culture at large rather than the teacher's responses. Responses, however, never become independent of consequences; rather, the nature, type, and frequency of the consequences change.

Similarly, the "self" in self-management becomes a progression along the continuum of verbal repertoires, the optimum use of responses vis-à-vis consequences, and the progressive ability to control more of the world in an abstract (e.g., verbally mediated manner) rather than through direct experience or direct manipulation of the contingencies. The advantage of the functional description over the tautological description is that the *functional one identifies what is to be taught and suggests how it can be taught (e.g., the pedagogical science)*.

Another important benefit of a functional curriculum is that instruction is identified based on what students can do now and what students need to do next, rather than on age, gender, ethnic status, disability descriptor, or physiological or genetic factors. Regardless of age, disability, or facility, the repertoires need to be learned, and at some stages they need to be learned in a certain order. A 15-year-old student and an 18-month-old student may require similar instruction, if their presenting repertoires of verbal behavior are the same.

Necessary distinctions in how students are to be instructed should be distinctions in pedagogy and distinctions in the students' location along the curricular continuum. Students without a speaker repertoire require more direct teacher-presented learn units, while students with a strong writer-as-reader repertoire can progress with carefully designed self-instructional materials; thus the number of teachers per student decreases with increases in the verbally governed repertoire of the students. The overall numbers of teachers to students across the entire schooling process may not need to change, but the particular distribution of professionals to students at particular stages of verbal behavior needs to be reorganized. Students who are have few verbal behavior repertoires more intense teacher-controlled instruction. They require more direct teacher-mediated instruction, while more verbally sophisticated students (and typically older students) will require fewer teacher-mediated learn units because the more sophisticated students can provide their own learn units. As soon as the student progress along the self-management or verbally governed curricular continuum, the ratio of teacher to student can increase.

It is suggested that we can determine what is instructionally and hence economically more feasible only by research that identifies relationships between learn units, objectives achieved, and subsequent habilitation. We can do this research by establishing research and demonstration schools to serve as prototypes. As the prototypes establish effective designs and cost/benefit relationship, communities can elect particular design arrangements consistent with the wishes of the community. This may involve more or less money initially, but at each stage the schools will be measurably accountable if the schools use comprehensive measures of learn units, criteria references object-ives consistent with good local and state standards, and standardized tests that use *validated measures*.

SUMMARY AND CONCLUSIONS

This chapter has proposed some of the unique and potentially powerful uses afforded by a behavioral selection perspective on curriculum analyses and design. Complex human behavior can be taught in curricula based on the concepts and distinctions associated with functional analyses of operations, verbal behavior, contingency-shaped behavior, and the distinction between learning to perform in a group versus the instructional processes needed to learn individual repertoires most efficiently. Many of these concepts remain to be tested. They do, however, offer testable and teachable objectives and subobjectives that meet the need of society for schools that teach students to perform complex human behavior independently. These concepts are viable alternatives to the constructs about curricula and schooling that have con-trolled much of what has served as the bases for curriculum design in the past. Functional approaches do not supersede some aspects of structural ap-proaches. Instead they can complement the functional approach in ways that are consistent with the kinds of educational outcomes that were envi-sioned by those associated with progressive education (Dewey, 1910). Thus, contemporary behaviorism has a unique contribution to make to curricular design in ways that we have only begun to realize.

References

Axelrod, S., & Greer, R. D. (1994). A commentary on cooperative learning. *Journal of Behavioral Education, 4*, 41–48.
Baer, D. M. (1987). Weak contingencies, strong contingencies, and many behaviors to change. *Journal of Behavior Analysis, 20*, 335–338.
Brophy, J. E. (January, 1983). If only it were true: A response to Greer. *Educational Researcher, 12(1)*, 10–11.
Buskist, W., Cush, D., & de Grandpre, R. J. (1991). The life and times of PSI. *Journal of Behavioral Education, 1*, 215–234.
Carr, E. G., & Durand, V. M. (1985). Reducing behavior problems through functional communication training. *Journal of Applied Behavior Analysis, 16*, 111–126.
Catania, A. C. (1998). *Learning* (fourth ed.). Upper Saddle River, NJ: Prentice Hall.
Chomsky, N. (1959). A review of B. F. Skinner's Verbal Behavior. *Language, 35*, 26–58.
Chu, H.-C. (1998). *Functional relations between verbal behavior or social skills training, and aberrant behaviors of young autistic children*. Unpublished Ph.D. dissertation, Columbia University, New York.

Cooper, J., Heron, T., & Heward, W. (1987). *Applied behavior analysis*. Columbus, OH: Merrill.

Dewey, J. (1910). How we think. Boston: Heath.

Donahoe, J. W., Burgos, J. E., & Palmer, D. C. (1993). Selectionist approach to reinforcement. *Journal of the Experimental Analysis of Behavior, 58*, 17–40.

Donahoe, J. W., & Palmer, D. C. (1994). *Learning and complex behavior*. Boston, MA: Allyn and Bacon.

Donley, C. R., & Greer, R. D. (1993). Setting events controlling social verbal exchanges between students with developmental delays. *Journal of Behavioral Education, 3*(4), 387–401.

Engelmann, S., & Carnine, D. (1982). *Theory of instruction: Principles and applications*. New York: Irvington.

Greer, R. D. (1989). A pedagogy for survival. In A. Brown (Ed.), *Progress in behavioral studies* (pp. 7–44). Hillsdale, NJ: Erlbaum.

Greer, R. D. (Ed.). (1996a). *Acting to save our schools (1984–1994)*. New York: Plenum.

Greer, R. D. (1996b). The educational crisis. In M. A. Mattaini and B. A. Thyer (Eds.), *Finding solutions to social problems: Behavioral strategies for change* (pp. 113–146). Washington, DC: American Psychological Association

Hawkins, R. P. (Ed.). (1986). *Selection of target behaviors*. New York: Guilford Press.

Jadlowski, S. M. (2000). *Effects of serving as a reader on the self-editing repertoires of students with early self-editing repertoires*. Unpublished Ph.D. dissertation, Columbia University, New York.

Johnson, K. R., & Joe Layng, T. V. (1992). Breaking the structuralist barrier: Literacy and numeracy with fluency. *American Psychologist, 47*(11), 1475–1490.

Lamarre, J., & Holland, J. G. (1985). The functional independence of mands and tacts. *Journal of the Experimental Analysis of Behavior, 43*, 5–19.

Lodhi, S., & Greer, R. D. (1989). The speaker as listener. *Journal of the Experimental Analysis of Behavior, 51*, 353–359.

Lovaas, O. S. (1977). *The autistic child*. New York: Irvington.

MacCorquodale, K. (1969). B. F. Skinner's Verbal Behavior: A retrospective appreciation. *Journal of the Experimental Analysis of Behavior, 12*, 831–841.

Madho, V. (1997). *The effects of the response of a reader on the writing effectiveness of children with development delays*. Unpublished Ph.D. dissertation, Columbia University, New York.

Marsico, M. J. (1997). *The effect of teaching verbally governed responding to math and reading instruction on the accuracy of correct responding and independence*. Unpublished Ed.D. dissertation, Teachers College, Columbia University, New York.

Mill, J. S.(1950). A system of logic. In E. Wagel (Ed.), *John Stuart Mill's philosophy of scientific method* (pp. 20–105). New York: Harper.

Pierce, C. S. (Ed.). (1935). *Collected papers of Charles Saunders Pierce*. Cambridge, MA: Harvard University Press.

Skinner, B. F. (1957). *Verbal behavior*. Cambridge, MA: B. F. Skinner Foundation.

Skinner, B. F. (1984). The shame of American education. *The American Psychologist, 39*, 947–954.

Sulzer-Azaroff, B., & Mayer, G. R. (1986). *Achieving educational excellence using behavioral strategies*. New York: Holt, Rinehart & Winston.

Vargas, E. A. (1988). Event governed and verbally governed behavior. *The Analysis of Verbal Behavior, 6*, 11–22.

Vargas, J. S. (1978). A behavioral approach to the teaching of composition. *The Behavior Analyst, 1*, 16–24.

Whitehead, A. (1929). *The aims of education*. New York: Philosophy Library.

Williams, G., & Greer, R. D. (1993). A comparison of verbal and linguistic curricula. *Behaviorology, 1*(1), 31–46.

Zurcher, W. (1975). The effect of model supportive instruction on the performance achievement of beginning brass instrumentalists (pp. 131–138). In C. H. Madsen, C. K. Madsen, & R. D. Greer, *Research in music behavior*, New York: Teachers College Press.

CHAPTER

8

Writing and Designing Curricula

TERMS AND CONSTRUCTS TO MASTER

- Curricula as repertoires
- Habilitative repertoires
- Utilitarianism (individual and cultural)
- Automated Programmed Instruction
- Scripted instruction
- Terminal objectives (see also long-term objective)
- Short-term objectives (see also subterminal objectives)
- Three-term contingency as instructional objective
- Frame of automated instruction
- Teaching machines
- Prosthetic reinforcers
- Constructed response
- Learner controlled responding
- Successive approximation (see also progressive approximation and shaping)
- Steps of programming
- General case instruction

- Finest discrimination as it pertains to general case instruction
- Prompting and cueing tactics as they pertain to scripted or automated curricular design
- Fading tactics as they pertain to scripted or automated curriculum design
- Vanishing techniques as they pertain to scripted or automated curriculum design
- Stimulus prompts as they pertain to scripted or automated curriculum design
- Response prompts as they pertain to scripted or automated curriculum design
- Extrastimulus prompts as they pertain to scripted or automated curriculum design
- Extra response prompts as they pertain to scripted or automated curriculum design

- Intrastimulus prompts as they pertain to scripted or automated curriculum design
- Intraresponse prompts as they pertain to scripted or automated curriculum design
- Mastery criteria as they pertain to scripted or automated curriculum design
- Fluency (mastery with a rate of responding criterion)
- Errorless learning as it pertains to scripted or automated curriculum design
- Underwriting steps as they pertain to scripted or automated curriculum design
- Revision of frames as they pertain to scripted or automated curriculum design
- Modes of curriculum as they pertain to scripted or automated curriculum design
- Academic literacy as a curricular mode
- Self-instruction (self-management) as curricular mode
- Problem solving as a curricular mode
- Writing as a problem-solving repertoire (and as a goal of curriculum)
- Repertoire for solving problems as a goal of curricular design
- Writer as reader (characteristics and as a goals of curricular design)
- Independent reader (what it means for individualized instruction)
- Postindependent reader as a prerequisite for the comprehensively individualized classroom
- Self-contained classroom
- Subject-matter-specific classrooms
- Preindependent reader as a determinant of classroom organization

---

TOPICS

Curricula as Repertoires

Modes of Curriculum

Writing Programs of Instruction in Scripted or Automated Formats

Writing as a Problem-Solving Repertoire

Programming for Individualization and Independence with Groups of Students

Summary

CURRICULA AS REPERTOIRES

In this chapter we will describe procedures for designing, writing, and applying curricula based on research in behavior analysis, verbal behavior, and the epistemology of behavior selection. The prior chapter introduced a *functional approach* to curricula based on the epistemology of behavior selection and the application of the science of the behavior of the individual to pedagogy. Now we describe how to develop functional curricula and how to convert existing curricula to meet the functional repertoires of behavior.

MODES OF CURRICULUM

We organize the content of what students need to learn into four broad categories or *modes of curriculum*. Some of these modes were suggested in the writings of the Alfred Whitehead (1929) in his book, *The Aims of Education*. His use of the term *modes of curricula* suggested the kind of dynamic learning that education should provide. We can realize these dynamic and functional curricula now by drawing on the epistemology and pedagogical science described in earlier chapters. Interestingly, behavior selection has its roots in the pragmatic philosophy suggested by Whitehead (1929), Dewey (1916), and Pierce (1935). While the approach and these philosophers have been associated with "progressive education," we argue that progressive educators have not had the scientific underpinnings to realize the goals of education proposed by the instrumental pragmatists because they had no knowledge of the operant, repertoires, or the controlling variables for response classes and no science of individualized instruction. As we described in the prior chapter, the concept of *repertoires* and other key features of our behavioral philosophy and science provide the means to offset the approach to education that Whitehead characterized and criticized as the teaching of "inert ideas." The educational work of Whitehead and other pragmatists is important to our behavioral approach to education because that work contributes the breadth needed to provide a *comprehensive behavioral education*—education of the whole child, if you will. These modes of education provide the framework for the comprehensive behavioral approach to education introduced in this text (see Table 1). They serve to direct our curricular efforts toward ensuring a broadly based educational experience. The four modes of instruction include:

1. Instruction in basic academic literacy;
2. Instruction in "self-management" and "self-instruction," (Mithaugh, Martin, Agran, & Rusch, 1988) or self-regulation;
3. Instruction in problem solving; and
4. Instruction for, and enlargement of, the student's community of reinforcers.

Table 1 briefly outlines the four modes.

Basic Academic Literacy

This curricular mode is concerned with the development of mathematical, writing, and reading repertoires that are necessary for advanced study and habilitation. They include both concepts (e.g., identifying characteristics of species) and operations (e.g., performing steps in science, performing

TABLE 1
Summary and Outline of the Four Modes of Curriculum[a]

---

1. Basic academic literacy curricula
   (a) Writing, reading, speaking, computing, and categorizing literacy (include learn units for all of the independent responses of tacting, manding, multiple control involving verbal and nonverbal S^Ds, copying/echoing, responding to dictation, selecting/multiple choice/pointing to, matching, and dictating).
      (i) Concepts—Discrimination within and between stimuli all of the above learn unit types and vocabulary of disciplines (e.g., literature, science, mathematics, social studies, and arts).
      (ii) Operation—Writing, reading, speaking listening, and computing as *verbal behavior functions* (i.e., write/speak directions, follow written/spoken directions, choosing/selecting as part of performing operations). The student learns the behaviors of governing or directing others as well as acting under the verbal control of others).
      (iii) Progression through the repertoires of verbal behavior: listener, speaker, speaker–listener exchanges or conversational units, speaker as own listener (self-editing of nonwritten verbal behavior, reader, writer with teacher tutor as editor, writer as own reader (self-editing for the target audience of the writer), and extension of verbal behavior to mathematics.

2. Self-management and self-instruction curricula (repertoires of verbally governed behavior—Repertoires of saying and doing and reading and doing for setting and realizing goals)
   (a) Graduated reduction in programmed teaching or teacher-intensive instruction (learn units become operants); student acquires learner controlled responding.
   (b) Graduated increases in self-instructional responsibility (e.g., moving from automated and scripted programs to learner-controlled responding in PSI assignments).
   (c) Graduated increases in independence (e.g., increasing the number and duration of completion of assignments before learn units are consequated via expansion of number of PSI modules completed independently); independence tied to progression through the repertoires of verbal behavior.

3. Problem-solving curricula
   (a) Programming problem-solving instruction and each level of literacy and independence.
   (b) Programming functional operations for solving problems using the methods of fixing belief (e.g., predicting consequences of actions using the methods of authority, logic, and science).
   (c) Using the vocabulary of subject areas combined with the methods of authority, logic, and science (e.g., social sciences, mathematics, and literature).

4. Enlarged community of reinforcers
   (a) Acquisition of repertoires paired with thoroughgoing positive reinforcement operations as a critical component of all instruction (i.e., acquisition of "reinforcement value" for books as "learning readiness").
   (b) "Choosing" learned repertoires as reinforcers for other tasks (i.e., reading about social science or science in free time, reading literature in free time, listening to new musics (i.e., jazz, music classics), doing art activities, and engaging in science activities.
   (c) Behaviors increasingly matched or selected out by newly learned reinforcers, with less control by deleterious reinforcer/behaviors.

---

[a]The four modes are done as a spiral curriculum. That is, the basic literacy is applied to the other modes at each level of advancement.

mathematical operations). Some responses need to be controlled intraverbally (often imprecisely termed rote learning); some need to be rule governed or verbally mediated (e.g., avoid parallel motion in the voicing of four-part chorales in the common practice style). All will require mastery and some will require fluency criteria. We know a great deal about how to teach these provided that we apply what is known from the science of pedagogy as described in chapters 2 through 6. These repertoires can be delivered to students in curricula that use scripted or automated programs combined with teacher-controlled, tutor-controlled, and progressively learner-controlled learn units.

Self-Management/Self-Instruction

We also know a great deal about how to teach self-management, provided that we replace the "self" with what we know about reinforcement schedules, stimulus control, verbally governed behavior, research in applied behavior analysis, and tactics such as the Personalized System of Instruction (PSI), class-wide peer tutoring, self-management as verbally controlled responding, and point systems/token economies. In chapters 5–7, individualized instruction was described as both

(a) the preferred means of delivering instruction and
(b) a critical goal of instruction.

In the individualized classroom, automated instruction, scripted instruction, student- and tutor-controlled learn unit exercises, self-recording, self-evaluation, self-reinforcement, graduated independence instruction, and graduated self-instruction experiences are all used. The specifics of combining the above will be described in the section titled "Programming for Individualization and Independence with Groups of Students."

Problem-Solving Instruction

This mode of instruction was discussed in depth in chapter 7 and an extensive exemplar is found in the chapter on instructional analysis (chapter 3). Curricula designed to provide instruction in solving problems bring to bear repertoires associated with basic academic literacy and repertoires of self-instruction/self-management to provide solutions to problems. The majority of students will need to develop complex problem-solving repertoires. Whitehead (1929) and other philosophers described them early in the century, but without a science of pedagogy they were left to the improvisation of the teacher. Moreover, the repertoires of problem solving were not conceptually operationalized sufficiently to enable a curricula and pedagogy.

The thoroughgoing application of the science of pedagogy within behavioral systems will make teaching the repertoires of solving problems using the vocabularies of disciplines feasible goals of education. Our new challenge is to *design* problem-solving curricula across various problem-evoking settings. The portions or components of that design and the rationale for teaching these particular repertoires were presented in the previous chapter. Since the methods of fixing belief or "setting out to determine the best course of action" are at the heart of most problem-solving repertoires cultures have developed repertoires for doing so. The methods of authority and logic (Pierce, 1935) are those that are typically used when science is not available (or for those decisions for which science is not useful). Within science, the methods of agreement, disagreement, concomitant variance, residuals, and combinations are generic procedures.

The repertoires of solving problems (e.g., authority, logic, and scientific procedures) can be taught directly. The student requires both verbally governed (e.g., verbally mediated) and contingency-shaped components. In addition, these two types of response classes will need to be taught such that the settings that call for the repertoires evoke them (i.e., the "need" to know) along with the critical discriminative stimuli and consequences that follow. The difficulties and novelties of the problems will necessarily need to be programmed such that the student is successful, yet challenged. In other words, the difficulty and number of steps involved in obtaining solutions (e.g., reinforcement effects) will need to change commensurate with the expertise of the student.

Enlarged Community of Reinforcers

As we teach repertoires of behavior we can teach the reinforcers as well as the stimulus discrimination (S^ds and the responses). That process, if done with thoroughgoing positive reinforcement, leads to an expanded community of reinforcers. Students learn "to choose" to read, listen to a broad range of music, read about scientific advances with interest, and learn to enjoy different cultures and perspectives because of a history of having positive reinforcement paired with learning the subject matter of these communities. While it is true that one cannot speak of reinforcers independent of behaviors, it is the learned or conditioned reinforcers that will lead the way to further learning and an even broader range of interests—interests that reinforce behaviors. These learned reinforcers, that are part of a quality behavioral education, are necessary to compete with the destructive behaviors and short-term reinforcers that are plentiful in all societies, such as the consumption of drugs, overconsumption of food, and the conditioning of violence and violent behavior by mass media. Students who have habilitative communities of reinforcers have a better chance of withstanding the nonhabilitative reinforcers that are constantly conditioned by other forces in society (See the chapter on "A Behavioral Utilitarian Ethic" in Greer, 1980; and Hawkins, 1986).

The four modes of instruction and their curricula should be taught *concurrently*. That is, one need not, or should not, teach independence and problem solving after *all* of the basic academic literacy is acquired. The students should be taught instructional objectives for the self-management, problem solving, and enlarged community of reinforcers as they progress through the hierarchy of skills of academic literacy. Some have called this approach a spiral curriculum. Like the three components of teaching we described for teachers, each level of student expertise should provide instruction in all four modes. In fact, the problem-solving mode provides the context and, if well designed, the establishing operations for using basic academic literacy. This is not to deny that certain basic reading, writing, and self-editing repertoires are prerequisites for problem solving. The student must be fluent in certain mathematical operations and reading in order to apply the method of logic to solve word problems. But, as soon as basic mathematical components are fluent enough to be contingency shaped or automatic, we can proceed to teach the student to use the component skills in more complex problem-solving activities (Singer, 2000).

Each of the four modes of curricula is important. Currently the standards and curricula for teaching and assessing self-management, problem solving, and an enlarged community are not part of standard educational practice. For example well-designed standards for different grade levels such as those in the New York State Standards for Curricula and the Standards for Excellence from England include literacy and some components of problem solving and an enlarged community of reinforcers, but they do not include the self-management and the progression toward learner independence as goals for instruction. Yet, the lack of curricula and related pedagogy for teaching self-management and problem-solving repertoires are at the heart of complaints about what students are lacking when they graduate from schools. Thus, it is important to track instruction in all four modes of curricula throughout schooling. The following is an example of how these standards are expanded to include the self-management and enlarged community of reinforcers as curricular standards for CABAS schools.

WRITING PROGRAMS OF INSTRUCTION IN SCRIPTED OR AUTOMATED FORMATS

The micro- and macrotask analyses of curricula used in the teaching operations described in prior chapters are a process that has been described generically as the programming of instruction. The term *programmed instruction* was used by Skinner (1968) to describe the presentation of curriculum to students such that the teacher need not be present. He developed Programmed Instruction and Teaching Machines as a result of observing one of his daughter's classrooms. He observed the crowded classroom and the paucity of learning opportunities that students received given the number of students the teacher had to instruct. In the procedures that he developed, the student *constructed* responses on an automated teaching device he called a *teaching machine* (e.g., they emitted successively more complex responses leading to the emission of behaviors that were the short- and long-term objectives of a particular curriculum). By *constructed* we mean that the students write or constructs their response rather than choosing from an array of answers. Some programs presented frames of instruction (e.g., learn units) via teaching machines, others presented the frames via texts (e.g., programmed texts), and more recently, Programmed Instruction has been delivered through software developed for personal computers (Tudor & Bostow, 1991). We use the term *programmed* to refer to the generic process of arranging curricula in successively more complex steps leading to mastery. We use the term *automated* as the descriptor for any technology for delivering individualized instruction independent of the teacher. The process of programming found in automated instruction is the process of teaching that includes the antecedent, consequence, and some of the contextual control for behaviors as part of instruction, presented by a teaching device that performs with little or no intervention by the teacher. Of course, the term programming is also currently used to describe the process of developing software for computers when computers replace the teaching machine. In the latter process, software writers use special language to "instruct" the hardware of a computer to respond to certain "inputs." Interestingly, the origin of the term probably had to do with programming curricula for humans.

Teachers who are well trained present programs of instruction in scripted or improvised forms consistent with the behavioral principle of successive approximation and in learn units as we described in chapters 3–5. An automated program attempts to multiply the presence of the teacher. Thus, prescriptions for Programmed Instruction apply generically to scripted, automated, quasi-programmed material or even teacher-improvised programs.

First, the terminal and subterminal *behavioral objectives* for a target repertoire are identified and defined in behaviorally operational terms and categorized according to the four modes of curricula and the stages of verbal behavior. The procedures for doing so are described in basic texts on applied behavior analysis and are also covered in many texts devoted to curricular design, particularly in texts for teaching special education teachers. When we incorporate what we now know about our science, behavioral objectives are, of course, more than the specification of behaviors. They include the setting events, the target antecedents that evoke the behavior, the responses, and the postcedents or consequences for the behaviors. In other words, what are learned are the three-term contingency and the related setting events. In this manner the student is taught repertoires rather than inert ideas. The manner in which the teacher, or substitute teaching medium, leads the student to emit the target behavior in the target three-term contingency constitutes the process of programming. Scripted programs are ones that spell out the presentations of individual learn units for teachers such as those provided in Direct Instruction materials or in scripted programs for teaching individuals with certain developmental disabilities (e.g., the Edmark reading curriculum). When a mechanical or electronic device that requires the student to construct responses does the presentation the term used herein is automated instruction.

Automated Programmed Instruction

Skinner (1968) called each learn unit presentation by an automated device a *frame* because his teaching machine presents the material in a framed window. A frame included the target antecedent and all prompts and the opportunity or necessity for a response by the student. Once the student responded correctly another frame was exposed, which, along with prosthetic reinforcers (e.g., a token was dispensed) when necessary, served to reinforce a response. When the response was incorrect the next frame would not unlock until the student corrected the response as a correction operation (i.e., the student wrote the correct answer). In other words, in addition to the three-term contingency, Skinner included the basic interlocking behaviors of teachers in the design of the automated instruction—an unambiguous presentation of the S^d, the necessity for a response, and a reinforcing or correcting consequence. In its earlier stages of development, a machine called the teaching machine presented the program of instruction. The student constructed or wrote the correct answer. When the correct answer was written, the machine moved forward to a new frame. When the answer was incorrect, the student had to correct her response, before he or she could move forward to the next frame. Later versions supplied a token for correct responses, to ensure that the correct answer was reinforced (Skinner, 1968). The frame included the potential three-term contingency for the student plus the three-term contingencies of the teacher who was replaced by prompts built into the frame. That is, the teaching device reinforced correct responses, or provided correction operations for incorrect responses, and presented unambiguous and sequenced S^ds. The frame met the criteria for the learn unit as described in chapter 2. However, if the student required special interventions beyond learn units, such as the 180-plus tactics described in chapters 3–6, procedures were not available at that time to deliver the individualized tactics, nor were the tactics known. Bostow has used the principles of behavior and the tactics of Programmed Instruction in several research studies for teaching complex behaviors (Tudor & Bostow, 1991). Recently, four computer scientists have applied automated learn units to teaching complex computer programming repertoires (Emurian, Hu, Wang, & Durham, 2000). The wide use of behaviorally based instructional design for personal computers could realize the potential of computers to maximize effective instruction in classrooms and distance learning formats. The material presented in this chapter is basic to writing automated instructional programs that are consistent with the science of the behavior of the individual.

The automated presentation of instructional material has been accomplished by what is termed the *programmed text* format (see the programmed text by Holland and Skinner (1961) titled *The Analysis of Behavior* as an exemplar). In this text and hundreds of similar texts written in 1960s, programmed textbooks for the teaching machine were used to compensate for the lack of a teaching machine. These programmed texts required the students to write the response then check their response on the next page. If it was incorrect they corrected it and moved to the next frame. This procedure continued until the student mastered the material.

In automated Programmed Instruction, each of the frames or learn units is successively presented from simple to complex. The learn units are based on task analyses of the material and these are tested across many students, until the student masters construction or production of the target responses independent of prompts or supplemental reinforcement. Rather than the student reading and studying (e.g., memorizing) the material, he or she actually is shaped to construct the answer. When done well, it replaces teacher presentations, is more time efficient and thus cost-effective, and is maintained (see Stephens (1967) for a review of all educational research up to the late sixties, including numerous studies on programmed texts). Programmed Instruction was one of the few procedures that was found to be effective in all of the research conducted in education through the late 1960s.

The process of successive approximation or the "shaping" of the student's repertoires must be arranged such that the student can respond correctly to each learn unit moving the student as quickly as possible toward the terminal goal. For each learn unit, the student is required to do more than he or she was before and the student is provided with the material or antecedent prompts that shape the desired response. The prompts are faded as quickly as possible. Steps that are too small may bore the student or delay progress; steps that are too large are punishing to the student since he or she cannot perform correctly. A good program avoids both disadvantages. Vargas and Vargas (1991), who we draw on extensively for this presentation, suggested that the steps be written slightly larger at the outset than you think will be needed. This is also consistent with instructions in how to design general case instruction procedures (Engelmann & Carnine, 1982), where the scripted curriculum author is told to start with the finest discrimination that the student can perform reliably. Problems with steps that are too large can be spotted quickly and supplemental steps can be inserted to fix the problem. The author should observe the student in the earlier stages. Later, the rates of correct and incorrect learn units by several students will serve as a source of feedback to locate steps that are too large or that are poorly designed. In order for the program to be successful, the student using the program had to have specific entry-level skills or prerequisite repertoires.

If the program is designed to teach the repertoire needed in the world at large, the author will avoid the use of multiple choice responding (see chapters 3–5, for the differences between point to, matching, and production repertoires). A multiple-choice or matching frame may be used, however, as a *prompting frame* for evoking a written response on a subsequent frame. Some program designers may include the teaching of multiple choice and constructed responding with the same material. The program, however, should avoid teaching what Whitehead (1929) described as "inert ideas." Multiple-choice responding may in fact be the quintessential example of the teaching of inert ideas. The purpose of a good program of instruction, whether automated, scripted, or improvised, is to produce effective written or constructed responding that eventually or directly leads to problem-solving repertoires.

Some automated programs incorporate *branching programs*. That is, when a student makes one or more errors, the student is directed to a subprogram that teaches the missing repertoire before resuming the primary program. Of course, this is the same process that occurs in scripted programs wherein the objectives are broken down into even smaller steps. If the necessary prerequisites for the program are identified, however, the need for breaking down programs will be reduced. It is likely the authors of automated programs in the future will be able to draw on the behavior analysis tactics described in chapter 3–5 to use scientifically based operations to determine the particular students problem and provide the best tactic to teach that student—all in an automated fashion!

Various prompting or cueing techniques are used to evoke both stimulus attention and responding. Sulzer-Azaroff and Mayer (1986) and Cooper, Heron, and Heward (1987) provide excellent examples of the various cueing and prompting tactics. Also see the list of tested tactics from behavior analysis found in the chapter on the analysis of instruction. By carefully reviewing Holland and Skinner's programmed text tilted *The Analysis of Behavior* (Holland & Skinner, 1961), authors can induce tactics for providing verbal stimulus and response prompts. The tactics are described variously as stimulus shaping, vanishing techniques, and response priming. These intrastimulus, extrastimulus, and response prompts are designed to ease the student into emitting correct responses without errors.

The criterion for achieving objectives needs to include concerns of mastery with and without a rate of responding criterion (fluency). Sampling the responding of several adults and using their responses as the terminal rate of accurate responding goals may set the fluency goals for both. The purpose of the rate instruction is to provide the student with contingency-shaped responding in component skills in order that subsequent instruction in more complex skills will be mastered easily (Singer, 2000). Objectives can be added for teaching contingency-shaped responding after mastery is achieved. Alternately, those basic academic skills that require fluency (e.g., math facts, spelling, definitions, translations) can be taught using flash card or written list techniques after mastery of the operations is achieved or rate criteria can be built into the computer program itself. Having the student practice emitting the answers quickly under time constraints is one tactic to accomplish this objective. This procedure was developed in the precision teaching research and was described in Part I.

Vargas and Vargas (1991, p. 246) suggested four rules of thumb for writing Programmed Instruction. They are useful for writing either scripted (teacher delivered) or automated programs and are outlined below.

1. Use selection responses (e.g., multiple-choice frames or matching frames) to introduce or prompt the construction of responses. One may also provide a frame or learn unit that models or gives the correct response before presenting a learn unit that calls for a constructed response. For example "We say that a respondent is (a) triggered or (b) elicited" where the student responds by writing the word *elicited* as a function of selection. The subsequent frame would ask the following: "Therefore a respondent is ____ by an antecedent stimulus." The student writes the word "elicited." "When a selection response is used, it should lead, however, to a *constructed or written* response. The goal is writer behavior. The reader is cautioned that the response that are primed should be *production* rather than *selection* responses, because they are different repertoires. Eventually the student can describe all of the components of the respondent independent of prompts.

2. Begin with simple responses and move to more complex responses. In other words, apply good task analyses to the development of the program. Use the behavioral principle of successive approximation to arrange the order and complexity of responding. As an operant is mastered the number of operants increases for the student's response component of the learn unit.

3. Begin with verbal or nonverbal response prompts to ensure that the student emits a correct response and then fade those responses as quickly as possible. For example, in teaching simple addition operations, one can use marks over the numbers as in the following case:

$$5 + 6 = \_\_\_$$
iiiii iii iii
12345 678 91011 = 11.

4. Begin with easy or grosser discriminations and move to finer discriminations. But take into account the *learning history and verbal behavior level of your student*. The more fine-grained or subtle the discrimination that you obtain from the student at the outset, the fewer the number of learn units required to achieve the finer discriminations. For example, in teaching discriminations between shapes, the discrimination of a square from a circle is less fine than discriminating a square from a rectangle. If the student, however, can discriminate the square from the rectangle at the outset, fewer learn units will be required to learn to distinguish all of the basic shapes. Unfortunately, if the

student already discriminates finer steps than the program requires him or her to do, presentations that are not true learn units will occur and the more advanced student is penalized.

The program should be tested, revised, and retested repeatedly until the responses of each student who uses the program result in smooth, accurate, and fast acquisition of the desired objectives. It is better at the outset to underwrite or make the steps larger and use the pilot trials to determine when and how often to insert other frames or learn unit steps. In writing scripted programs for self-help skills, this principle is reflected in the motion of moving from least to more intrusive prompts. Table 2 presents a list of steps to follow as guidelines for writing automated programs. Table 3 provides some examples of terminal and short-term objectives. There are two maxims to follow that supersede all others. First, the tactics that are used must be designed to adhere to the basic strategies of the science. Second, pay close attention to the behavior of the students who try out the program as a basis for revision. Your early "trials" with students will help define the appropriateness of the steps in your program.

The reader will recognize that the principles of programming are synonymous with the principles of successive approximation described in applied behavior analysis texts. To repeat, the automated program substitutes for the teacher's presence. It allows the student to proceed through good learn unit presentations while the teacher is free to do other tasks. It is in a sense a fine-grained demonstration of the process involved in PSI also. In PSI, the objectives of units of programmed, scripted, or quasi-Programmed Instruction are arranged in a sort of macroprogram. Both procedures require labor-intensive preparation initially. The initially more labor-intensive process of writing automated instruction will require less teacher intensive intervention later, just as the design of PSI or macroprogramming will require less intensive teacher behavior at class time than will the lecture or transmission model later on. One hopes that those concerned with developing educational software will learn more about the science. The result would be instruction for mastery, rather than the use of the computer as a text.

Remember the term *programmed instruction* are applicable generically to the process of writing scripted instruction or to the writing of automated programs. Thus, what actually differentiates the traditional use of Programmed Instruction from other behaviorally based modes of presentation is the automation of the instruction. Table 2 provides a summary. Table 2 outlines the procedures.

It is now apparent that what was previously discussed as procedures for programming the presentation of curricula can be applied generically to any medium of presentation be it workbook, worksheet, teacher presentation, tutor presentation, automated text (e.g., programmed text), automated mechanical device (e.g., teaching machine), or automated electronic presentation (e.g., computer). Each of these mediums may be used singly, jointly, or in various combinations. Part of the tactics of organizing classrooms that consist

TABLE 2
Steps for Writing Automated Instruction Programs

1. Describe the terminal behaviors of the program in a thorough manner and relate them one of the four modes of curricula. The description should be operational in the sense that the responses and controlling stimuli are fully detailed. The description should include the setting that occasions the behavior, the antecedent stimuli that are to evoke the behavior, the response, and the resulting effect of the response (e.g., consequences).

2. Describe the entry behaviors and stimulus controls necessary for the student to begin and complete the program. This will require several trials with several students. Pay careful attention to the student's existing repertoires of verbal behavior.

3. Break down the terminal goals, specified in Step 1, into a series of logical subobjectives. The subobjectives or component discriminations should be fairly large initially. Include the antecedent and contextual control for the response in all goals. The control is the "need to know," even if the need to know is initially the receipt of a prosthetic reinforcer. That is, the establishing operation or motivation must be part of the operant control if the response is to result in repertoires that result when needed.

4. Begin with the initial subobjective, and break it down into a series of learn units. Include the antecedent and contextual controls in your steps. In other words, be sure that each frame is a true learn unit.

5. Sketch an outline of the learn units.

6. Write the antecedent stimuli and the answers or verifications that will be presented to the student after his or her response.

7. Do Steps 5 and 6 for all of the subobjectives.

8. Write the additional explanatory verbal prompts for each frame or, if the response is nonverbal, write the extrastimulus prompts.

9. The verbal prompting material should provide the salient stimulus control and provide a form of the desired response but not the desired response (e.g., "eliciting" is used as a prompt for "elicits" in the example above). Fade the prompts quickly in successive learn units.

10. Observe a student run through the program. Time and record correct responses/incorrect responses by learn units, and when the student has diculty, change the program while the student is present after you have recorded the error and stopped your clock. If you are doing a computer-based program, build in data collection and timing.

11. Revise the program, and add prompts and additional learn units as the first run through indicated. Replace stimulus and response prompts that do not work with ones that do work.

12. Run several students through the program, making further revisions as necessary. At this point, corrections are made in the program after students finish the entire program. The data should result in displays of number per minute of correct and incorrect responses (rate) to learn unit frames.

13. Determine the mean and range of several students' performance on the revised version.

14. Record and graph rate correct and incorrect and achievement of objectives for subsequent students. Use a cumulative graph. Each frame and with the student's response is a learn unit.

15. Problems encountered by subsequent students call for (a) changing a description of prerequisite, (b) teaching prerequisites that were not there, (c) fixing inadequate reinforcement operations, (d) determining the need for a new set of frames or prompts, or (e) determining the need for a branching program.

TABLE 3
Standards, Modes of Curriculum, and Curriculum Design and Pedagogy Sources

Standards for Grades 1–2	Mode of curriculum	Curricula design and pedagogy
Student reads 10 first-grade-level fiction books and retells story orally (first objective), and student writes simple summary for second objective for 10 additional books	CABAS academic literacy	Pedagogical and curricular operations as specified in chapters 6–8
Student "chooses", to read for up to 20 min fiction at first grade level during free time (student purchases reading time with tokens)	CABAS enlarged community of reinforcers aesthetic reading	Pedagogical and curricular operations as specified in chapters 6–8
Student "chooses" to read for up to 20 min fiction at second grade level during free time (student purchases reading time with tokens)	CABAS enlarged community of reinforcers aesthetic reading	Pedagogical and curricular operations as specified in chapters 6–8
Student "chooses" to read for up to 20 min nonfiction at first grade level during free time (student purchases reading time with tokens)	CABAS enlarged community of reinforcers technical reading	Pedagogical and curricular operations as specified in chapters 6–8
Student "chooses" to read for up to 20 min nonfiction at second grade level during free time (student purchases reading time with tokens)	CABAS enlarged community of reinforcers technical reading	Chapters 6–8
Student checks own and other student's answers using answer sheet with 100% accuracy two successive occasions each	CABAS self-management objectives	Chapters 6–8
Student plots own performance accurately across two literacy areas with 100% accuracy	CABAS self-management objectives	Pedagogical and curricular operations as specified in chapters 6–8
Student sets own STO objectives and achieves them for four LTO academic literacy objectives	CABAS self-management objectives	Pedagogical and curricular operations as specified in chapters 6–8
Student determines tokens for achievement of objectives for four academic literacy STOs	CABAS self-management objectives	Pedagogical and curricular operations as specified in chapters 6–8

of students who engage entirely in individualized learning activities involve how the teacher juggles the multiple uses of mediums (e.g., automated, tutor-presented scripts, teacher-presented scripts). The design and organization of such classes are based on the research findings and practices contributed by the PSI model (Keller, 1968).

The obtaining of the terminal objective should provide the prerequisite repertoire for a subsequent program of study. Also, the maintenance of the repertoire needs to be built into the subsequent instructional requirements. These can be independent drill and practice exercises with timed goals. The student can check his or responses on a calculator or an answer form (see the self-management curriculum guide). In addition, if the setting events for the stimulus–response–consequence repertoire are not programmed, this will need to be done. For example, using our addition example, the setting for the responses and stimulus control is a problem involving the counting of several sets of a phenomenon followed by the *need to sum those sets*. The student is given a mixture of coins, bills, and checks. The sum is needed to determine the final cashier total. Optimally, this and other relevant setting events (e.g., measurements of rooms) can be built into the initial program, thus programming the setting stimuli and the subsequent three-term contingency.

The author of automated instruction would do well to draw also on Engelmann and Carnine's (1982) procedures for teaching the general case. These programming operations are described at length in the chapters on teaching (chapters 5 and 6). In fact, the operations described in that chapter in detail should be incorporated in the process of designing automated instruction. These include considerations of the context or setting events, the target antecedent, the target behaviors, the prosthetic, and eventually the natural consequence.

As students become more adept, they will need to learn how to teach themselves new material. In essence, the student moves from carefully programmed materials (automated or scripted) to materials that require the student to teach him- or herself (see chapter 7). That process is built into the description of PSI that was described in chapter 6 and in the following sections devoted to delivery of curricula. However, before doing so, the reader should be informed of the hierarchy of types or modes of learning that must be designed into the schooling process if we are to produce the types of individuals that contemporary and future societies will require and reward. The verbal repertoires of writing are critical.

WRITING AS A PROBLEM-SOLVING REPERTOIRE

Curriculum reforms that attempt to insert opportunities for students to grapple with "real world" problems move in the right direction but need to be aligned with the tenets of our science. The critical function of writing is that the written text of the writer should function *to affect the behavior of a reader as the writer intends*.

This holds whether the writers and readers are engineers, poets, teachers, scientists, friends, lawyers, businesspersons, or family members. Of course, the standards of usage (e.g., form and structure) are critical ingredients of writer repertoires, but the *function is the purpose and takes precedence.* As such, it seems likely that if this function is taught at every level of expertise, the functional goals of writing are more likely to accrue. Currently, the functional effects of writing are not taught directly as content. Students now learn the effects of writing as an accidental by-product of learning linguistic repertoires. The following outline is provided as an initial scheme to design instruction that teaches the functions of writing. The linguistic components of writing are to be taught simultaneously with and in the service of the verbal functions.

In the early stages, each product of writing should be read immediately and responded to immediately by the teacher functioning as reader or the tutor or automated device functioning as reader. The reader responds to what was written. The earliest form may be tacts or forms of simple mands such as "book," "play," "read," "draw," "color." These, in turn, are elaborated into mands such as "May I exchange my tokens?" "Give me the ball," "I want the puzzle," "I want the 8½-inch by 11-inch construction paper and the large box of crayons," and "Check my math paper, please." As writers become more effective, they affect their reader's behavior with minimal numbers of revisions, and the teacher may demand more complexity and greater attention to structure. One of the useful setting events for such behaviors is a period of time in the daily schedule in which *all communication occurs in written form* (e.g., the reader responds to the initial writer as a writer also). Such a setting establishes the necessary context or establishing operations (e.g., deprivation conditions) necessary to evoke the behavior and creates *the need to know.* Classroom materials and activities are obtained with written forms of verbal behavior as the student becomes progressively more proficient. Teacher consequation to instructional responses in reading, math, and social studies repertoires, for example, are communicated in written forms. Queries from the student regarding problems encountered in drill and practice exercises, automated, or tutored programs are made in written form. Subsequent responses from the teacher are also in written form. We described the pedagogical operations and the research basis in detail in chapter 6.

The functions of writing require writers to *tact* their environment such that the reader can make vicarious contact with that which was described by the writer. Writers describe simple phenomena such as "We have no milk in the refrigerator" or elaborate phenomena such as the description of photosynthesis. At still another level, the writer describes what he/she read about the verbal content made by others to verbal or nonverbal phenomena. Thus, the writer's verbal responses may be related to verbal or nonverbal events. The more removed the writer is from the nonverbal event, the more "abstract" are the nonverbal controls. The controls become the behavior of others per se. One sees the evolution of what Skinner called "ruled governed behavior" or what Vargas and Vargas (1991) called verbally governed behavior. That is, as verbal behavior substitutes for direct *contact* with the nonverbal environment the behavior of writing may become more abstract. Writing that is directed toward describing real phenomena in real time (e.g., biology, behavior analysis, organic chemistry) must be checked against the nonverbal events or verbal effects on verbal behavior. As students become more sophisticated, that contact need occur less frequently. In the early stages, the verbal behavior must be checked against its descriptions repeatedly. For example, in order for the description of photosynthesis to be an authentic scientific tact, photosynthesis must be observed; without the observation, the student is simply emitting an intraverbal. Of course the intraverbal definition of the process of photosynthesis is important too.

Esthetic forms of verbal behavior (e.g., poetry, fiction) may contact verbal or nonverbal relations or events. The writer may describe his or her own contact with the nonverbal environment (e.g., "The sky shouted blue"). The writer emits interesting forms of intraverbals that are reinforced by their effects on the reader (e.g., fiction and poetry). This esthetic repertoire requires instruction directed to that end. It is a different writing repertoire. That is, the two major repertoires are scientific or technical writing and esthetic writing. The audiences for each repertoire are different and the writing repertoires needed to affect the different audiences are different also.

Repertoires that have functional effects other than esthetic ones can be related to other needed consequences. The subject matter of history calls for verbal behavior that is tied to the use of the *method of authority* as the critical mode of inquiry. A historian presents the burden of proof by authenticating his or her source. Literary criticisms about almost all topics require the use of the method of logic tied to the use of the method of authority. Given the existence of certain events (e.g., authoritative source), writers present their verbal and written arguments such that readers can confirm or deny the writer's logic. In the initial stages, the writer must receive confirmation (e.g., reinforcement) immediately for certain statements and corrections that function as prompts for improved written verbal effects. As the repertoire is strengthened, the assignments are lengthened and the contact from a reader other than the writer him- or herself (self-editing) may be less frequent.

Initially, words, sentences, or short paragraphs require immediate teacher or peer editor consequences followed by revision. At advanced stages, entire chapters or books are written before feedback is given. *The writer is not, however, bypassing a reader, even at the most advanced stages*. What happens is that the writer becomes more adept at functioning as *her or his own reader* and can do so for longer periods of independence. Skinner has described this as the self-editing function of writing.

The Writer as Reader: A Self-Editing Function

As the student of writing requires fewer editing responses from another person who functions as editor, the writer must become his or her own reader. In effect, the writer responds to his or her own words, as would the target reader. When the "self" functions as the reader, the self must be as effective for the writer's purpose as was the teacher/reader initially. Lodhi and Greer (1989) found that young children functioned as both speaker and listener in certain settings. This function must be nurtured and reinforced in the more abstract repertoires of writing. Traditionally, the development of self-editing expertise has occurred only by chance with some students. By making the effects more obvious in the process of curriculum design and programming, the teacher or program designer will be more successful with more students. Jadlowski (2000) found that when students functioned as editors for other students efforts, the target students became more proficient in self-editing their own writing; that is, they decreased the number of rewrites needed. Table 4 provides a brief summary of the process of teaching self-editing in a systematic manner.

The rough outline of this programming procedure requires that the program designer fill in extensive details. Simply initiating and revising the program of writing instruction, however, can provide such details. The effects on the writer's behavior and the ensuing data must shape the components of the

TABLE 4
Self-Editing Function

1. The student writes instructions for one or two-step operations:

    (a) Another reader responds to written text (serving as a reader or editor of other students' texts is part of the curricula for teaching self-editing) (Jadlowski, 2000).

    (b) The is text revised until reader responds correctly per the writer's script (i.e., the editor is given a script of what he or she is to look for as effective writing or the reader must perform according to the instructions. When the instructions are unclear the draft is returned to the initial writer *until* the editor can perform the task effectively. Each draft or each correction within the rewrite is a learn unit, dependent on the level of the student.

2. The student writes two or more instructions for two-step operations:

    (A) Serves as own reader until the self-editing results in an effect is consistent with what a teacher or student editor would have done (e.g., history of several responses with initial success in having the teacher respond accurately)

    (B) Receives feedback (from another reader) on an FR 2 and then VR 2 accomplishment/product gradually thins the learn-unit schedule consistent with development of operants.

       (i) Continue until mastery

       (ii) Progressively extend responding before someone else (teacher or tutor) functions as a reader.

       (iii) Count the number revisions to criterion as a measure of self-editing.

       (iv) The long-term goal, for example, might be the production of two successive writing assignments that require no revisions.

3. Expand the writing assignments to include increasingly more complex goals. Writers must have the experience of editing the writing of their peers because this experience is a critical tactic for teaching self-editing (Jadlowski, 2000).

program. By setting writing objectives that teach the functional effects of writing to the writer, we teach the by-product of independent writing and independent thinking. Thus, we have identified the effects of the repertoire and the component behaviors for direct instruction. They are made observable, thus measurable, and therefore teachable.

The self-editing repertoires of writing are then tied to the methods of authority, logic, and science as these latter methods affect the traditional subject matter of history, science, and literature. Thus, the repertoires traditionally associated with thinking or problem solving are developed as writing behaviors, converting the nonoperational process of thinking into overt operations that can be taught as learn units. Researchers concerned with artificial intelligence or designers of computer programs follow a similar process but may not operationalize the process accurately because they do not possess the vocabulary of the science. Combining the functions of verbal behavior with the functions of methods of determining beliefs (e.g., what to do) builds writing skills and problem-solving skills simultaneously. The problem-solving and self-editing functions can be taught as three-term contingencies and embedded into learn units (Table 4).

The Learn Units of Writing

As described in chapter 5 on the learn unit, the responses that constitute the writer's learn unit depend on the existing level of the student's performance and the occurrence of verbal feedback by the teacher who functions as a reader. Written learn units occur in those instances when the teacher has returned verbal comments specific to certain written units. The effects of the learn units depend on convincing the reader. The teacher functions as an editor also, suggesting changes (e.g., corrections that function as prompts that change writing) and providing numerous reinforcers for increasingly effective writing responses. Learn units for teaching self-editing repertoires, as is the cases with other learn units, are counted in terms of time and at specific levels of functioning. If the teacher supplies 20 separate comments (e.g., corrections and reinforcements) for a 20-page paper, 20 learn units were provided in the time period involving the assignment period and the return of the paper to the student (see the chapter on learn units).

Each writing assignment should specify the objectives of the assignment and other criteria. For example:

- The authority for style (e.g., American Psychological Association style manual, Modern Language Association style manual);
- The authority for spelling (e.g., *Webster's New World Dictionary*, United Kingdom or United States versions); and
- The authority for grammar or the structural objectives (e.g., Strunk and White's manual on writing).

The objectives for substance can be specified according to the methodology used (e.g., methods of authority, logic, or science as well as the subject matter objectives of the assignment). The student responses are then compared to these criteria. Incorrect responses are corrected or noted and adequate or excellent responses also are noted. Each comment constitutes a learn unit. Reliable scorers will perform consistently with the assignment objectives and the criteria. In addition, effective teachers will return the papers ensuring that the papers are rewritten until they are consistent with criteria. Careful attention should be given to reinforcing variety in the responses that meet the criteria if one of the goals of the instruction is creative responding.

The following section provides some generic guidelines to use in establishing learning objectives or criteria and their relation to learn units for both scripted and unscripted curricula.

Generic Measurement Criteria for Establishing the Mastery of Instructional Objectives

A. Existing Scripted Curricula

1. Scripted Curricula typically specify 80 or 90% as the criterion for two successive instructional sessions. Note, however, that the particular number of sessions may involve empirical observations for individual students and that observation may suggest different criteria for different students for different instructional goals.

2. Some curricula have unit tests. Your students should meet at least 90% criterion on those tests. If you measure responses of students in all instructional sessions prior to the test, there is no reason to require two successive tests at the criterion. These guidelines apply to both the reading and the math curricula. *Of course, mastery without a rate of responding should be followed with an objective for a rate criterion for important component skills.*

3. Competent adults rates of responding may be used to set rate criteria (Lindhardt-Kelly & Greer, 1997). However, Johnson and Layng (1992) recommend that the rate be based on the rate that results in retention, stability of accurate responding, and endurance or the operant control of the task such that the student will perform for long periods of time without the need for learn units (e.g., the results of the behavior reinforce the student's responding accurately and quickly). However, in all cases the performance of the individual student should specify the number of sessions that are best for that student, regardless of the curricular materials used. Let the data of your students guide instruction.

B. Guidelines for Adapting Nonscripted Curricula

1. Use the smallest subunit objective that is detectable. The student should meet a 90% criterion for successive testing of the material that works for

retention for the student. For example, a unit test following a chapter in a social studies text (e.g., global studies) should be answered on at the 90% level and at a rate of responding that shows that the student is fluent. The addition of single digits from 1 to 5 would require two sessions in which the student was correct at the 90 or 100% followed by meeting a rate criterion for two sessions, before moving the student to single digit addition from 5 to 10. The particular math problems used should differ, at least in the order.

2. For work sheets, arrange their sequence according to some instructional objective—such as "same with same" for some visual stimulus dimension. Keep in mind that if they are to serve a useful end, they should prepare the student for sameness capacity followed by discrimination instruction programs. That is, first the student shows matching, then discrimination (e.g., not same). This process leads to academic repertoires such as the early matching skills for Edmark or other reading instructional materials or one-to-one correspondence or the use of manipulatives to teach operations in addition and subtraction.

3. The student's individual and independent work should serve two purposes:
   (a) teaching *the particular objective that is explicit* and
   (b) teaching *implicit self-management objectives* as spelled out in the self-monitoring curricular guidelines described in chapter 6.

4. Problem-solving objectives are met when the student solves an academic problem for the first time using repertoires that they have learned. For example, the solution constitutes an objective if the student learns some rules (verbal behavior about the subject) about seeds and then uses those to satisfactorily solve (i.e., be verbally guided by the rules) in an experiment. If the student does not solve the problem, teach him or her to master the component rules/operations and have the student repeat the experiment until he or she is successful. The control of the verbally written rules can be observed by watching the student refer to the written directions step by step. Marsico (1999) found that the process of the student moving from verbally mediated responding to the solution of problems by contingency-shaped behavior was reliable and that her students developed self-editing independence by using these verbal scripts to guide their behavior. The guidelines are the same as those for a teacher. That is, when a teacher uses written directions to implement a continuous or progressive delay procedure after reviewing the student's likelihood of benefiting from the procedure, applies it, and is successful, it is an instance of verbally mediated behavior or problem solving. After that particular application, the use of the procedure should be taught until it is contingency shaped (e.g., the student performs correctly without recourse to the written instructions).

C. Setting Fluency or Rate of Responding Criteria

Clearly, you are going to need to make decisions about the addition mastery plus fluency guidelines consistent with the research. It is likely that basic discriminations and repertoires such as those associated with math, phonetics, and vocal reading fluency will need rate criteria. The question of whether the rate criterion needs to be extended to the performance of operations such as those involved in subtraction of double digits, division, or the following of guidelines to solve word problems needs research. Until the research produces hard answers, these decisions must be based on each student's individual performance.

PROGRAMMING FOR INDIVIDUALIZATION AND INDEPENDENCE WITH GROUPS OF STUDENTS

The most prevalent objection about the viability of instruction that is entirely individualized is that it is not feasible in large classes. However, the approach to teaching as an activity involving the teacher leading a group through the same material is actually a relatively new phenomenon. The one-room schools of early American communities actually predated the group ensemble approach. Yet, the concern that the individualization in behavioral models is impractical is the most often voiced criticism. Some say that the use of behavioral models of education in special education was made feasible by the staff-to-student ratio found in classes for students in special education.

In fact, early forms of comprehensively individualized classrooms (Greer, 1980) were done in music classes often with as many as 150 students and one teacher. In these music performance groups, the students also individually:

- Learned to perform on a major instrument;
- Learned a second ensemble instrument;
- Learned basic chords on the piano or guitar;
- Learned to write simple compositions; and
- Learned to improvise on a simple blues chord progression.

Each of the groups of instruments (e.g., brass, woodwinds, percussion) requires different instruction. There are often subgroups also (e.g., first, second, and third stands) as well as different grade levels (e.g., sixth, seventh, and eight grades). The levels of expertise in such ensembles can range from students who can barely play a chromatic scale to those that can perform concertos with advanced expertise. Yet the data on these programs for several years showed that the individualization was feasible and worked with in the confines of a 45-min class session.

Thus, the task of individualizing instruction in a classroom of 25 to 50 students seems less difficult. However, in order to do so, the way in which the class is organized and the planning necessary to arrange individualization as a means to learning and as a learning goal is radically different from the standard *modus operandi*. Furthermore, the expertise of the teacher is critical and the students' require certain basic levels of verbal behavior as described in earlier chapters.

For an individual teacher, who does not have the benefit of the kind of systems support found in a behavioral systems school, the task of curriculum design is more difficult, but it can be, and has been, done. In this case, the teacher can approach the task one subject matter at a time, gradually adding to the range of individualized curricula term by term. The initial planning and curricular design and the subsequent modifications needed are labor-intensive. However, as the curriculum options grow and the organizational and behavioral expertise of the teacher grows, the teacher's role shifts dramatically. The teacher becomes a designer of instruction, a motivator, and a strategic scientist of instruction, not a purveyor of information (Keller, 1968).

There are sufficient amounts of written material available for almost all subject matter and almost all levels of ability. The task of locating them and procuring them is not easy, of course. Most of the materials are not adequately programmed or scripted, but over time the teacher can modify the material. Successive experience with the different levels of materials for different students offers the teacher the opportunity to modify the content and the programming of the material. General case teaching approaches can be inserted, as can prompting and fading (e.g., vanishing procedures). Functional content (see chapter 7) can be inserted along with the subject matter content (e.g., writer–reader exercises) and units of Programmed Instruction (designed by the teacher or purchased) can be inserted. However, the above approaches presume that the students taught by the teacher are independent readers. Some generic categories for programming individualization are *preindependent reader* and *postindependent reader* categories. Each shall be dealt with separately.

Independent Reader Classroom Design

Once the students have achieved a basic reading repertoire that allows them to follow written instructions for instructional purposes, the design for the independent reader classroom is appropriate. In essence, the classroom design can be based on the concepts of the Personalized System of Instruction (PSI) developed by Keller for large college classes. Within this category, however, are two subcategories as the self-contained classroom and the subject matter classroom. In the former instance, the same teacher teaches many or most subject matter and related repertoires within one classroom. In the latter instance, a particular subject or topical area (e.g., history, algebra, English composition) is taught by a single teacher often to different groups of students.

The Self-Contained Classroom

In the self-contained classroom, teachers are responsible for most of the repertoires taught to a single group of students. Thus, curricular materials are located for the range of abilities (e.g., baseline repertoires of the students in the class). Once existing materials are located for each level, they can be arranged in an initial sequence. Each level becomes an objective with sub-objective for each objective. At this point, the teacher needs to apply the concepts described in chapters 6 and 7. Are the objectives valid? Is the sequence valid? That is, does the acquisition of the different levels of skills lead to useful ends? Will the repertoires be functional in the subsequent school experience and in the world at large? Are they writer or reader repertoires or exemplar identification or exemplar production repertoires? Is the context (e.g., need for the repertoire) built into the curriculum?

The teacher or school embarking on this path will have to make do initially with only successive approximations of what is best. Gradually, with successive individual students, new repertoires and better-programmed approximations can be inserted. As the material becomes better programmed, the need for individualized and teacher-intensive learn unit presentations will be replaced by automated or tutor-scripted learn units. The teacher will initially need to invest many hours preparing the individualized material and locating and modifying the individualized material. Over time, the modifications and newly written material will develop into an extensive corpus of programmed materials that will be on hand for the students. Of course, if teachers share material across different self-contained classrooms, the corpus will grow faster. Better yet, in a comprehensive behavioral school program the supervisor facilitates the development of curricula for teachers in a unit of the school or an entire school just as was done with the music programs and the CABAS schools (Greer, 1980).

Delivery

The next step involves organizing the delivery of the various units (e.g., ability levels or the of subject matters) to the students such that they proceed through the material at optimum individual rates.

The classroom for the independent reader should be arranged with desks such that the teacher and tutors can quickly reach students. Students may be faced away from their peers, preferably in screened carrels with shelves for holding books and materials and workspace for writing and the use of computers or other equipment. Such an arrangement allows students to have their own "office area" as is the case in the workplace. An area of the room should be arranged also as a reinforcement activity corner also.

Each student should have his or her own individualized PSI folder with work for the day or period inserted. There may be an *in* basket (for work to be done) and an *out* basket for work that awaits comment from the teacher. Visual display (graphs) of each subject matter/repertoire are available either in loose-leaf binders or posted or produced on the computer monitor. Each desk or computer should also have a timer for timing responses to instructional tasks (i.e., self-monitoring). Dictionaries and texts should be located at individual carrels or in a central location for ready access. The order of instructional activities (e.g., math, English) can vary from student to student based on reinforcement value in some cases or for self-management objectives in other cases (see the description of point systems described in chapter 6).

Clear, concise, and positive rules for engaging in independent work are established at the outset. These are accomplished by direct reinforcement for engaging in independent work by tokens or point systems. Initially, engagement and accomplishment are reinforced. Successively fade engagement reinforcements until only accomplishment is reinforced. Subsequently only mastery and fluency are reinforced. These stages vary for individual students based on their own individual progress. Exchanges for backup reinforcement are individualized also. Some students will need to exchange points after short periods of work or achievement. Subsequently, these are extended and the student eventually selects his or her own exchange time. In time, the accomplishment of tasks will reinforce responding.

Students go to their desk and begin working as soon as they enter the classroom. If students come before class, they may begin work or they may stay after class at the convenience of teachers and parents. Eventually, continuing to work after class or before class can be an earned privilege (see the PSI system designed by William Zurcher in Greer, 1980).

The goal is for all students in the class to receive most of their instruction in automated, scripted, or quasi-programmed form. These various formats of instruction are delivered by automated devices (e.g., computers, programmed texts), texts, teachers, peer or cross-age tutors, and teacher assistants. All students will be engaged in curricula for each mode of learning (e.g., basic academic literacy, self-management, problem solving) *at his or her appropriate level*. Measurement is continuously done and variously done by the student, teacher, tutor, and teacher assistant. Much of the communication that occurs is done in written form with written responses. The classroom can be individualized in a comprehensive manner and the relevant repertoires needed by students as self-learners are enhanced are taught by the structure of the classroom. The repertoires being learned are similar to those that individuals engage in as adult members of society. That is, the students go about their tasks, reinforced by the effects of their effort or their self-reinforcement, and they manage the order and succession of tasks that they need to accomplish. They also learn to monitor how well they are doing and how to work to improve their own performance. They learn to arrange the day, week, and term to accomplish those goals.

Subject-Matter-Specific Classrooms

Teachers who teach the same subject matter (e.g., math, English literature, second languages, music) but to different groups of students throughout the day can do so as described in the section on the self-contained classroom. The subject matter teacher can prepare material that is more comprehensively programmed; however, such a teacher must deliver the curricula to a larger number of students. The advantage is that more automated instruction can be developed in less time. The disadvantage is the limited amount of time that is available.

Wide ranges in baseline levels of performance in a specific subject area are expected, but they are typically narrower than the range of skills in a self-contained classroom *provided* that there are certain entry-level skills required for class membership. If entry-level skills are not met, the provisions described for the self-contained classroom will need to be replicated in a subject matter classroom.

As in the case of the self-contained classroom, the responding is all individualized, self-managed, continuously monitored, and revised and modified repeatedly based on the data. Students begin work on entering the classroom and leave or discontinue when the bell rings. Access to the material may be possible before or after school or during free periods or lunch. In time, students may learn to move between subject matter classes based on their arrangements and accomplishments rather than a bell schedule. Different ratios of students to teachers will result from an analysis of the attainment of objectives and costs of objectives and learn units.

Classes devoted to teaching other languages will need to add frequent class-wide tutoring and other characteristics of prereader/writer classrooms since the student are at that stage in the new language. However, they will have self-management repertoires driven by their native language repertoires to assist them in delivering learn units to each other and to themselves.

Preindependent Reader Classrooms

During the preschool, kindergarten, and early primary grades, students do not have the reading skills required to function in the independent reader classes. This holds also for some children who have severe handicaps for extended periods or throughout their school years. Initially, instruction for these children must be scripted and delivered as described in the CABAS literature presented in chapter 5. Teachers and teacher assistants must rotate students delivering individualized scripted curricula. Our current experience suggests that the ratio of instructional personnel to students should be small (e.g., 3 to 1 or 4 to 1). However, with children who have no significant verbal handicaps, this ratio may be as high as 1 to 20 (see the description of categories of students in chapter 5–7).

Ensemble instruction in reading or math may be arranged consistent with the procedures outlined in Distar Reading and Math (Engelmann, & Carnine, 1982). Early on, the students should be taught to monitor their own correct and incorrect responses during ensemble instruction. This is done by providing frequent probes that measure the students' progress in order to individualize tactic that are required when the students are having difficulty. As the student acquires independent reading and writing skills, the classroom is shifted to the independent reader design. Procedures for observing individual responding in groups can be used by the teacher, teacher assistant, and in some cases tutors. As automated or computerized programmed instruction becomes more available, the individualization of preindependent reader classrooms will become less teacher-intensive in terms of delivery and more design intensive as in the case of the independent reader classrooms. The pedagogical operations for doing so are described in detail in chapter 5.

Group Instruction as Goals

There are instructional goals that are group-like in nature, such as working together to accomplish joint goals, performing in musical groups, or engaging in athletic tasks. These too should be programmed in terms of *individual objectives, subobjectives, and the contingencies*. Procedures for doing so are described at length in chapter 5. These repertoires are reinforced and programmed as group contingencies as a necessary condition. The goal of such instruction is cooperative rather than competitive behavior. In the case of all instruction, the goal is to teach all of the students, not to select out individuals who have already learned the instructional repertoire being taught (Axelrod & Greer, 1994).

In short, the purpose of such instruction is to develop cooperative problem-solving repertoires in which methods of solving problems are combined with verbal behavior repertoires such that the reciprocal exchange of behavior pays off for the joint reinforcement. When the purposes of group instruction concern the development of cooperative social skills, such purposes can be isolated from the utility of group instruction as a means of instructing students in social skills. The pedagogical skills and related social skills curricula are treated at length in chapters 5 and 6.

Progressive Independence Hierarchy

Throughout this chapter and in chapters 5 and 6, we proposed a scheme for teaching the student to become progressively independent of teacher. However, the student is not described as becoming independent from his or her environment; rather, the actual controlling variables (e.g., antecedent and postcedent stimuli change) change for the student. The student shifts from functioning as both speaker and listener to writer as reader. The student shifts from having completed a few learn units in short periods of time to having completed many more learn units in a more extended time with less direct teacher supervision. Initially, the teacher collects all data, while at the other end of the progression the student collects almost all of the data on his or her responding. Students also progressively learn to arrange the order of work and reinforcement periods and set short-term and long-term objectives. The student moves from automated frames or scripted learn units to coping with lengthy assignments in which he or she must read and teach him- or herself material before taking quizzes (e.g., characteristic PSI units). In short, the progressive independent hierarchy outlines the independence or self-management curricular mode as shown in Table 5.

TABLE 5
An Outline of a Curriculum to Teach Self-Regulation, Self-Management, Time Management, and Independence: A Progressive Independence Hierarchy

(A) The comprehensively individualized classroom as the setting:

  (1) Assignments for all subjects, areas, and modes of curricula available at the outset of the lessons.

    (a) Learn units are consequated per schedule described below and returned immediately or in delayed fashion.

    (b) Student objectives for self-regulation (e.g., one of the curricular modes) are incorporated in the PSI arrangement of instruction.

(B) Independence of self-regulatory hierarchy in the comprehensively individualized classroom:

  (1) Young and severely delayed children (preindependent readers) receive scripted and automated programs of instruction with immediate correction or reinforcement following responding.

    (a) Prereading, speaker, and listener instruction must continue until the repertoires are mastered; thus the student has the prerequisite repertoires to move to greater independence.

    (b) The instruction introduces systematic thinning of teacher controlled learn units leading to next level.

  (2) Early reader/writer period

    (a) Functional reader/writer behavior, rather than structural behavior, is reinforced by teacher or automated device. Students must affect the behavior of a reader or follow the directions of a writer fluently.

    (b) Written responses are used for the students' written responses; that is, written communication replaces vocal communication for classroom requests as well as teacher responses to written learn units.

    (c) Instructional antecedents are increasingly shifted from vocal to written form. Thus, when students encounter new concepts associated with subject matter and other curricular modes (e.g., basic academic literacy, problem solving), the instruction is progressively in written form.

    (d) The gradual shift from vocal/vocal learn units to read/write learn units is done via successive approximation. That is, the shift in stimulus and response control occurs at the speed that works for each individual child.

    (e) Repertoires for seeking information from dictionaries, encyclopedias, and systematic observations are taught commensurate with the development of read/write repertoires.

Table 5 (*continued*)

(f) Teach self-editing repertoire (e.g., writer as own reader) systematically. Successively larger numbers of write learn units are completed before the responses are consequated by teacher, tutor, or automated consequences occur.

(g) The number and length of assignments are progressively extended based on the students' success (e.g., one assignment, two assignments, multiple assignments). Similarly, periods for the exchange of points and tokens for backup reinforcement are progressively delayed.

(h) Students participate in determining their movement through the progressive independence hierarchy described above. Initially, students make choices (e.g., acquire mand repertoires) for immediately delivered reinforcement. Gradually, students select backup reinforcers, points to token relationships, opportunities to exchange tokens, and number and extent of learn units to be completed before tokens are dispensed, before the consequences for the complete learn units are received.

(3) Intermediate and advanced reader/writer period

(a) Each of the components listed under No.2 above are extended until students complete extended projects (e.g., solving of problems) independently before interacting with a teacher or teaching device.

(b) Students progressively use skills for teaching themselves new material derived from the progressive instruction toward independence.

(c) Teachers need to continuously evaluate the repertoires needed for the workplace or graduate or professional school.

(4) The progressive independence hierarchy and data collection

(a) Data collected entirely by teachers, tutors, and automated devices at early schooling levels.

(b) Students progressively collect data on their own responses contingent on:

(i) The reliability of their data and

(ii) Their progress through the progressive independence hierarchy.

(5) Students increasingly design data collection procedures and objectives for their own goals.

When the students have achieved these latter repertoires, they are capable of designing their own instruction. They can flourish in classes that provide lectures without instruction and they can solve problems through individual research and self-instruction.

SUMMARY

This chapter has provided an overview of how curricula are organized, written, task analyzed, and delivered as a result of the contributions of the science of behavior and the epistemology of behavior selection. Procedures for programming instruction in scripted or automated forms were described along with how these programs are introduced. The four modes of curriculum (e.g., basic academic literacy, self-management, and problem-solving repertoires) were described and incorporated into the curricula.

The critical roles of writing and writer/reader repertoires were described along with a hierarchy for developing self-editing repertoires for problem solving. The conceptions drew on the analysis of verbal behavior that evolved in turn from the science of behavior.

The process of designing a classroom that comprehensively individualizes curriculum was described as an outgrowth out of the PSI model. This progressive curriculum serves both (a) to allow individualization to occur and (b) to teach independence. Finally, we outlined a progressive independence hierarchy outlining the progression that leads the student from moment-to-moment dependence to the independent repertoires that will allow her or him to flourish in graduate or professional schools and the workplace. Taken together, chapters 7 and 8 provide the framework to integrate the science of pedagogy with curricula that consist of functional repertoires.

References

Axelrod, S., & Greer, R. D. (1994). A commentary on cooperative learning. *Journal of Behavioral Education, 4*, 41–48.

Cooper, J. O., Heron, T. E., & Heward, W. (1987). *Applied Behavior Analysis*. Columbus, OH: Merrill.

Dewey, J. (1916). *Democracy and education: An introduction to the philosophy of education*. New York: MacMillan.

Emurian, H. H., Hu, X., Wang, J., and, Durham, A. G. (2000). Learning Java: A Programmed Instruction approach using applets. *Computers in Human Behavior, 16*, 395–422.

Englemann, S., & Carnine, D. (1982). *Theory of instruction*. New York: Irvington.

Greer, R. D. (1980). *Design for music learning*. New York: Teachers College Press.

Holland, J. G., & Skinner, B. F. (1961). *The analysis of behavior*. New York: McGraw–Hill.

Hawkins, R. P. (1986). Selection of target behaviors. In R. O. Nelson & S. C. Hayes (Eds.), *Conceptual foundations of behavioral assessment* (pp. 331–385). New York: Guilford Press.

Johnson, K., & Layng, T. V. J. (1992). Breaking the structuralist barrier: Literacy and numeracy with fluency. *American Psychologist, 47*, 1475–1490.

Jadlowski, S. M. (2000). *Effects of serving as a reader on the self-editing repertoires of students with early self-editing repertoires*. Unpublished Ph.D. dissertation, Columbia University, New York.

Keller, F. S. (1968). Good-bye teacher . . . *Journal of Applied Behavior Analysis, 1*, 79–90.

Lindhart-Kelly, R., & Greer, R. D. (1997). *A functional relationship between mastery with a rate requirement and maintenance of learning*. Unpublished paper, Columbia University Teachers College.

Lodhi, S., & Greer, R. D. (1989). The speaker as listener, *Journal of the Experimental Analysis of Behavior, 51*, 353–359.

Marsico, M. J. (1999). *Textual stimulus control of math performance and generalization to reading*. Unpublished Ed.D. dissertation, Columbia University Teachers College, New York.

Mithaug, D. E., Martin, J. E., Agran, M., & Rusch, F. R. (1988). *Why special education graduates fail: How to teach them to succeed*. Colorado Springs, CO: Ascend.

Pierce, C. S. (1935). In C. Harsthorne & P. Weiss (Eds.), *Collected papers of Charles Saunders Pierce* Harvard University Press.

Singer, J. (2000). *Rate, contingency-shaped, and verbally governed instruction in component math skills and acquisition of complex math operations*. Unpublished Ph.D. dissertation, New York.

Skinner, B. F. (1968). *The technology of teaching*. New York: Appleton-Century-Crofts.

Stephen, J. M. (1967). *The process of schooling*. New York: Holt, Rinehart, & Winston.

Sulzer-Azaroff, B., & Mayer, R. G. (1986). *Achieving educational excellence*. Columbia, OH: Merrill.

Tudor, R. M.,& Bostow, D. E. (1991). Computer-programmed instruction: The relation of required interaction to practical application. *Journal of Applied Behavior Analysis, 24,* 361–368.

Vargas, E. A., & Vargas, J. S. (1991). Programmed Instruction: What it is and how to do it. *Journal of Behavioral Education, 1,* 235–252.

Whitehead, A. N. (1929). *The aims of education.* New York: Mentor.

PART III

Organizational Behavior Analysis: A Support System for Expert Pedagogy and Curricular Design

Teaching and
Mentoring Teachers

TERMS AND CONSTRUCTS TO MASTER

- Cybernetic system
- Instructional variables
- Teacher accuracy
- Supportive supervision
- Self-monitoring by teacher coach/supervisor
- Teacher performance rate/accuracy (TPRA) observation
- Reliability observer
- Transducer of behavior
- TPRA for individual students
- Faultless antecedent stimulus presentation
- TPRA score or teacher performance rate/accuracy score
- Teacher errors
- Verbally mediated strategies
- TPRA observations for group instruction
- Personalized System of Instruction for teacher training (PSI)
- Modules for teacher education
- Tactics of instruction
- Principles of behavior and their relationship to instruction
- *In situ* training
- Supervisory repertoire
- Administrative repertoire
- Administrative rate
- Supervisory tasks
- Supervisor rate per hour
- Cyclical changes associated with visual displays of data
- Supervisor modules for continuing education
- Teacher modules for continuing education
- Organizational behavior analysis
- Participatory management
- Consulting teacher

TOPICS

Supportive Training and Supervision of Teachers in the Classroom

Teaching, Maintaining, and Expanding Teacher Repertoires through
PSI-Based Procedures

This chapter describes the procedures for:

(a) *Motivating* and teaching teachers in the classroom as they perform
    teaching functions;
(b) Using the *Personalized System of Instruction* to teach the *three functions of
    teaching* as a strategic science of instruction;
(c) *Monitoring* teacher performance and the performance of the professional
    who are responsible for teaching teachers; and
(d) *Developing, maintaining*, and using school and classroom level data
    management to *ensure high-quality instruction*.

The role of teacher educator may be performed by individuals who have
position responsibilities associated with consulting special educator, school
principal, school psychologist, supervisor, curriculum coordinator, director
of special education, head teacher, or professor of education and teacher
training. We shall use the terms supervisor, teacher educator, or teacher coach
interchangeably. Regardless of the position title, the procedures treated in this
chapter provide the wherewithal to teach and maintain a high level of instructional quality and productivity in a positive working environment.

SUPPORTIVE TRAINING AND SUPERVISION OF
TEACHERS IN THE CLASSROOM

The supervisor, like the university teacher trainer and the consulting special
education teacher, is a teacher of teachers. He or she functions as a teacher of
the science and practice of effective pedagogy for less experienced teachers
and as a collegial consultant in pedagogy for the more advanced teachers. The
individual doing the teacher training must have the repertoires of an advanced
strategic scientist of instruction. Her or his expertise in the science provides a
natural rather than forced authority. Thus, the power implicit in the role will be
effective *only* if expertise in the science and practice of teaching as behavior
analysis is the source of the power. Unlike traditional approaches that *are not
based on research* and *extensive application of research in practice*, much of the training
and consultation that is based on a science of teaching require the supervisor's
presence in the classroom with the teacher. That is, the training is done *in situ*.
According to the research on behavior analytic supervision and teacher
training, the presence of the supervisors and the role that they play in the

classroom is one that is not only desirable for student and teacher learning, but one that teachers come to appreciate (Ingham & Greer, 1992). In order for a mutually positive relationship to occur, the supervisor must help the teacher, by teaching new strategies and tactics to the novice teacher. For more advanced teachers, supervisors provide an additional set of eyes and ears to help in the objective location and solution of difficult instructional problem. A mutually beneficial relationship occurs as a function of the careful arrangement and monitoring of the contingencies of supervision and teacher education. The principal controlling variable for a positive symbiotic relationship between the teacher coach and the teacher is the progress of the individual student as revealed by the data at the individual student level.

Training New and Less Experienced Teachers

The supervisor trains the new or less advanced teacher in the classroom by several routes. These include:

- Demonstration of the application of behavior analysis to all aspects of teaching and curricular analysis;
- Use of the teacher performance rate/accuracy observation procedure (TPRA) to develop behavior analytic teaching repertoires;
- Teaching the teacher to monitor graphic presentations of each individual student and using the graphs for teaching decisions that are implied by those visual displays (via strategic and tactical decision analysis); and
- Teaching the teacher to monitor the summary graphs for all the students in the teacher's classroom (classroom summary graphs) and the related decisions implied by those graphs.

Demonstration

Beginning teachers or teacher interns begin their training for a particular classroom by first serving as a second or "*reliability*" observer for the supervisor while he/she or another advanced teacher demonstrates instructional procedures. This continues until the teacher trainee can collect data that are in agreement with the model teacher and the supervisor. This step also requires that the teacher intern be familiar with the scripted teaching procedures and other curricula used by the teacher along with the instructional objectives set for each student. When the trainee acquires accuracy as a *transducer of student* behavior and is familiar with the curriculum and objectives, the trainee begins teaching portions of the class or individual students under continuous supervision.

The supervision at this point is aided by using the teacher performance rate/ accuracy (TPRA) observation procedure. The TPRA procedure focuses the new teacher on his or her students and the teacher contingency-shaped teaching responses of the teacher at the level of each learn unit presentation. The TPRA observation procedure is used to decrease teacher errors (i.e., replace teacher presentations that are not learn units, or that are flawed learn units, with presentations of learn units that are rapid and flawless). Our research and experience allow us to predict that as the teachers' presentation of instructions shifts to accurate learn units the students learn as a direct result of the teachers' instruction (Ingham & Greer, 1992; Selinske, Greer, & Lodhi, 1991). These observations and the supervisor consequence to the teacher are critical. Even expert teachers profit from them; however, for the novice teacher, they are the key means for developing effective behavior analytic teaching practices in the classroom. The TPRA is used to teach teacher assistants and parents also. In our research and school-wide applications over the past 20 years, the number of teacher observations has been a key predictor of effective schooling and pedagogy. It is not unusual for the novice teacher to receive 20 of 30 TPRA observations in the first week of teaching.

TPRA Observation Procedures

We will describe a TPRA observation at the level of a teacher and one student (individual instruction) and at the level of all or several students involved in individualized instruction.

TPRA for Individual Students

When the teacher is involved with tutoring an individual student, two forms of data are collected: (1) the rate (number per minute) and accuracy of teacher presentation of learn units per the scripted program and (2) the rate and accuracy of student performance (number per minute correct and incorrect) for the learn units presented.

The supervisor and teacher identify the student, the particular scripted program to be used, the type and schedule of reinforcement used, the antecedents involved, the instructional context, and prerequisites. For the novice teacher, the supervisor will have the teacher read and practice the procedures in the classroom under the supervisor's observation. The teacher trainer and the teacher trainee record the start and stop times as well as the elapsed times for the instructional period or program. The presence and absence of the appropriate instructional materials are also noted, such as are the exemplar and nonexemplars *faultless* or are the data collection materials present (e.g., stopwatch or timer, data collection forms)? In addition, the supervisor reviews the student's graphs with the trainee and determines the trend in the instructional sessions immediately preceding the presentation to be done. This step is key in the process of training accurate decisions and instructional analysis. The data points for previous sessions must be accurate data, up to date, and displayed correctly.

The start time of the session coincides with the teacher's presentation of the antecedent(s) for the first learn unit in a block of a minimum of 20 learn units or a block of learn units constituting a lesson or instructional session. If the instruction is interrupted (e.g., fire drills, classroom management problems, knocks on the teacher's door, intercom messages), the elapsed time clock is stopped until learn unit presentations are resumed at which time the clock is restarted. The clock for elapsed time is stopped when the last learn unit presentation is consequated or the consequation is omitted by the teacher.

Both the teacher and the supervisor simultaneously, but independently, record whether or not the student's response is correct or incorrect with regard to the prompt level or subobjective identified as appropriate for the student prior to the observation (e.g., the behavioral objective specified in the scripted curriculum or individual program for the student). Incorrect responses of the student are recorded as a minus, as are the lack of a response, and correct responses are recorded as a plus. The teacher should typically record the student's response after consequating the student response (e.g., a reinforcement operation or correction operation immediately after the student's response or lack of a response). Responses or their lack thereof may be recorded via pencil and paper (on a clipboard), with an automated data collection device, or with two or more mechanized counters for some programs. When the teacher records student responses textual passages that they read aloud, either photocopies of the passage to be read or transparencies placed over a copy of the reading passage are used by the teacher and supervisor to collect the data. In the latter case, only individual incorrect responses are marked on the transparency or the photocopy. Subsequently, the correct and incorrect responses are totaled and converted to number minute *correct and incorrect*. Initially, the particular incorrect response must be identifiable, not simply the ratio of correct to incorrect, in order for point to point agreement to be ascertained between the supervisor and the teacher. As the trainee becomes reliable, simple subtractions of incorrect counts can be made from the total words in a passage.

The supervisor also collects data on the teacher's performance as the instruction unfolds. Some supervisors use a TPRA checklist while others use a code. Either way the teacher is recorded as having

- Correctly reinforced the student's correct responses, if the student responded correctly in the predetermined intraresponse period allotted (e.g., 1, 5, or 10 s);
- Incorrectly reinforced an incorrect response,
- Omitted a reinforcement operation for a correct response or did not respond quickly enough (e.g., another response by the student intervened),
- Reinforced a student's response before the student had completed the expected response, and
- Appropriately corrected the student (e.g., correction done as prompt).

In the latter case, the student's response was, indeed, incorrect and the teacher's correction procedure included the student's emitting an imitative correct response while attending to the relevant antecedent stimulus. Typically, the correct response emitted during the correction subroutine is *not* reinforced (for exceptions, see the chapters devoted to teaching practices for students with particular verbal repertoires).

If the correction is incorrectly omitted, done incorrectly, too delayed, or emitted before the designated intraresponse period or the student is not required emit a correct response in the correction subroutine under instructional or target stimulus control, the correction is recorded as incorrect.

Table 1 summarizes the basic options and the code used. A check mark is used to indicate that both the teacher and the target stimulus were correctly presented. That is, the forms of each were accurate (e.g., the teacher

TABLE 1
TPRA Code for Teacher Consequences during One-to-One Instruction

1. R = Reinforcement operations were performed accurately. That is, they were done immediately, an apparently effective reinforcer was used (determined prior to observation typically), and the teacher did not satiate the student (the amount or time was appropriate). The reinforcer may include or be limited to verbal (vocal or nonvocal) praise, along with another consequence (e.g., token, point, opportunity to request or mand an activity).

2. ®a circled R = A reinforcement error occurred or a reinforcer was omitted when it should not have been, or the delay between the response and the reinforcement operation was too prolonged. If a consequence was not a designated or apparent reinforcer, the reinforcement operation is also noted as incorrect. The same holds true for inappropriate amounts or duration. If the reinforcement operation was delivered in any way that serves to obviate potential reinforcing effects, a reinforcement error is noted.

3. C = The predetermined correction procedures were performed accurately and timely (no delay). The student was appropriately attending to the teacher and target stimulus antecedent, and the student emitted a correction response. In some cases, for some programs, the appropriate correction procedure is planned ignoring. That is, no correction is performed and the next learn unit is presented. In this latter case, the teacher is scored as a C if ignoring occurred.

4. ©= The correction was omitted or performed incorrectly by the teacher, or the student did not emit a corrected response. If a response of a particular student is to be reinforced, and that reinforcement was omitted, a (R) is recorded also. Record reinforcement errors as (R), along with the (C) or C, depending on whether the other portions of the correction procedure were done accurately. Typically, if a corrected response is reinforced, an R is recorded again with a C or (C) depending on the accuracy of the other components of the operation.

5. P = The response was punished appropriately (e.g., verbal disapproval, a token removed) as designated by the instructional program for the particular student. This is a rare designation, since punishment operations are rarely designated. However, in the rare case, the P is recorded when the teacher punishment operation (e.g., removal of good behavior ribbon) was correctly done by the teacher.

6. ℗= A student response was disapproved either contingently (the student was incorrect) or noncontingently (the student was correct) by the teacher, and there was no predetermined punishment operation specified. This designation occurs frequently with teachers not trained in behavior analytic tactics.

7. If the teacher presented the antecedent correctly a check mark is recorded; if the antecedent is incorrect a circled checkmark is recorded.

antecedent was unambiguous and consistent with the script and the target stimulus was faultless). If either or both are in error, the observer draws a circle around the check mark.

The TPRA observations identify the teacher's rate of accuracy (antecedent presentation and postcedent presentation to the student's response or lack of response) and rate of errors, or inaccuracies, in presenting learn units (circled antecedents or consequences). The observation also results in two data points for the student or students taught (number per minute correct and incorrect) and the percentage of agreement between the supervisor and teacher concerning the student's performance (point to point agreement plus disagreement divided into agreement). The number per minute correct and incorrect (rate) for the student is determined by dividing the elapsed time of the observation into the student's correct and incorrect performance, respectively (e.g., separate calculations for each), and plotted graphically. The teacher's rate correct and incorrect is calculated and plotted also to provide a visual picture of the teacher's performance. These data are used primarily to teach the teacher or to eliminate teacher errors as a possible source of a student's difficulty. The teacher coach or supervisor reviews each correct and incorrect learn unit with the teacher using the form described. The teacher convinces the supervisor that he/she, the teacher, discriminates the problem and can correct the problem, during the review of the observation with the supervisor. As the teacher becomes familiar with the procedure, he or she simply reviews the written comments and data on the observation summary and asks questions of the supervisor.

Many supervisors stop observations and correct the teacher's behavior immediately during the observation. This provides an immediate learn unit for the teacher and also provides additional immediate opportunities to determine if the teacher has modified the error. We have not yet done the research to determine whether stopping immediately or waiting until the end of the program is more effective.

Rather than plotting the teacher's rates of correct and incorrect performance, a single score for the teacher can be calculated; this is the rate/accuracy of the teacher in an algebraic score. It is obtained by subtracting the errors (all circled codes) from all accurate teacher responses (all encircled codes). The score may result in a negative score (e.g., -5.2 or -1.2), when inaccuracies exceed accuracies. The result of the subtraction is divided by the elapsed time (e.g., elapsed time into correct minus errors). The teacher's performance and the performance of the teacher (TPRA score) are graphed on a single display. We have used this algebraic calculation for statistical analyses of large groups of data in order to compare a single measure of the teachers' performances with other measures of teacher, supervisor, student, and parent behaviors in the CABAS schools (Greer, McCorkle, & Williams, 1989). However, use of *the rate correct and incorrect is the appropriate display of data for teaching teachers* to perform effectively because the algebraic display does not reveal accuracies and inaccuracies independently.

The speed or rate of presentation (more learn units in less time) is important, thus the importance of time in the calculation (see chapter 3 for the research supporting this statement). Initially, supervisors will concentrate on eliminating teacher errors and increasing accurate performance. As the errors decline or are eliminated (mastery), they concentrate on increasing the rate of errorless performance (fluency).

As errors decrease and accuracy increases, the students' accurate rates should increase and the inaccurate rates decrease. If this does not occur for specific students, the student and context of the learn unit presentations should be examined through the verbally mediated procedures outlined in the chapters on teaching and the analysis of instruction. However, if the teacher's TPRA scores and student data across numerous observations of the teacher are summed and divided by the number of observations conducted, the mean accurate TPRA score should increase along with appropriate changes in the mean correct and incorrect student responses. The teacher's errorless performance will agree with errorless student performance. Trends in these data will show whether or not teacher's contingency-shaped skills are increasingly more accurate, inaccurate, or not changing. The supervisor is responsible for obtaining improvements. Just as the teacher must learn to fit tactics to individual students, the supervisor will need to fit tactics from applied behavior analysis to improving the teacher's performance (see the list of tactics in chapters 5 and 6).

Subsequently, as teacher accuracy becomes less of a concern, the teaching operations per specific students may be examined by verbally mediated strategies. Such strategies will result in changes in programming for specific students as described in the chapter on teaching and instructional analysis. Problems in curriculum sequences are also identified by the process of elimination as described in the teacher and instructional decision chapters.

The information obtained by the TPRA is used in several ways. It is used to instruct the teacher in the classroom during or immediately after the observation. The TPRA data are also used at the teachers' regular individual weekly conference with the supervisor to assist the teacher and supervisor to identify instructional objectives for students, teachers, and supervisors. Finally, the mean scores for the teacher for all observations for each week become part of the composite mean for all teachers in the school for the week in question. These latter statistics function to assist the supervisor, teachers, and consultants who review supervisor performance. The composite data identify improvements or difficulties and can be used to *set school-wide objectives based on school-wide teacher performance*. When the teachers have rare incorrect presentations, the data are summarized as cumulative number of observations with and without errors. The cumulative graphs are useful to review the supervisors as well as the teachers' performance.

For supervisors and teachers working in some settings (e.g., preschool, primary grades, or schools for students with developmental disabilities), the supervisor may stop the observation (and the timer) after a learn unit for which the supervisor wishes to make an instructional assist to the teacher. The supervisor may, at that moment, demonstrate an instructional procedure, correct or clarify an instructional procedure, or revise the instruction or simply provide a verbal prompt. The clock is restarted and the teacher has the opportunity to correct his or her performance. Alternately, the teacher is given feedback after the session and subsequently teaches a second session incorporating the feedback. For teachers working with students for whom the interruption would be detrimental or create difficulty, the feedback is given after class and a subsequent observation is scheduled.

For novice and less advanced teachers or for advanced teachers using or developing new scripted or programmed instruction, the observation serves to teach the teacher to emit accurate operations associated with the learn unit. Once the teacher's *delivery* of the scripted procedures is ruled out as a variable for a particular student's difficulty, the teacher and supervisor can pursue problems associated with inappropriate fit of the scripted or programmed instruction to the individual student. If the teacher's presentation is errorless, then the locus of the problem resides with the instructional program per se, the context or motivational setting, or the student's history and rarely phylogenic variables (e.g., the student has a hearing impairment).

At this point, the supervisors and in time the teacher's uses of verbally mediated strategies come into play. The experienced supervisor will draw on thousands of prior experiences to predict more likely sources for the difficulty. He or she can then suggest procedures from the literature or prior experiments or, better yet, prompt the teacher to suggest possible strategies and tactics to rectify the problem. If there are no identifiable operations from the literature to solve the problem, solutions that are more likely to work may be tested directly through experimentation with the student in question. The accumulated revisions in teacher operations, scripts, or programs provide a repertoire of effective tactics that advance the expertise of both the supervisor and teacher and ultimately the performance or the student.

The *in situ* training process should consist of complete learn units between the supervisor and the teacher. The fact that the supervisor and teacher work jointly on solving the instructional problems serves to foster a cooperative relationship, if the supervisor focuses on reinforcing correct or progressively correct performance. Also, the fact that the supervisor too will encounter difficulties that are resolved through researching the literature or experimentation emphasizes the necessity of teachers and supervisors being involved continuously in learning about better pedagogy. As the teacher becomes more sophisticated, it is not unusual for the teacher to provide an instructional learn unit for the supervisor. This latter issue emphasizes the necessity of a continuing education program devoted to expanding both the teacher and the supervisor's expertise in the sciences of behavior and pedagogy.

TPRA Observations for Group Instruction

The TPRA observation for the teacher working with the total class provides information on the teacher's performance in managing the performance of all students. It is most valuable for students who are able to work on individually programmed instruction, while the teacher, tutors, and other instructional personnel function to tutor students. These students are the students who are performing at the level of reader/writer and the teacher uses the tactics described in the chapter devoted to teaching students with this level of verbal behavior. In this instructional setting, many of the learn units are presented in textual form (work sheets, workbooks, computer programs) to students according to their individual achievement levels at the moment. However, vocally presented learn units are also incorporated. This occurs when a teacher encounters a student problem and provides brief tutorial learn units. This is easily done because most of the learn units between the students and the teachers are in written form and are collected after the session for the period of time the observation is done, while the tutorial presentation data are collected in the class as they occur.

Students or instructional personnel draw a line or place a mark on all students' individual papers to *designate the onset of an observation period*. They also place a mark after their last response on a learn unit when the *end of the observation period occurs*. During the observation, the supervisor records vocal antecedents and postcedents (e.g., teacher consequences) and later sums the written learn units that have all components (antecedent, response, and postcedent). Those that are accurate are tallied and the inaccurate ones are subtracted. Similarly, student accurate and inaccurate responses are tallied. Both student and teacher performance are converted to rate as in the case of the individual teacher and student observation. In the latter case the performance of the teacher involves all students' responses. The more learn units that are consequated accurately, the higher the number per minute and the lower the intraresponse times between learn units.

The total for the written learn units performed by students and the teacher from different time periods may be used also. In this case, the complete learn units may include corrected responses returned to each student at the beginning of the next day. More immediate consequation of learn units is desirable, but in classrooms with large numbers of students, it is often necessary to include overnight learn units. That is, papers returned and seen on the day of the class are also counted in a 24-hr time frame. These become number of written learn units per day for all instruction or for a particular subject area. However, the completed learn units are counted for the time period in which the student reviews the teacher's written comments and must include the students' corrections. The learn unit is complete *only when the student views the corrected paper and completes the corrections*. Written learn units that are complete include items with correct, incorrect, or corrected marks. The feedback must specify whether the response was correct or incorrect in order to be complete and the student must attend to the correct mark and perform corrections for the incorrect marks. The data show *that it is necessary for the student to perform corrections before the learn unit is completed* (Hogin-McDonough, 1996).

When doing an observation, determine the correct and incorrect responses associated with the particular instruction in process from the material or from the teacher. If the instruction involves individualized seatwork, count teacher student vocal interactions or written vocal interactions that occur at the time. Return the next day or after a period set for the feedback to the students to complete the observation. Total the completed learn units for vocal performance of the teacher and the number of written learn units that were corrected from the prior assignment. You may separate the data also into *in situ* learn units and written learn units. Determine the correct and incorrect learn unit responses before corrections. Divide the total learn units received by the correct and incorrect responses respectively for a given time period. The time period is the total time for the initial work on the lesson plus the time involved for the students to do the corrections. For example, if you observed a lesson on 1 day for 20 min record that time. When you return the next day, or later on the same day, check the papers to determine the completed learn units. Use the 20-min time to determine number per minute correct and incorrect for students. Responses that have not been marked by the teacher are errors and those that have been marked and corrected by the teacher are accurate learn units. In some classrooms the teacher will be correcting and completing written learn units as the students perform individually. In the latter case you can determine the number of completed learn units and correct/incorrect learn units at the time of the observation. Be sure to include tutor-controlled learn units and student self-completed learn units. Thus, a teacher who is running a classroom that used tutor, teacher, and computer learn units will have significantly more learn units.

Alternatively, you can observe *in situ* learn units done by the teacher/tutor/teaching device at only the time of the 20-min observation, in which case you also want to determine the number of written learn units on a daily basis for the teacher, by totaling all of the completed written learn unit(s) per day. Better yet, advanced teachers will total the learn units that they teach each day and the correct and incorrect responses of the students. Typically this is done by subject areas such as science, math, reading, and other subject areas. These are then summarized for a total for the day. As the students increase their self-management accuracy, they will total their own learn units for each subject area as part of their instruction in self-management.

The individual responses of students are placed in the students' folder. The teacher or student also graphs the prior performance. The overall correct and incorrect response rate of the student results in: (a) more advanced material for the student when criterion is achieved, (b) new material, (c) new tactics or material designed to correct the students' problems, or (d) brief vocal tutorial interactions followed by more independent responding by the student.

The individual student and classroom (all students') performances on learn units resulting in written responses with textual antecedents are classified by function (see chapter 6 on teaching students with advanced verbal repertoires). Learn units may also be classified by subject matter. Curriculum functions are particular response repertoires. These include essay responding, report writing, exemplar production (e.g., fill in the blank, complete the sentence, or construct the exemplar), and exemplar identification (e.g., finding the correct exemplar as in multiple-choice responding). For mathematics, the repertoire includes: identifying the correct answer or operation, performing the operation and obtaining the correct answer, identifying the correct operation for a problem, doing the operation correctly and producing the correct response, and identifying the problem and converting it to operations and solutions. Similarly, functional repertoires are identified and classified for music, art, and vocational instruction. These functions and what they consist of are described in the chapters on curriculum and teaching students with advanced verbal repertoires (chapter 6).

Learn unit rates are compared within categories, since *the opportunity for rates varies with the function of the repertoire and the component behaviors for each learn unit in the repertoire*. For example, the behaviors involved in writing a report are extensive. A learn unit consists of a written comment that specifies how the student is to correct the response or responses. Each counts as a unit at the time that the student receives the comment. Learn units involving exemplary identification involve fewer student and teacher responses and are thus more likely to be more frequent.

Teacher rate and accuracy for written responses with textual antecedents are based on the numbers of correct approvals, inappropriate disapprovals, correct corrections, or incorrect corrections using the formula (correct minus incorrect divided by the time frame involved). For units that involve long periods of time (e.g., essay writing), the time frame is the in class time that is involved in writing and correcting responses. When the assignment is done as homework, the time is the number of minutes on the day that the responses are reviewed and corrected by the student. For example if the school day is 5 hr then the number of minutes is 300 divided into all of the learn units (total, correct, incorrect). The homework learn units are then added to the totals for all other instruction for that day.

Refer to the chapter on the measure of teaching for the designation of response criteria for various steps. The supervisor must be familiar with the learn unit and how students' graduate to larger numbers of responses in order to identify the correct learn unit for the students and teacher who are the target of the observations (see chapter 2).

The supervisor modifies the TPRA to include specific targets, such as the measure of the teacher's reinforcement of students following or not following classroom rules while the teacher simultaneously provides learn units in the class. This is a critical repertoire for the teacher as we described in chapters 5 and 6. Once the teacher is fluent with individual instruction, he or she is simultaneously observed for both individual instruction and control of the entire classroom while providing individual instruction. The teacher's accurate reinforcement of the behaviors of other students who are appropriately engaged in individual tasks is counted in the teachers' total consequation tally along with the teacher's consequation of the individual student's behaviors that she is teaching. These types of TPRAs are graphed separately. The TPRA should be directed at the specific repertoires such as the written responses to textual learn units or teacher-presented learn units across students, with or without classroom management data. When collecting data on presentation of both learn units and classroom management reinforcement, record learn units as usual and record accurate reinforcement of following rules as an *A* and an error in reinforcement, such as paying attention to an infraction of rules as a circled *A*; record any disapprovals as a circled *D*. Use this information to reinforce and correct teacher behavior following the observation. Include learn units and reinforcement of rule following in your calculation of the teacher's accuracy.

Objectives and Learn Units

There are two measures of contingency-shaped performance that are important for the teacher: First the TPRA is used to assess fine-grained instruction with one student and, second, the numbers of learn units presented and the number correct per day/week for all students in the classroom. This latter measure is a classroom-wide measure of both productivity and quality (Greer et al., 1989; Dorow, McCorkle, Williams, & Greer, 1989). It is important to subcategorize learn units by functional repertoires (e.g., essay writing) and subject matter in order to identify areas that need assistance or commendation and to measure overall change per category.

Both the TPRA and classroom summary data must be judged in light of the attainment of valid and functionally defined objectives and the category of objectives achieved. A teacher of a 12th grade English composition class that is devoted to teaching students to write college-level papers may generate low rates of learn units compared to a high-school biology class emphasizing the identification of botanical exemplars. In the case of the English class, the objectives need to be defined and measured such that the objectives identify the functional components of the writing of compositions (e.g., their effect on a reader—see the chapters on curriculum). Subobjectives that are tied to the terminal or long-term objectives must also be defined. *The subobjectives must be capable of division into levels that provide good and functionally useful training and measurement levels commensurate with the baseline repertoires of those students encountered.*

The role of the supervisor is to support the teacher in effective design of instruction and its measurement. General and specific strategies and tasks from the science guide both teacher and supervisor. On occasion the supervisor will have authority by virtue of her expertise, while on other occasions the teacher will have the authoritative expertise. In the latter case, the supervisor's work in the classroom with the expert teacher serves two purposes. First, the supervisor will acquire expertise to teach new or intermediate level teachers. Second, the supervisor can assist the master teacher by providing feedback to the teacher such that the teacher herself may identify and strive for her own goals. The goals may range from maintenance skills to doing research on improving the general design of instructional arrangements in order to contribute to more effective instruction at the classroom level or at the level of the science of pedagogy.

This scheme provides a set of general operations that will allow the redeployment of teachers to supervising roles, based on natural fractures that are related to engendering more effective classroom instruction. In the English teacher example, expertise in the design and instruction of composition may change the deployment of staff. Expert teachers who are strategic scientists will become the subject matter supervisors. This need not change the number of teachers. Rather, the number of students assigned to teachers may vary according to the utility of the design. All may be involved in reading compositions, while the new supervisor may devote time to changing teacher behavior and improving the curriculum in concert with the teachers. Students will spend more time writing and rewriting papers with increased numbers of teacher consequences to student responses and an increase in the effectiveness of those consequences as judged by the achievement of objectives. Allowing the data of the individual student and the replication across collective students will result in the continuous modification of the deployment of instructional and supervisory staff. The key role of supervision is to maintain instructional design such that the learning behavior of students controls instructional deployment.

Costs can be related to such deployment and student effects within the constraints of a fixed budget. For example, design X may result in lower costs per objective than design Y. Alternately, the importance of the objectives may justify additional budgetary resources. The decision as to whether or not English composition is more important than another subject area will call for value judgments and the setting of priorities. Providing measures that identify instructional effectiveness vis-à-vis the objectives identified makes those decisions more accountable. In a community in which the students are most likely to attend college, the importance of English composition objectives will differ from schools in which students will enter more direct vocational training schools. Decisions about budgetary allotments can be tied more directly to the community in terms of educational performance, not simply the provision of more services that may or may not result in more effective attainment of the community and schools' values.

The supervisor tools discussed thus far are related to direct classroom skills and productivity. They must be related to the *vocabulary of the science* and *analytic repertoires* by direct instruction from the supervisor. The analytic repertoires are three-term contingencies also. We teach them through the measurement of decisions that the teachers make about the visual displays of their students' progress. The particular protocol that we use was first introduced in the chapter devoted to teachers' analytic repertoires ("The Strategic Analysis of Instruction and Learning," chapter 4). We repeat that protocol, but with emphasis on the role of the teacher coach in the analysis of the decision-making responses of the teacher as a tool to teach the decision-making repertoire to teachers.

Analysis of Instructional Decisions

One of the most important components of *high-quality teaching consists of the teacher's use of the visual displays of student learning to make instructional decisions.* While the use of the data to make decisions about instruction is only one of the uses of data, it may be the most important one. When the teachers' decisions are accurate, the students' make optimum progress. When the data are not used to make decisions, the student's progress is thwarted. There are several steps in the process.

The teacher must know:

1. *When a decision is needed* (i.e., when to continue with the current operations and when a change is required);
2. *What kind of decision is needed* (i.e., move the student to a new objective or change the instructional tactics);
3. When new tactics are needed, *what strategic questions must the teacher ask* about the data and the performance of the student (i.e., where in the learn unit context is the problem most likely located)?
4. When a likely source of the problem is determined, *what existing research tactics are likely to fix the problem* (i.e., the use of interspersal of known items, changes in reinforcement schedules, changes in antecedent or consequences delivered by the teacher, changes in the display of instructional material)?
5. After the tactic is implemented, *do the data show that it is working or not working* (i.e., continue with the new procedure or find a better tactic or ask more strategic questions about the effects of current instruction or the student's learning history)?
6. To *continue* the above sequence *until a solution is found.*

In order to teach this repertoire to teachers, first the decision-making response of the teacher must be measured. The following section defines the responses and the contingencies as well as the measurement of decision-making responses.

A. Criteria for Decisions and Their Measurement

1. A single graph check should include a standard number of instructional programs for a teacher (i.e., five different curricular programs such as math, composition, reading, global studies, reading, self-monitoring, identification of own goals) with a single student that can be blocked in numbers of decision opportunities (usually 20, across five instructional lessons or programs). The reviews should be done with programs that the supervisor has checked for reliability of the teachers' instruction using TPRA observations. The TPRA observation eliminates many possible errors in instruction that are attributable to the contingency-shaped repertoire of the teacher. Once the teachers errors are eliminated as a source of the problems, the analysis can focus on the other components of the learn unit context. Each group of five programs for each student results in a separate error score. The programs should include 20 decision opportunities. Of course, the numbers of learn the level of verbal behavior of the students and the curricula involved determines units in any given program. For example, many prelistener to speaker programs include 20 learn unit sessions while math lessons for self-editors will vary in the number of learn units per instructional session or assignment.

2. Use the decision criteria listed below to determine the correct and incorrect decisions for each program within the sessions described above.

   • Determine the *correct percentage of decision errors and the total number of decision opportunities*. Divide total opportunities into correct decisions to obtain the percentage of correct decisions. This result should be graphed on a separate graph for the teachers, as are their TPRA observations. For example, if the teacher had 20 opportunities for decisions and was correct for 15 of those opportunities, the correct decisions score for the teacher would be 20 divided into 15 or 75%. Do this calculation for each program or calculate from all decisions opportunities and responses for all programs depending on the focus of the teacher training at the time. The goal is 100% accuracy in decision-making responses. However, there may be several subobjectives associated with teaching decision responses. Initially the target may be less than 25% errors, followed by progressive decreases in errors until decisions are errorless. The decision goals like TPRA goals are set for successive module components until the performance is errorless.

B. Decision Opportunities

The protocol that we describe is based on the research done by Keohane (1997). She found that the decision accuracy of teachers predicted the immediate and long-term outcomes for the teachers' students. We have subsequently replicated the effects of using the protocol across several schools. While the number of data paths that should be considered for a decision opportunity may change with future research, the one we specify below has a scientific basis in terms of student outcomes.

A count of three *data paths* is the earliest opportunity for a decision. The first data point is a point of origin. Thus, from the first data point of the onset of teaching a new objective to the second data point is 1, to the third data point is 2, and to the fourth data point is 3. At this point determine whether the trend is descending or ascending or there is no trend. If a trend, or lack of, or the achievement of an objective or criterion, is present a decision should be made whether to continue, change tactics, or move to a new objective. If the direction changes for one or more data paths, a judgment of trend across five data paths should be made. If no clear trend is apparent at five data paths, a decision opportunity is at hand. Of course the achievement of a criterion involves only the number of consecutive sessions and data points per the program under review. In other words you do not count data paths for criteria achievement decisions; rather the number of sessions that constitute the achievement of instructional objectives is the rule. In those cases in which a student performs on the first session with 100% accuracy, do not treat this as the achievement of an objective since such a high initial level of performance indicates that the student has already mastered the repertoire prior to instruction. That is, the subobjective step is too small for the student and a more advanced subobjective needs to be taught.

C. Trend Determination

Ascending trends should result in continuation of the current teaching operations in most cases. A descending trend or no trend should result in a decision to change tactics and a horizontal broken line should be present. If criterion was achieved the broken line should appear also. Each decision opportunity that is consistent with these definitions is a correct one, while a decision that is not or the lack of a decision is an error. *Each session run without change is another error*, and when a decision for change is made after these errors it is not counted as a correct decision because it is, at this point, a correction. The correction is an instructional opportunity for the teacher coach with the teacher. Prompt decision-making questions until the teacher comes up with a viable decision.

D. Learn Unit Analysis Decision

Once a decision for change is made, the next step is the analysis of the particular decision. The strategic choice is counted as a correct decision, if the description of the potential problem by the teacher includes:

- A *defensible reference to the learn unit* (i.e., the relationship between the teacher and student's three-term contingencies and the three-term contingency of the student);
- The *three-term contingency of the student;*
- The *instructional history* of the student; or
- The *setting events.*

Of course the terms used to characterize the decision must be accurate and reasonably viable scientific tacts (e.g., tactics selected from the lists in chapters 5 and 6 or from a research paper). The teacher has many alternatives, however, since the systematic process of fitting scientific tactic to instructional problems has been introduced only recently in the research literature. Thus, correct responses consist of any reasonable reference to the problem that draws on scientific tactics and scientific verbal mediations. However, if the analysis of the potential problem draws on prescientific rationales or hunches not grounded in the science, the strategy analysis is counted as an error. Once again the teacher coach uses the response of the teacher as an opportunity to reinforce or correct the behavior of the teacher. Another correct decision, at this point, may be to do a correlation (AB design) or functional analysis drawing on the above components as the basis for the analysis. Each correct or incorrect decision for doing the experiment, of course, constitutes a decision point, for which correct or incorrect tallies may be made and each decision an opportunity for the teacher coach to teach the teacher.

E. Selection of the Tactic

If the tactic selected for the solution to the student's problem is data based from the literature, another correct decision is counted. If the tactic is not from the literature and the student's instructional history, then the decision is counted as an error. The teacher may skip the above, item D, and jump directly to the tactical decision, without an error if the tactic is related to a probable strategy. In the latter case a correct decision would be counted and the correct decision would be reinforced by the teacher coach. An inappropriate decision is an error and creates an instructional opportunity for the teacher coach. That is, prompt the teacher through a strategic analysis and obtain the response from the teacher; do not simply provide an answer. The teacher coach is of course providing learn units to the teacher in the process of analyzing instructional problems and arriving at solutions.

F. Implementation of the Tactic

If the tactic is implemented correctly according to research-based procedures, the teacher receives another count of correct. If the tactic is not implemented accurately, the count is an error. Following the implementation, return to the trend analysis.

The TPRA observation procedure is used to teach classroom practices that should be the initial and standard teaching procedures that a teacher should use in the classroom. The goal is for the tested contingency-shaped teacher behaviors to become automatic and fluent. The teacher coach must teach them *in situ*, if the contingency-shaped behaviors are to be consistent with behavior principles and tactics.

The procedure for analyzing instructional decisions provides tools to teach/ measure the verbally mediated repertoire of the teacher. Using the operations described, the teacher coach can prepare the teacher to make accurate decisions about what should be done when the basic contingency-shaped operations that are taught via TPRAs are not effective with a particular student. Obviously, the teacher must have fluency with the vocabulary of the science in order to do instructional analyses. The procedures we have described provide the basic operations to connect classroom practice to the science in both common practice and in the process of analyzing and solving instructional problems. All of these repertoires need to be coordinated and organized into a systematic curriculum for the teacher. In addition, the sciences of behavior and pedagogy need to be applied to the instruction of the teacher just as the teacher applies them to the teaching of the student. If the teacher is not learning, it is the function of the teacher coach to provide the strategies and tactics that result in the teacher learning all three major repertoires at increasingly more sophisticated levels of performance. Fortunately, there is research literature that provides the necessary expertise. In addition, the CABAS system has used the procedures from this literature to train over 500 teachers in CABAS schools. The literature associated with the three teacher repertoires is termed Personalized System of Instruction (PSI) (Buskist, Cush, & de Grandpre, 1992). There are over 100 studies in university settings that have consistently found that PSI is the best set of procedures to use to teach mastery in undergraduate and postgraduate subject matter, ranging from behavior analysis and physics to music history. It replaces the traditional lecture and workshop model in the continuing education of professionals. All research comparisons between PSI approaches versus lecture/workshop approaches have shown that the PSI approach is more successful (Buskist et al., 1992). Thus, we incorporate PSI in our continuing education for all professionals.

Please note that the use of PSI and the teaching of the three repertoires of teaching as behavior analysis are completely compatible with advanced undergraduate and graduate degree programs. They have been used at Teachers College Columbia University to train many masters-level students and Ph.D.-level students to apply advanced behavior analysis in a comprehensive manner for all types of instruction for students with the complete range of verbal repertoires. Thus, the procedures we describe are applicable to university courses, classroom instruction, and practica, as well as to in-service professional training.

TEACHING, MAINTAINING, AND EXPANDING
TEACHER REPERTOIRES THROUGH
PSI-BASED PROCEDURES

The Personalized System of Instruction developed by Fred S. Keller has several key features that are incorporated into both classroom instruction for students with reader writer repertoires (see chapter 10) and the *in-service training of teachers* in the science of behavior and the science of pedagogy (Keller, 1968). *Modules* of in-service training are established for each teacher. Typically these modules are grouped in clusters of 5 or 10. Meeting the criteria for a cluster of modules in CABAS schools results in one or more of the following for the teacher:

- A salary increase (usually in addition to annual cost of living increments);
- An increment in ranks (Teacher I, Teacher II, Master Teacher, Assistant, Associate, or Senior Behavior Analyst Teacher, or Supervisor in CABAS Schools); and
- Commendations and awards as well as CABAS Board Certification of expertise.

In addition, weekly performance efforts result in weekly commendations (e.g., most improvement in learn units, most improvement in objectives achieved with students, most quizzes at criterion). However, the progress of the teachers *needs not be tied to salary and advancement*. In many schools that use only components of CABAS, the PSI module system has been used with success to provide an excellent data-based continuing education program for teachers. In the latter case the teacher coach must be ingenuous in the development of motivational setting events and the delivery of effective forms of reinforcement. It is well documented in the behavior analysis research that simply posting teachers' progress and reviewing that progress on a regular basis is effective reinforcement for many teachers. The TPRA and decision analysis protocols have built-in learn unit capabilities that typically reinforce teacher behavior, and the natural effects of these latter repertoires soon reinforce their use. Nevertheless, the teacher coach must always be alert to the arrangement of contingencies to motivate teacher learning.

Each module has three components. They are the three key repertoires of teaching as advanced applied behavior analysis that we described in detail in earlier chapters. One is devoted to *contingency-shaped objectives* (i.e., automatic behavioral teaching practices). One is devoted to advances in *verbal behavior about the science* (i.e., the vocabulary and scientific tacting repertoire). One is devoted to *verbally mediated strategies* (i.e., the scientific analytical repertoire). Each component of each module is adjusted to the level of experience and prior achievement of the teacher along with the learning objectives and performance levels of the teacher's students (e.g., grade/age levels, subject and curricular functions). In CABAS schools we use standard criteria set for each collection of modules that constitute a level of teacher expertise or rank. Examples of these modules are provided in the study guide available from the author. But the use of standard criteria is optional. While the basic rudiments of the science must be mastered by all, the order in which those are mastered are best arranged to meet the immediate needs of the teacher and his or her students. The implicit reinforcement effects of using the latter procedure are apparent.

The verbal repertoire associated with the science is a key one because it must be used to discuss student performance, devise new strategies at the level of instructional analysis, and serve as the means of focusing efforts on the relevant instructional variables. The repertoire involves two response classes: (a) providing accurate and rapid written definitions of principles and tactics of the science and correct classroom exemplars of those terms and (b) providing vocal descriptions of the science per the student and the visual displays of student and teacher performance (i.e., scientific tacts). In both cases, the objective is for the teacher to describe the principles, strategies, and tactics per the precision of the science and to describe his or her own performance and that of the student precisely according to the science.

The same repertoires are taught to teacher assistants based on their role in the classroom and their educational background. In some cases, teacher assistant repertoires are tested vocally (sometimes with audiotaped readings of the assignments). Of course they are to receive extensive training via TPRA observations in order that they learn to use effective contingency-shaped classroom practices. Teacher assistants are not typically involved in acquiring verbally mediated strategies, although some teacher assistants do develop expertise in the analysis of instructional variables because they become interested in doing so when the teacher coach contingencies and the coaching repertoires of the teacher are adequately arranged.

Teaching Verbal Behavior about the Science

The modules for the verbal repertoire are arranged, like the other repertoires, according to levels of sophistication. The early levels are concerned with basic principles and tactics and include topics such as the following:

1. Rules, praise, and ignoring;
2. Principles and tactics of verbal (vocal and written) reinforcement;
3. Principles and tactics of nonverbal prosthetic reinforcement (e.g., token systems, PSI hierarchies, edible reinforcement; activity reinforcement);
4. Principles and tactics of direct measurement of learn units;
5. Principles and tactics of social behavior measurement;
6. Principles and tactics of visual display of learn units;
7. Principles and tactics of the measurement of social behavior and their use in visual displays; and
8. Principles and tactics of evaluation or research design (e.g., multiple baseline, reversals, changing criteria designs, probe designs).

More intermediate levels involve more sophisticated treatment of each of the above tactics and in addition include the following topics:

1. Reading and summarizing (research summary forms) basic and applied research;
2. Providing accurate visual displays of all students individually;
3. Describing, using, and taking advantage of TPRA observations;
4. Summarizing the research in subcomponents of the science (e.g., findings in math teaching research, music education research, reading research, writing research, verbal behavior research);
5. Principles and tactics of motivation (establishing operation);
6. Principles and tactics of reinforcement schedules; and
7. Principles and tactics of stimulus generalization; principles and tactics of general case instruction.

Each topic for the teacher's module (e.g., "basic reinforcement operation") is assigned a reading or readings. For inexperienced teachers or teacher assistants, the readings might include portions of pamphlets covering subtopics in behavior analysis tactics (e.g., "how to use reinforcement"); at the next level the teacher might read a chapter in an intermediate-level text such as the one by Sulzer-Azaroff and Mayer (1986), while at still another level he or she might read Catania's (1998) coverage of reinforcement and/or Cooper, Heron, and Heward's (1987) treatment. Each successive module that addresses "reinforcement operations" can treat the subject in greater depth or from a more advanced perspective. Typically in CABAS schools, initial modules address tactics alone (e.g., basic reinforcement operations); the next level addresses both principles and tactics (e.g., reinforcement principles, reinforcement operations, relating the two).

Ideally, each reading should identify the objectives that the teacher is to master by providing practice with short answer or short essay questions that are treated by the supervisor as learn units for teachers. Tutorial assistance is usually identified (e.g., periods in the week when an advanced teacher or the supervisor is available to answer questions and provide tutorial drill). Teachers may or may not avail themselves of this opportunity. Many simply read/study the material and request a quiz when they are ready to take the first test trial. If teachers do not meet criterion on the first quiz trial, they may use the feedback on their quiz from the supervisor and avail themselves of tutorial assistance (at their discretion). Typically, after teachers have taken two or three quizzes, they are able to achieve criterion on their first trials at taking further quizzes. They learn what to study, how to practice the response, and whether or not they need to seek tutorial assistance. They also learn more about how to use PSI with their own students.

Subsequent topics in the science typically involve the use of prior material. Thus maintenance of prior repertoires is programmed by virtue of the interdependence of the subject matter. The order of introduction of subject matter is typically based on (1) what is needed by the teacher to operate effectively in his or her classroom and (2) traditional treatments of the science in beginning, intermediate, and advanced texts.

Teachers proceed at their own pace consistent with the basic science and the tenants of PSI. However, it is important for new teachers, particularly, to master the basic and intermediate material rather quickly because the progress of their students is tied to the teacher's facility with the precision of the language of the science as that language pertains to contingency-shaped and verbally mediated repertoires. The completion of five modules, for example, may result in a salary increment, an increase in rank, and official commendation. We use the completion of modules as a "portfolio assessment" of teacher progress and as an inventory of what the teacher has learned and needs to learn. The collection of modules also functions as syllabi for university-level instruction. The availability of tutors and the frequency with which other teachers complete their modules are often setting events that enhance the reinforcement effects of completing quizzes and modules. Weekly meetings between individual teachers and supervisors are devoted to the progress of teachers and their students. Assistance for both the teachers' study and the teachers functioning with their students is offered at those meetings. Supervisors provide feedback on all components of teacher performance concentrating on positive reinforcement of improvement and corrections (e.g., suggestions) proposed in prompt forms in meetings in and outside of the classroom.

If the system is running appropriately, teachers will take quizzes frequently, will appreciate TPRA feedback, and will seek assistance with the development of new instructional tactics. Once the teacher is engaged in learning the science and receiving supportive instruction and assistance from the supervisors, the resulting morale of the instructional staff is high. Although initially in the CABAS schools teachers were mandated to complete modules within a generous grace period, in more recent years the rate of completion has been left entirely to the individual teacher. This latter revision is more consistent with the self-paced component of PSI.

The same basic principles for setting up PSI modules for teachers are in operation for students in classrooms in which each student proceeds entirely at his/her own pace. If particular teachers or particular modules are causing undue delays, break down the topic more and increase the reinforcement possibilities as described in the chapter on teaching.

Components of Modules for Classroom Performance

The component of each module concerned with the contingency-shaped behavior of the teacher (e.g., actual classroom performance) has two components. The first area concerns the observed *in situ* performance of the teacher that is tapped by the TPRA or other observation procedures. The second area concerns the daily performance of all students in the teacher's class. This is taught via the use of summary graphs of the performance of the teacher (i.e., learn units to criterion, taught and correct learn units, objectives achieved) and the performance of individual students as well as the conglomerate performance of all of the students in the classroom.

In Situ Observations

The *in situ* objectives include the following teacher repertoires and each requires:

(a) Setting objectives based on the teacher's initial performance and agreed on goals with the teacher and
(b) Using appropriate observations to provide feedback and to shape the teaching behavior of the teacher to achieve the objective specified.

The target objective should exceed the mean of the baseline performance, but only to the extent that the objective is attainable in a week or two. Typically, the objective must be achieved for two or three successive observations.

The response classes within this repertoire include but are not limited to the following:

A. Teacher performance rate/accuracy:
 1. Teacher performance rate/accuracy with individual students (or pairs of students in which learn units are alternately presented to each student in the pair).
 2. Teacher performance rate/accuracy with group choral responding.
 3. Teacher performance rate/accuracy in PSI settings with textual antecedents and responses. This is a setting in which the students are working at individual tasks following written instructions and responding mostly in written form. The teacher circulates among the students dispensing vocal learn units as needed or the students come to the teacher for learn units. This latter TPRA is used in classes for students who are learning advanced verbal repertoires including advanced self-management.
B. Instructional control of social behavior and student attention:
 1. Group engagement and rule-following responses;
 2. Group engagement coupled with one-to-one instruction; and
 3. Behavioral observations of aberrant behaviors.

The specific instructions and coding procedures for all observations were provided earlier in this chapter. The discussion at present concerns how they are used to set performance objectives for the classroom performance component for each module. The observations and supervisor learn units for the observations provide the basic teaching tool for assisting the teacher to achieve objectives in contingency-shaped instructional operations.

Initially, new teachers will need assistance with both instructional control of the group and their performance with academic responses as tapped by the various forms of the TPRA. While instructional control (e.g., all students following class rules and engaged) is a necessary prerequisite, it is only a *prerequisite* for academic responding. Once the rules are stated positively and are sufficient, yet few in numbers, the teacher teaches them by frequent reinforcement of rule-following behavior. It is critical, however, to shift the reinforcement to academic responding (i.e., learn units) as soon as possible because in well-designed classrooms, the social behavior of students or instructional control (by the teacher) is a by-product of academic reinforcement. As the teacher becomes more sophisticated and as the diversity of individualized and well-programmed material becomes available, the reinforcing consequences of academic responding will increasingly obviate the need for reinforcement of the topography of engagement (see chapters 5 and 6). As the teacher progresses through modules, the emphasis on academic instruction will eventually replace the emphasis in instructional control per se.

Teachers may progress from module to module on any component of a module (e.g., contingency-shaped behavior) before completing all components of a module. That is, progression on contingency-shaped behavior should not be held up because the teacher has not completed all three components. Thus, a teacher may be working on the sixth module for a contingency-shaped objective, the fifth vocabulary component, and the third module component devoted to analyzing instruction. However, the teacher should not fall behind extensively in one of the three components. Thus, proceeding with a new component may at times require an individual *contingency contract* (see Sulzer-Azaroff & Mayer, 1991) specifying the completion of all components at a given level. Much of this falls under the responsibility of the instructional skills of the supervisor or professor.

Daily Performance of All Students

The establishments of target objectives for learn unit presentations and the achievement of objectives by the teacher's students are based on the baseline performance of the teacher and classroom in question. The target objectives for learn unit presentations (with or without correct responses) and objectives attained must be slightly higher than the baseline (typically across 2 weeks) but achievable in a 2- or 3-week period. In preschool programs, increases of 50 to 100 learn units per day are typical goals. However, the specific objectives must be based on the functional areas taught, the subject matter, and the grade and verbal levels of the pupils in the class. Like the performance of the teacher with the student, the performance of the supervisor with the teacher is strategically governed by initial performances and success, not a blanket level of performance. *It is critical to encourage the teacher to set his or her own goals.* The research shows that gaining the teacher's input in setting her own goals is an effective tactic to motivate study.

For example, if the initial learn unit presentation rate for all students daily is 500 units and the correct rate is 300, an initial and reasonable level might be 2 weeks of 500 unit daily presentations with 350 correct responses. More advanced teachers would teach 2000 learn units with more than 1800 correct responses. The achievement of instructional objective targets should be arranged similarly. The subject matter, age, and repertoires of the students in the class and the functional repertoire (e.g., writing, reading) determine the initial goals for the teacher and the classroom. The teacher, however, must take a major role in setting the objectives.

Supervisory Repertoire: Critical Operations and Self-Monitoring

The driving force for the behavior of both teachers and supervisors is the progress of students. The objectives and measures of supervision and instruction reflect the importance of the individual student. Supervisory responses function to promote:

(a) Learning for students;
(b) Instructional skills of teachers, which in turn increases the probable learning of students; or
(c) The care and well-being of students.

The procedures used by supervisors to monitor their own performance that have been developed and used in CABAS schools have been subjected to experimental tests and found to be valid predictors of teacher performance and student learning (Greer et al., 1989). Specific repertoires (e.g., TPRA observation, visual feedback, vocal feedback) have been demonstrated to affect both teacher and student behavior. The effects of other repertoires (e.g., completing payroll, ordering supplies, producing reports) on student and teacher behavior are less clear. However, the completion of these tasks is necessary to the functioning of the school.

The *supervisory tasks* that are counted vary in the effort, skill, and time required producing the effect desired. Some require several responses in a short period of time while others require numerous responses over a considerable period of time (e.g., production of components of an annual report). Each supervisor's rate is useful for that particular supervisor and is not comparable with that of another supervisor unless both supervisors have the same responsibilities. For example, it is not unusual for the number per hour of the senior administrative officer to be less than that of an associate or assistant administrator. This is due to several factors. Administrative chores are often time-consuming before a product (effect) occurs in terms of student/teacher behavior change. Senior supervisors frequently complete more administrative chores than do junior supervisors. Senior administrators are more easily distracted from classroom/instructional activities for both necessary and unnecessary reasons. The usefulness of the assessment concerns the effects on students relative to performance of each supervisor separately. The collective rates of all the supervisors combined, like the collective rates of TPRA by the teachers, are, however, critical to the instructional effects of the entire school.

Typically, each supervisor keeps an individual and daily tally or log of accomplishments that meet the criteria described above. Also, the job profile for each supervisor is operationally defined with specific assignments to specific responsibilities that meet the generic criteria. Supervisors are assigned classrooms and teachers, administrative functions (e.g., staffing, personnel chores), daily data summary chores, data plotting and teacher decision reviews, and preparation of teacher modules. The ultimate responsibility for the school resides with the senior administrator and in CABAS schools equally with the university consultant. It does seem to be critical from our experience that all supervisors need to be involved in pedagogy and curricular issues at the level of the individual teacher and student. Attempts to separate supervisory responsibilities (e.g., pedagogically related assignments) from administrative responsibilities by assigning each to two different individuals, respectively, have led to difficulties. All supervisors share administrative and pedagogically related responsibilities in CABAS schools. This serves several purposes. The system allows individuals to rotate responsibilities. Supervisory assignments do not take a back seat to administrative chores. But, supervisory responsibilities cannot be done without concern for administrative issues. Everyone must attend to the instruction of individual students. The result is that each supervisor has critical roles and responsibilities in both the administration and the supervision of the school, and those roles are monitored by the supervisors and made public. Exemplars of supervisor logs and data collection are available from the author.

Supervisors and peer reviewers of school performance (e.g., in CABAS schools this is done by university or peer CABAS consultants from outside of the school) and superintendents of school districts use the data in numerous ways. Each supervisor checks off her accomplishments of tasks daily on a supervisor checklist. This, in turn, is divided by the number of hours actually spent by the supervisor (in and out of school) to determine weekly number per hour data points which are in turn plotted on each supervisor's graph and displayed along with the graphs of teacher performance. Both the checklists and the graphs serve useful purposes. When rates fluctuate, often a review of the checklist determines the reason. This review is then used to determine answers to questions such as was the division of labor useful vis-à-vis goals for the school, the performance level of teachers and students, or impending deadlines (e.g., site reviews by state officials, grant application deadlines). The checklist also functions as a log of activities that is useful for a variety of reasons such as distribution of time with regard to school goals and personal professional goals. When supervisors suggest that they are harassed or underutilized, a review of the checklist results in useful ideas for locating the source of the difficulty (e.g., too many administrative chores or too many responsibilities that are not personally satisfying). When numbers per hour measures of supervisors are not consistent with the general overall school data, something is typically awry. For example, observations are not being done reliably or the effects of the observations are not valid. Over time, cyclical relationships between the school calendar and supervisor rates become clear. This can lead to reassignment of responsibilities or changes in goals or simply the acceptance of certain *cyclical changes*. In all cases, there must be a direct or indirect benefit for a student or students. There must be a product or a by-product of the accomplishment of a task. The task must result in an effect on the school or community or have the potential to do so. Reports, grants, or articles are broken down (albeit arbitrarily) into per page counts. This latter decision is an arbitrary decision in that the impact on the environment may actually result in only one effect. However, such projects involve extensive periods of time and a more protracted response time. Other criteria could be used (e.g., references, data citations, analyses); however, these latter efforts are too time-consuming to tally on a practical basis and may interfere rather than assist the supervisor with the performance. Some writers (e.g., Skinner, 1979) actually count words and this indeed may be a more natural environmental fracture.

Numerous decisions are made about criteria surrounding each task in the process of operationalizing and clarifying the objectives of each task. This has a utility beyond the measurement system—to wit, clarification of objectives and operations.

Each supervisor has a list of weekly tasks (agreed upon with cosupervisors) with criteria for completion operationalized. These tasks become both the behavioral objectives and the measures of job performance. Each supervisor is also expected to identify other tasks that they perform either occasionally or frequently that contribute to the generic objectives. However, part of the responsibility of the supervisor is to identify and operationalize objectives and tasks that constitute effective supervision. As the objectives of the school evolve, so do the objectives and tasks of the supervisor. Typically, the behavior analytic supervisor or other supervisors collaborate on identifying the tasks and objectives. Frequently, more advanced teachers supply advice. It is critical for the supervisors and colleagues to arrive at definitions and objectives on which all agree.

The supervisor data serve to provide:

(a) An *accountability* measure;
(b) A *log of activities;*
(c) *Clarification of roles*, responsibilities, and objectives;
(d) A way of making the *power of the administrator responsible;*
(e) A *means of changing roles* to more closely realize the objectives of the school;
(f) An objective means of *assessing the need for more or fewer supervisors;*

(g) A set of *contingencies* to ensure a *nonadversarial relationship* between teachers and supervisors;

(h) A *form of reinforcement for supervisors* to accomplish steps toward goals; and

(i) *A tool for discovering relationships* between supervisory accomplishments and teacher or student behavior.

Supervisor Modules

Supervisors continue their in-service education through PSI modules in a fashion similar to that of teachers. In some cases, the modules will be very similar to those of advanced teachers. Again the key to what the supervisors learns is based on what the supervisor needs to contribute more immediately and what the supervisor knows about the science of pedagogy and curriculum. Some supervisors have managed to do quite well with only a few years of teaching experience in a CABAS school. However, these individuals have often had to have basic teaching competencies as part of their own modules. The greater the repertoire of skills and knowledge in the science, the more effective and efficient the supervisor is in assisting the teacher and assisting in curriculum related activities.

Supervisors who have advanced skills and considerable experience in the science continue to advance these repertoires as they encounter problems in classrooms and help find solutions with whom they assist. Often the modules of the supervisor identify objectives related to stronger effects on student and teacher behavior that result from:

1. Observation and demonstration;
2. Experiments done with teachers;
3. Increased numbers of quizzes in which teachers achieve objectives; and
4. Implementation of organizational modifications that improve measures of efficiency and cost-effectiveness.

Other modules are devoted to supervisor-established long-term objectives. These include effects such as:

1. Expanding the number of research-based classrooms;
2. Designing and implementing clerical management systems that show improved efficiency;
3. Producing a grant application or producing an annual report;
4. Developing a new parent training thrust;
5. Changing the deployment of instructional staff based on the findings of a behavioral analysis;
6. Developing any of many projects that benefit the effectiveness of the school; and
7. Completing research studies that benefit students and teachers in the school and their publication and presentation at professional conferences to benefit other schools.

The supervisor uses documentation of the school effectiveness, as shown by the data, and achievement of modules as the grounds for salary increases rather than relying on years of service alone. Those superintendents of school systems who have the behavioral expertise will want to systematically reinforce the efforts of those supervisors who demonstrate measurable increases in school effectiveness. Perhaps, some well-trained supervisors will eventually develop entire school systems in which behavior analysis is applied to a comprehensive manner to all of the schools in the system.

Collegial Relationships through Organizational Behavior Analysis

In some schools, the use of behavior analysis is extended to the secretarial, clerical, maintenance, transportation, and social work departments. In each case, the components of each job are identified per the objectives of the school. These objectives are then operationalized into tasks that produce desired effects and products or by-products (e.g., minutes of meetings). In each case, an attempt is made to tie counts to the objectives and priorities of the department and the school. What maintenance or transportation object-ives contribute to the production of learn units or a safe environment for students? Clerical staff who are involved in reproducing data forms (photocopy or ditto), scripted curricula, and other clerical tasks are typically asked and taught to record counts of their products (e.g., pages typed, photocopied) according to specific criteria related to the needs of students and teachers. Clerical staff can assume these chores and teachers can produce more learn units. The more junior-level clerical tasks are converted to number per minute rather than number per hour. Only within-job roles are comparable. However, over time, necessary or sufficient rates and exemplary rates are identified. This leads to a more objective assessment of how staff should be deployed.

As in the case with all other school staff, the data on performance serve as a source for the collegial analysis of roles and sources for reinforcement of staff productivity and quality. It is important to enlist the assistance of each staff member in identifying measures and objectives of each person's position. It is equally critical to have all individuals understand the objectives and measures of all of the positions in the school. Individuals in each position define and specify their role in achieving the goals of the school. No one should simply be given a list of chores and an assignment unless and until that person has had an active role in defining his or her tasks. The evidence suggests that when students set their own objectives (for example) learning is increased.

Periodically, the responsibilities of individuals need to be reviewed. The supervisor or the staff person can initiate this review. As the special skills and interests of staff members evolve, they should be encouraged to contribute these skills and interests to the degree that they contribute to the objectives of the school. The self-monitoring procedures become an integral and useful part of the participants' responsibilities when the procedures are monitored, introduced, and reinforced appropriately.

This notion of *participatory management* must pervade the management of instructional and administrative personnel. In addition, it is important for each member of the organization to suggest innovations. Such suggestions should result in experimental analyses to determine their utility and other benefits. These analyses are a joint venture of the staff member(s) and their administrative supervisors. It is not unusual to discover that office assistants are very effective administrative supervisors. They can often suggest and implement more effective management procedures. When office assistants assume such administrative tasks for supervisory staff, their other tasks must be assumed by others (e.g., the supervisor him- or herself, other secretaries, student assistants, parent volunteers). Developing high rates of accurate performance should result in assignment of more desirable tasks, not the addition of tasks that are chores for the individual. The latter will simply punish good work. It is as important to provide "job satisfaction" as it is to increase quality and productivity.

While senior supervisors will often need to work beyond the normal work hours, the goal is to achieve what is needed within the 40-hr week. Thus, we emphasize rate of quality performance rather than total tasks done.

The evidence indicates that time away from a responsibility is critical to the quality of performance and to the individual's satisfaction with his or her work environment. Ideally, individuals will acquire the proficiency, expertise, and fluency to complete the work in prescribed periods of time. If the number per hour of a staff member is associated with the frequent use of overtime, either the responsibilities must be reduced for that individual or the contingencies need to be invoked to increase the fluency of the staff member while maintaining the quality of work.

It is important to provide reinforcement in each position that results from the performance of the task in a competent manner. Self-monitoring assists each person in doing so. Often an analysis of the manner in which an individual engages in certain tasks shows preferences. These preferences can be used in determining reassignment of responsibilities when the task that is highly preferred by one person is a low preference task for another position. In other words, responsibilities can be exchanged when the reassignment results in higher reinforcement value. When responsibilities cannot be reassigned, they can be rearranged according to the Premack Principle, with high preference activities following the satisfactory accomplishment of low preference tasks.

It is also important to provide reinforcement from administrative superiors in the form of written memos, weekly announcements of improvements in performance that the individual and supervisor have both identified to be important objectives for the individual and the organization. (Note that the preparation and distribution of memos that praise performance or that function as corrective prompts are included in supervisor counts.) To some degree, the performance of administrative staff must be the responsibility of the supervisor just as the performance of the teacher must be the responsibility of the supervisor. The use of clear operational definitions of tasks, reliable self-monitoring, analyses of rates of behavior vis-à-vis preference, supervisor reinforcement, and participatory management are all key ingredients in an effective organization in which each individual takes pride in his or her role and the overall performance of the school. Aversive management procedures should be avoided and replaced with positive approaches based on good and participatory behavior analysis.

Department heads maintain visual displays of their department. The measurement and its analysis should work to assist individuals not only to perform more efficiently but also to enjoy their work. If the data are made public and reviewed in a participatory, collegial, and *noncompetitive* manner, the result will be a nonadversarial relationship between administration and other staff. These are the contingencies of reinforcement that will avoid an adversarial relationship and foster a collegial and positive work environment.

Deployment of Supervisory Staff

The extent to which administrators devote their time and attention to pedagogy and organizational behavior analysis in the model we described departs radically from the deployment of administrators in many schools. However, applications of this model of supervision have proved viable and have led to learning rates from four to seven times the rates found in the classrooms prior to the implementation of CABAS. Not only is the deployment of time and activities radically different, but also the expected expertise of supervisors is even more a departure from the norm. Of course, traditional administrative activities can be continued and additional behavior analytically trained supervisors can be employed as additional staff. There are two potential problems with this approach, however. First, simply adding additional supervisors to the administrative budget makes the model less cost-effective and cost-effectiveness is a key issue. Second, if the titular administrative officers are not involved in pedagogy, the financial and political processes will *not* be tied to the learning progress of individual children. In short, the contingencies developed in CABAS to make learning the central business of schooling will be short circuited.

Provided that the contingencies are arranged such that student learning is always the controlling variable, there are a variety of arrangements in staffing that can be tested. In each case, the design of the administrative and supervisory model must be decided based on student learning effects rather than the inertia of traditional top-down approaches to administration.

The most obvious way in which the data may serve to redeploy teacher, supervisor, and student relationships is in terms of the manner in which teachers are assigned students. The comprehensive measurement of student, teacher, supervisor, and support personnel makes redesign of staffing arrangements based on student benefits feasible for the first time. Moreover, one does not have to wait until the end of the year or term. The use of reversal, multiple baseline, changing criterion, and alternating treatment designs (e.g., multiple schedule, multiple treatment) allow changes to be made on a day-to-day (or at least weekly) basis.

The deployment of teacher and teacher roles can be related to functional curricular categories (see the curriculum chapter) as well as subject matter areas. Moreover, arranging for "team" teaching can be tested in terms of the teaching arrangement on learn units and instructional objective achieved. Members of the team may rotate instructional design and feedback (scoring papers) roles with presentation or instructional design roles. The analysis can be extended from learn units and objectives to the costs. For example, arrangement of the staffing for classes of 60 with two teachers may be more pedagogically effective (or less) with equal cost benefits. The use of electronically mediated instruction (e.g., computers) that is programmed for good behaviorally based instruction *and measurement* can result new ways in which teachers administer, manage, and mediate learn units. For example, PSI can be managed over class E-mail in classrooms with adequate numbers of computers.

Such analyses will likely reveal that different arrangements are more or less effective depending on the particular team involved. Other findings will suggest more generic principles to guide the profession in general. If teachers learn the verbal behavior about the science and the scientific process (e.g., experimentation, replication, redesign, report results), they will contribute to the science of pedagogy and schooling and indeed the science of behavior. The process of functioning as a strategic scientist is another source of professional reinforcement in addition to the effects on students, the appreciation of parents, and the economic advantages that will accrue from the effects on the economy of effective instruction.

In summary, if there is a problem, apply "good behavior analysis" to the problem. Maintain comprehensive measurement of real behavior effects, and the analyses and needed changes will suggest themselves. If the progress of the student is tied to teacher behavior and teacher behavior to supervisor responses, effective education will result provided the data and the epistemology is thoroughly behavior analytic.

The Politics of Change

There are numerous threats to the implementation of the Comprehensive Application of Behavior Analysis to Schooling. However, if the system is truly driven by the performance of students and the supervisor provides the necessary expertise and support for teachers, the effects of the model will result in effective change. In some cases, the initial efforts may only be done with one or a handful of receptive teachers. New teachers or teachers having difficulty with their classes are eager for assistance. If the supervisor introduces behavior analysis to these classrooms and the necessary expertise is provided, the evidence suggests that the approach will prevail.

It is a good rule of thumb to continue with the expansion based on requests at least until the approach becomes the *modus operandi* for several classrooms. Teachers who want to be maximally effective will adopt the procedures as they see them work. Moreover, they will go on to improve on them. This is more likely to occur if the performance of the teachers is posted for view by colleagues and parents. Perhaps there are a handful of teachers who will resist, but they would probably resist change of any kind.

Reasonable doubts and reasonable objections from teachers should be approached as potentially useful modifications. However, *the pervasive uses of measures meeting the characteristics of learn units cannot be rejected*. If the changes that occur do so as a result of power based on knowledge (e.g., scientific applications that work or lead to workable solutions), the effects of the change will reinforce the continuance. The keys to this control for the system are the TPRA observations and instructional decision analyses with knowledgeable feedback from the senior colleagues along with the maintenance of graphs at the level of student, teacher, classroom, and school.

Philosophical objections need to be countered with empirical tests in the classroom. Are the philosophical objections viable? If so, redesign the objectives taught. If the ensuing data on students, teachers, supervisors, and parents' effects on the system suggest that the philosophical objections are groundless, then discard the objections. In a thoroughgoing learner-driven system, the children and parents are the primary citizens as well as the customers. Moreover, if the school leaders are enthusiastic, a positive effect on teachers and supervisors will follow. In a comprehensive behavioral school, each learn unit that is successful is a vote and each objective obtained is a vote, when they are beneficial to the student, parent, and community at large. Moreover, the election occurs constantly, not simply once every 6 weeks.

One support system that has been useful to the CABAS effort is the provision of a behavior analytic consultant to assist all of the school personnel. This person provides a fresh look at the system and typically has several years of experience in education as behavior analysis. Such a person must, of course, be accountable at the level of the data on the individual student and on the summary measures of each component of the school. Typically, this person reports to both the supervisor and the board and is held accountable to both. They are senior behavior analysts from other behavioral schools or are university professors who have the necessary expertise.

The redesign or initial design of a thoroughly effective school is a long-term effort. Design or redesign must be followed by empirical tests that lead to revision after revision. Education has not been accustomed to such systematic, systemic, and prolonged efforts. The politics of fad and fashion have prevailed. However, if the designed system is tied continuously to the measurement of student learning as the prevalent controlling variable, the acceptance of good procedures and the rejection of bad ones will prevail.

In an effective school where parents see the effects on their children and the continuous display of teaching and learning, good procedures will prevail because of political pressure. In the worst-case scenario, when parents are prevented from access to the data of schooling effects, or their lack thereof, the interests of the student will be lost to the politics of fad, fashion, and lethargy. *The worst-case scenario cannot be allowed to prevail if we are to provide voices for all children.*

References

Albers, A., & Greer, R. D. (1991). Is the three-term contingency trial a predictor of effective instruction. *Journal of Behavioral Education, 1*(3), 337–354.

Buskist, W., Cush, D., & de Grandpre, R. J. (1992). The life and times of PSI. *Journal of Behavioral Education, 1*, 215–234.

Catania, A. C. (1998). *Learning* (fourth ed.). Englewood Cliffs, NJ: Prentice Hall.

Cooper, J. O., Heron, T. E., & Heward, W. L. (1987). *Applied behavior analysis.* Columbus, OH: Merrill.

Dorow, L. G., McCorkle, N., Williams, G., & Greer, R. D. (1989). *Effects of setting performance criteria on the productivity and effectiveness of teachers with students.* Paper presented at the International Conference of the Association for Behavior Analysis, Nashville, TN.

Greer, R. D. (1992). *L'enfant terrible* meets the educational crisis. *Journal of Applied Behavior Analysis, 23*, 65–69.

Greer, R. D., McCorkle, N., & Williams, G. (1989). A sustained analysis of the behaviors of schooling. *Behavioral Residential Treatment, 4*, 113–141.

Hogin-McDonough, S. (1996). *Essential contingencies in correction procedures for increased learning in the context of the learn unit.* Unpublished doctoral dissertation, Columbia University, New York.

Ingham, P., & Greer, R. D. (1992). Changes in student and teacher responses in observed and generalized settings as a function of supervisor observations. *Journal of Applied Behavior Analysis, 25*, 153–164.

Keller, F. S. (1968). Good-bye teacher . . . *Journal of Applied Behavior Analysis, 1*, 79–89.

Keohane, D. (1997). *A functional relationship between teachers use of scientific rule governed strategies and student learning.* Unpublished Ph.D. dissertation, Columbia University, New York.

Selinske, J., Greer, R. D., & Lodhi, S. (1991). A functional analysis of the comprehensive application of behavior analysis to schooling. *Journal of Applied Behavior Analysis, 13*, 645–654.

Skinner, B. F. (1979). *The shaping of a behaviorist.* New York: Knopf.

Sulzer-Azaroff, B., & Mayer, G. R. (1986). *Achieving educational excellence.* New York: Holt, Rinehart, & Winston.

Sulzer-Azaroff, B., & Mayer, G. R. (1991). *Behavior analysis for lasting change.* San Francisco, CA: Holt, Rinehart, & Winston.

The School Psychologist and Other Supportive Personnel: A Contemporary Behavioral Perspective

TERMS AND CONSTRUCTS TO MASTER

- Projective tests
- Archival records
- Standardized achievement tests
- Bad behavior
- Portfolios of achievement
- Social skills instruction
- School survival repertoire
- Instructionally based collaboration
- Emotional or affective repertoire
- Nomologically derived constructs nomological networks
- Ecobehavioral analysis
- Biospecification of behaviors
- Independence of behaviors
- Pedagogical expertise
- Correspondence teaching
- Behavior selection

- Behavioral disorders (emotional disturbance)
- Scientifically questionable labels
- Self-injurious behaviors
- Assaultive behavior (aggressive behaviors)
- Environmental adjustments
- Social Reinforcement (peer and teacher)
- Individualized curriculum
- Individualized pedagogy
- Social skill deficit
- Inventories of student repertoires
- Behavioral excess
- Aberrant behavior
- Academic repertoires
- Social repertoires
- School survival repertoires

- Parent education and parent educators
- Parenting inventory
- Classroom inventory for inclusion
- PSI modules for parents
- Parent performance rate/accuracy observation
- Parenting groups
- Coercive traps
- Mutually reinforcing relationships
- Portfolio and inventory assessment

TOPICS

Treatment of "Bad Behavior"

Parent Education

Teaching Students Behavioral Skills

Assisting Related Services Personnel

Assisting Teachers to Become Strategic Scientists of Instruction

Portfolio and Inventory Assessments: Archival Records

Norm-Referenced and Projective Tests

Summary

In the past decade school psychologists have shifted to a data-based, proactive, collaborative, and student-centered professional role (Task Force on Comprehensive and Coordinated Psychological Services for Children, 1994; Yesseldyke, Dawson, Lehr, Reynolds, &Telzow, 1997). The new functions of school psychologists, working with parents, teachers, school counselors, and social workers, can be realized or enhanced by the science and technology of pedagogy and schooling. Today, school psychology can offer schools much more than the traditional role that evolved in the normative practice of schools. In fact, the systems applied behavior analysis that we have introduced in this book owes its origins to behavioral psychologists, many of whom were instrumental in the development of strategies and tactics. Schools that provide system-wide coordinated applications of behavior analysis to the whole student, and all of the members of the school community, call for school psychologists to play a key role. *All of the chapters in this book are for the school psychologist;* however, this chapter is devoted to roles that may be realized uniquely by school psychologists. Our school psychologist is part behavior analyst teacher, part supervisor, parent educator, and *full-time strategic scientist of instruction.*

Traditionally, schools have relegated psychologists to two functions—(a) administering standardized and projective tests for classification purposes and (b) fixing or passing opinions on "bad behavior." Tests are important, but they take on different functions from both a behavioral selection perspective and the new role for school psychologist—the same holds true for the analysis and treatment of bad or "aberrant" behavior. In addition to changes in these two functions, the behavioral school psychologist has a host of new roles that result from the science (Shapiro, 1987).

These new roles embrace: (a) social skills instruction (preventive and remedial); (b) professional collaboration with teachers, community services, and parents that is instructionally-based; (c) the insistence on data-driven procedures; and (d) the need to serve the range of needs of children. The new role calls for professionals who are data-driven purveyors of research-based preventative and remedial services. Education is broadly construed to include all of the needs of the child from academic services to healthy lifestyles and well-being (see Greer, Dorow, Williams, & Asnes, 1991). The objective of this chapter is to show how these new roles for school psychologists are served through comprehensive and systemic applications of behavior analysis to the range of the needs of students and fellow professionals.

TREATMENT OF "BAD BEHAVIOR"

The use of the term "bad behavior" as opposed to terms such as behavioral disorders, emotional disturbances, hyperactivity, and attention deficit syndrome, among other common professional terms, is only partially tongue-in-cheek. Behavioral selection has a long history of eschewing such labels (Kazdin, 1988; Skinner, 1956). Those whose scholarship is devoted to the education of children with behavioral disorders or emotional disturbances echo the same caution about labels and the need to use data based procedures to solve clearly defined problems (Kauffman, 1989; Paul & Epanchin, 1991).

When the bad behavior is viewed from an instructional perspective—one grounded in a strategic science of pedagogy and behavior analysis—we can proceed with systemic and data-driven solutions and preventative actions. When special services result in the use of professional expertise to fix the problem, the process *is not only be appropriate for the student in question, it is also in the best interests of all involved*. Corrective tactics from the literature can be used after the psychologist has identified the controlling variables through a functional analysis. In other words, the steps are those identified in chapter 4 on the analysis of the source of instructional problems as well as other procedures to be identified in this chapter. We use the term *bad behavior* in order to suggest something that can be fixed with a little or a lot of instruction and analysis. The term is used generically to apply to the range of behavior problems and environmental sources. The research in applied behavior analysis since 1968 (Sulzer-Azaroff, Drabman, Greer, Hall, Iwata, & O'Leary, 1988) has shown how school psychologists can assist teachers and parents to fix bad behavior and avoid the potential for damage that results from the use of professionally acceptable but scientifically questionable labels. We describe how that research can be incorporated into a new and critically important role for the school psychologist to play in arranging the schooling environment to fix existing bad behavior and more importantly how to avoid problems by providing schools that are effective.

Psychologists who function as behavior analysts, using the tools of ecobehavioral analysis and pedagogical expertise, can observe students in classrooms, analyze the controlling variables or existing classroom contingencies for their role in the bad behavior, and assist the teacher in replacing the bad behavior with useful behavior (Greenwood, Carta, & Atwater, 1991). Even when the teacher is not trained in the science of pedagogy, a behaviorally trained psychologist using the procedures outlined in the chapter on mentoring teachers can obtain useful results. As teachers become strategic scientists of instruction, the usefulness of behavior analysis for the school psychologists is multiplied. When teachers and psychologists use the same verbal community of the science of behavior, joint applications of the sciences will replace bad behavior with useful repertoires.

What Is Bad Behavior?

From the perspective of the student and behavior selection, the student's bad behavior has momentary utility. Bad behavior results in short-term gains for the students and long-term losses for the student and the community. The community legitimately identifies the behavior as bad for the community and not in the best long-term interest of the student. But an echobehavioral and functional analyses determines that the behavior is being reinforced by the contingencies in existence at the moment, regardless of the behavior's origin. For the student, and at the moment, the bad behavior has some momentarily adaptive purpose (e.g., reinforcement results for the student). This perspective shifts the locus of the solution from treating behavioral disorders to the redesign of the instructional environments—the ecology of the contingencies of reinforcement and punishment in the classroom. The goal is for the environment to produce both short- and long-term gains for the student.

A classroom that is based on an individualized behavioral instruction allows professionals *to replace the bad behavior with a repertoire of acceptable behavior* following the installation of the appropriate instructional contingencies that were selected following the analysis of the source of the behaviors. Instruction is designed to replace assaultive or self-injurious behaviors with repertoires that are functionally appropriate and less costly in terms of responding. These repertoires, in turn, lead to acceptable forms of reinforcement. The student is taught to emit behaviors that lead to frequent reinforcement without resource to bad behavior. In the novel *Iilywacker* (Carey, 1991), the protagonist describes, after having learned to read at the late age of 50 years, how he was amazed at how much energy he saved by no longer having to manufacture lies. The new reading repertoire obtained the reinforcement that was obtainable formerly only by lying. In the same vein, Sulzer-Azaroff and Mayer (1986) found that occurrences of truancy and vandalism decreased in classrooms and schools when the better instruction resulted in better academic skills. Carr and Durand (1985) found that students who had emerging speaker repertoires stopped assaulting their teachers and peers when they were taught to substitute communicative behaviors. Kelly and Greer (1992) and Martinez and Greer (1997) found that multiplying the daily learn unit presentations received by students who had listener repertoires only resulted in the elimination of assaultive and self-injurious behaviors. Chu (1998) found that teaching a mand repertoire in a social skills setting replaced aberrant responding of autistic preschoolers with social interaction. All of these studies built on Ayllon and Roberts (1974) research that showed that reinforcing correct academic responding acted to decrease inappropriate classroom behaviors in classrooms for children who where performing on grade level.

In the long term the real source of the bad behavior is an educational deficit, and we argue that the most efficient way to fix the problem in an educational setting is to teach more effectively. The use of the science of pedagogy and schooling provides the wherewithal to provide effective instruction.

One of the prevailing approaches to bad behavior in the behavior analysis literature prescribes the use of a functional analysis of the *immediate* controlling variables of the bad behavior (Iwata, Dorsey, Slifer, Bauman, & Richman, 1982). This particular approach suggests that the analysis should determine whether the source of the behavior is attention, escape, or automatic reinforcement of the bad behavior itself. In the last case the behavior has acquired inherent reinforcing properties. After the general function of the aberrant behavior is identified, treatment involves providing a behavioral tactic that eliminates the reinforcement effect for the behavior. For example, providing frequent availability of a preferred item/food/activity on a noncontingent response basis can reduce or eliminate the behavior. While this is approach can serve as a first aid in the short term (most of these studies involve short sessions rather than day-long sessions), it raises problems on a long-term basis. That is, the real source of the problem is the lack of repertoires that recruit reinforcement in a response-cost-efficient manner. Learning to read obviates the need to lie, learning to mand or tact provides an easier and more accessible means to reinforcement than temper tantrums or self-injury, learning to mand in a play setting results in more conversational units or real socialization and elimination of assaults by preschoolers, and the provision of frequent learn units obviates the need for differential reinforcement of incompatible behavior. In these latter cases the long-term solution to the problem is teaching the repertoires that were educational deficits. Thus, in the recent mandates of the IDEA law to provide functional assessments of bad behavior there are at least two approaches—(a) an assessment of the educational repertoires and deficits of the student and teaching the repertoires as we describe in Part I or (b) the use of the functional analysis of the immediate controlling variables. Of course, if the psychologist does not have access to the educational system in the school, an immediate analysis of the controlling variables for the bad behavior is an essential step.

In still another example of the importance of educational treatment, children or adults with autism spectrum disorders or pervasive developmental disabilities engage in repetitive behaviors or stereotypy. While there is a plethora of behavioral tactics that momentarily reduce the stereotypy (e.g., self-stimulation) ranging from DRO to mildly aversive punishment procedures, Greer, Becker, Saxe, and Mirabella (1985) found that conditioning play related stimuli such as puzzles, games, or toys as newly conditioned reinforcers had a generalized and long-term effect on eliminating stereotypy as automatic reinforcement. Nuzzola-Gomez, Leonard, Ortiz, Rivera, and Greer (2002) replicated this finding and also found that conditioning books as reinforcers also eliminated stereotypy. These data support the belief that the root solution is effective instruction to eliminate deficit repertoires. The problem is not an excess of bad behavior; it is the lack of alternative behaviors that are more easily emitted and a paucity of conditioned reinforcers. Thus in a learn unit context analysis, the source of the problem is a deficit in the instructional history of the individual. This is not to suggest that in certain clinical situations the use of short-term behavioral tactic related to the existing reinforcers does not have short-term utility. However, in educational settings the real solution is effective instruction.

The psychologist who deals with *the educational deficit associated with the bad behavior* must be sophisticated in the science of pedagogy and supervision discussed throughout the book if she or he is to fix the problems. When the controlling variables are isolated, they are probably going to result in a need for significant changes in instruction. This will require the psychologist to draw on the expertise described in the chapters both on teaching and on supervision. The psychologist can provide the missing instructional expertise for the teacher, when that expertise is lacking.

The analysis needs to go beyond fixing the bad behavior. Regardless of how effective a treatment for a particular bad behavior may be, it is likely that a more thoroughgoing analysis will require educational environmental adjustments (e.g., changes in instructional operations), if new repertoires and their new controlling variable are to replace the function of the bad behavior. Simply eliminating the behavior by punishment or continuous access to preferred activities is not a long-term or acceptable solution.

From a contemporary behavioral perspective, the student, as is the case with all living organisms, behaves as a function of the selection of behavior by the existing reinforcers or punishers, the student's history of instruction at home and in school, and the setting events in the environment—*his/her* environment, *his/her* punishers, and *his/her* reinforcers. Bad behavior results in short-term reinforcement for the student, but acts to his or her long-term detriment. The various behaviors may momentarily function to obtain peer or teacher attention in cases where the student was not receiving reinforcement for academic responding. In curriculum and pedagogy geared to the group mean, it is inevitable that some students will be unable to obtain social reinforcement from peers or the teacher for the average standard of academic performances. The student cannot meet the average criterion at least at that moment. On the other end of the scale, an advanced student will not receive instruction that is interesting (e.g., reinforces attention and other appropriate responding) and will increase the number of nonacademic responses that are reinforced. If left alone, he or she will read or engage in other appropriate activity associated with conditioned reinforcers resulting form thoroughgoing behavioral educational practices. Again the source of the problem is an inadequate individualized curriculum and pedagogy.

When bad behavior is treated as social skill deficits or in some cases emotional deficits, the deficit becomes the impetus for specifying curricula and pedagogy to teach a useful repertoire. Often the reinforcement of academic skills (Ayllon & Roberts, 1974) eliminates the problem if the level of difficulty and related learn units are appropriate for the student. A prudent approach is to obtain a baseline assessment of what that student can do across academic, social, emotional, problem-solving, and self-management repertoires. In the CABAS schools, the curricula and pedagogy for students are based primarily on *inventories of the students' existing repertoires*. While standardized tests are considered also, they are secondary. The student is observed in a constant environment on repeated occasions for each response class associated with the domain in question (e.g., social skills or their deficits). If the student emits verbal or physical assaults, the problem will show up in the inventory as a deficit in one or more response classes associated with the inventory. The teacher and psychologist locate or write scripted or programmed curricula to remediate the deficit. Instruction is then designed to rearrange the environment such that the student receives adequate amounts of reinforcement. Even if the instruction takes the form of planned ignoring across professionals, the intervention is an instructional one.

If the bad behavior in question is behavioral excess, there is often an accompanying deficit. That is, while all students argue and even assault others from time to time, students who are seen as "overly aggressive" can be identified as also having a behavioral deficit. Rather than only identifying the problem as a behavior that needs to be decreased or eliminated, *a thoroughgoing behavioral and environmental analysis will identify the contingencies in effect and the kinds of, and frequency of, reinforcements associated with the setting events, setting stimuli, discriminative stimuli, and the behavior.*

The rationale underlying this approach includes the following beliefs.

- Organisms continuously emit behaviors associated with reinforcement (e.g., to allay momentary or long-term deprivation or contingencies that evoke escape or avoidance). These range from stereotypy to thrill seeking.

- Behavior is not maladaptive for the *individual who is engaging in the behavior at the moment* (e.g., the fight obtained peer attention when no attention occurred without it).
- Individuals receiving reinforcement for, and adequate instruction in, behaviors that lead to either natural or prosthetic reinforcement will engage in minimal numbers of inappropriate behaviors. Behaviors will match the contingencies in the environment. Behaviors distribute themselves in environments consistent with the individual's community of reinforcers that exist in the current environment.
- Social repertoires function under the same lawful principles, as do academic responses; therefore, social repertoires are subject to the same contingency-shaped and verbally mediated instruction described in detail in the chapters on teaching and supervision.
- Behaviors are dependent on environments.

Does this eliminate psychological constructs such as IQ, temperament, or cognitive deficits as sources for the problem behaviors? Of course not, but these psychological constructs are nomological clusters of behaviors that summarize repertoires; they are not the sources of behaviors. On close inspection, we argue that these psychological measures, which are based on nomological constructs, are intimately tied to instructional histories and phylogenetic variables for the individual. As we pointed out in the chapters on using behavioral tactics for instructional purposes, there are tested instructional interventions that alter the psychological predispositions of the individual.

The bad news is that this means that the appropriate behaviors learned for school do not appear necessarily in homes because the environments (e.g., setting, antecedent, and consequences) are different. It is also quite possible that *behaviors that are inappropriate in school have survival value on the street*. The good news is that the student may learn "appropriate" behavior in school and have concomitant success, while continuing to engage in "bad behavior" in the home because the latter is successful momentarily in the home. Parents who want the bad behavior to go away at home can learn to change the home environment through instruction from the school psychologist or other professional who functions as a *parent educator*. That is, the elimination of reinforcement for bad behavior in school and the establishment of high rates of correct academic responding for individually appropriate curricula occurs as a result of rearranging the school environment. When the parent elects to extend the behavior change to the home, the psychologist or other appropriately trained professional can do so by drawing an operations described in the literature on behavior analytic parenting repertoires (Forehand & McMahon, 1981; Patterson, 1982; Wahler & Dumas, 1986). The psychologist replaces the illness or socially deviant label with an instructional deficit—one that can be remedied instructionally by the parent.

PARENT EDUCATION

The literature on teaching parents to use tactics from applied behavior analysis fits into an instructionally based approach. In CABAS schools or other schools based on a thoroughgoing application of behavior analysis to instruction, the echobehavioral inventory can be extended to parent–child interactions in a kind of curriculum for behaving at home. The inventory or baseline assessment across the home environment identifies strong and weak parental instructional strategies and ambiguous or vague objectives. That is, like the comprehensive inventory in the school, a comparable inventory in the home can identify parental skills as instructional objectives to teach parents to use in the home.

While the parent typically pinpoints some of the problems, a complete inventory identifies potential future problems. The training is then not only for remediation but for prevention. In the CABAS schools, we provide instruction in

1. How parents can teach their children more effectively;
2. What they need to teach; and
3. How to advocate effectively for their children.

A behavioral home assessment assists the parent to identify what it is that they want to teach and how pedagogy is used to teach each of the target domains.

If compliance is the issue, compliance is operationalized across home settings and pedagogy taught for those settings. We do not expect that pedagogy to generalize necessarily to teaching the child to enjoy being read to or to follow rules by the parent in the home. This requires further objectives and further pedagogical applications. We should not assume that because the reinforcement operations for one set of parent–child interactions appear similar to the psychologist it does not mean that the parent will emit them for different situations. We then design PSI modules to teach parents basic vocabulary about the science and contingency shaped skills to teach and shape behavior.

We identify:

- The pretraining and posttraining teaching skills needed across various home and school environments;
- The goals that parents are to set, including priorities for their children and themselves; and
- The advocacy skills needed by the parent to monitor school placement and the instruction received by their children.

When psychologists have time built into their schedule for parent education, some of the training can occur in the home with the parent and child. A first-level vocabulary of the sciences is taught by readings, discussions, and quizzes and the actual applications are taught (a) *in situ* in the home or (b) in the school.

In the latter case, the parents, *mom and dad*, learn skills first with children other than their own and when they have achieved acceptable levels of performance, the parents apply the skills to their own child. Videotapes are made of parents with their children and used for both data collection and instruction. The training is outlined in the section on teacher training through PSI. Typically, first levels of contingency-shaped and verbal vocabularies about the science are taught. The goal is application but with accurate use of the terminology. We do not substitute nonscientific terms for scientific terms as, for example, "reward" for reinforcement; rather, we teach the scientific term and its utility but at a beginning level.

Of course there are many problems in the homes that are tied to the success of our parent education effort. Are the parents in agreement? Are there marital or other difficulties that hamper successful parent education? Clearly, these problems call for other professional interventions and we would suggest good behavioral interventions for these too. Teaching parents to arrive at consistent rules and teaching parents to be consistent or contingent can eliminate or ameliorate at least some of the marital difficulties. When children behave well as a result of their parents using good parenting skills, at least some of the sources of marital difficulties can be relieved.

Parents are taught basic teaching repertoires through the teacher rate/accuracy observations described in the supervisor chapter. The particular skills that are taught are adapted to the particular parent needs, but all parents should be taught to use skills for the following functions or settings.

1. Leisure settings, (e.g., using positive procedures watching television, playing with toys, playing with siblings, setting and teaching children to follow home rules, how to travel with children);
2. Instruction of self-help or daily living skills (i.e., the parent teaches the child to tie his/her shoes, pick up his/her toys, take the trash out using positive reinforcement operations); and
3. Doing homework with the student (e.g., praise, corrections, shaping, and the avoidance of coercive interactions; see the tutor training procedures described in chapter 6); and
4. How to use community settings as opportunities for teaching.

The parent learns to use effective pedagogy across different settings and for different repertoires. The parent educator can teach only a few of those basic tactics described in chapters 5 and 6; thus the psychologists will want to teach those tactics that are most helpful to the parents. Parents need to learn also how to identify objectives and subobjectives and to respond to the behavior of their child to guide parenting operations. The parents learn to provide learn units for social and cognitive skills and to identify the goals of instruction including the appropriate social behavior of their children. In many cases the psychologist will need to help parents establish rules such as those associated with bedtime, sibling interaction, compliance, and home responsibilities. There are several excellent and reader-friendly texts devoted to teaching parents to use basic behavioral parenting skills that can assist psychologists to teach parents the basic verbal behavior about the science and, most importantly, in home contingency-shaped repertoires.

The development of good parenting skills and individualized curricula for the children facilitates collaboration between parents and teachers in terms of specific teaching operations. Parents who are well trained can assist teachers, and the common vocabulary allows parents to specify how they wish teachers to help them. The psychologist who functions as the parent educator can serve as the locus for home to school and school to home instruction. The basic belief of the CABAS model of schooling and of the Association for Behavior Analysis is that the parents are not to be blamed for the problems; rather, the problems are instructionally derived and instructionally fixable.

Not all parents avail themselves of instruction in schools that apply behavior analysis in a comprehensive manner (between 30 and 100% do in any given school year). Participation is, of course, the prerogative of the parent who is always viewed as the client (Donley & Greer, 1993). However, one of the psychologist's roles is to recruit parents for parent education.

Academic Weaknesses: The Parents' Role

Parents can have significant effects on the academic skills of their children, if they are taught how to present the academic antecedent and how to correct and reinforce student responses (Duvall, Delquardi, Elliot, & Hall, 1992; Trovato & Bucher, 1980). The parent educator or the teacher can specify the material to be taught and then teach the parent how to teach the target curricula using learn units. *It is important to script or to program curriculum and teaching operations for parents. It is equally important to observe and provide TPRAs with the parents.*

Teaching parents how to tutor their children using sound pedagogical procedures can prevent or replace the occurrence of a punitive or coercive relationship between parents and students. From time to time, parents and students report that homework sessions develop into heated exchanges when parents become involved. Parents who are taught to use scripted procedures and learn units avoid this pitfall. Tutoring sessions become enjoyable times that occasion assistance for the student and structure mutually and positively reinforcing exchanges between parents and children.

The parent trainer teaches parents to use scripted or programmed material and reinforce using consequences that are probable reinforcers (e.g., television time, game time, appropriate praise). They are taught to move at the child's speed (i.e., use successive approximation) and to *appreciate* the individualization of instruction; they learn to pay attention to the performance of their child to guide their parenting actions. When parents are brought under the control of scripted procedures or to use learn units in many applications, they learn also how to avoid punishment. Part of the process of teaching the parent to pay attention to the performance of their children results from collecting data on correct and incorrect performance. The student and parent plot the data together on graphs that serve to: (a) reinforce successive increments in student performance, (b) keep the parent focused on the goal at hand, and (c) communicate progress or difficulties to teachers and the parent trainer.

Reading and verbal behavior repertoires are emphasized, particularly with young children who are learning to become independent readers (see Parts I and II for classification of stages of verbal behavior mastery). Parents of young children are, however, not told simply to read to their child. They are shown how to reinforce the child to participate and how to pair that reinforcement with reading stimuli in order to *condition books and reading stimuli such that they acquire strong reinforcement value for children*. Young students can learn to read one or two words as part of the reading sessions. As their skills increase, the parent and child can learn to "trade" words, sentences, and then paragraphs. Most importantly, reading sessions become pleasant periods because parents avoid coercive traps as a result of knowing what to do, how to do it, and what to avoid. The real goal of these reading sessions is to condition books and reading stimuli as reinforcers.

Parenting Groups

The maintenance of newly acquired parenting skills is enhanced by the formation of small parent groups that meet with the parent educator, if the contingencies of these meetings are arranged appropriately. The grouping can be based on the interests of the parents (e.g., social skills training or academic skills), the age or characteristics of the children, or the language spoken in the home (e.g., Spanish, Chinese). Meetings focus on sharing data, sharing tactics that work, or introducing new tactics when the existing ones are not working. The sessions also provide the parents with a support group (i.e., a source of peer reinforcement). Parents telephone each other to discuss assignments and to assist each other in advocacy efforts. Parenting groups also provide parents with the contacts who can also assist them in advocating for quality education for their children. Because they learn to identify the components of a quality education together, they can draw on the group when increased numbers are needed to lobby for effective schools. The formation of parent groups establishes the conditions for the school to become a community devoted to the well-being of students.

Psychologists play a critical role in parent education in behavioral schools. Another and related role is providing individual students with skills to deal with parents or teachers in cases where the parents or teachers are not accessible. This role is also critical in the inclusion process for children with disabilities. One of the tactics that has proven useful for some students is to teach the students themselves to use behavioral skills.

TEACHING STUDENTS BEHAVIORAL SKILLS

The teaching of simple behavioral skills is a key component of teaching students self-management in the classroom setting by teachers who have the necessary expertise. The psychologist can teach the student these skills also for specific problems that arise. *This strategy is most useful when either teachers or parents are not accessible to the school psychologist and the student is accessible.* It is also a basic repertoire that is taught to students to assist an educationally effective inclusion effort.

Students, who habitually become embroiled in coercive traps with parents or teachers, can be taught to turn what can be characterized as *mutually aversive relationships* into mutually and positively reinforcing relationships (Polirstok & Greer, 1977; Sherman & Cormier, 1974). The student is taught to identify and record disapprovals and approvals emitted by the student to the teacher and by the teacher to the student. This is done by role-playing activities in which the student acts out disapproving and approving teacher behavior and doing the same with the student's disapproving behaviors to the teacher. As the student learns to identify the acts (i.e., approval and disapproval) reliably, he or she is taught and reinforced to collect data on the exchanges in the classroom or the home. Often this monitoring procedure is effective in itself when the behaviors are graphed. When necessary, contracts and token economies with backup reinforcers are designed for goals that the student sets with assistance from the psychologist. The student reports regularly to the psychologist. In these sessions, the data on student, teacher, and parent approval/disapproval are shared. Points or tokens that are earned are exchanged for activities or events and new goals are set.

When the home setting is a problem, the student who is old enough (pre-adolescent or adolescent) can serve as the change agent when a good program is written and when the student is adequately trained. The student learns to avoid emitting disapprovals, not to respond to parent disapproval, and how to approve positive parent behavior (e.g., smiles, attention, compliance). Of course, this process also teachers the student him- or herself how to behave appropriately. That is, in order to reinforce the parent behavior, the offspring's behavior becomes that which the parent approves, just as when the student reinforces certain behaviors of teachers the student becomes socially appropriate (Polirstok & Greer, 1977). If the parent or teacher is not accessible, the psychologist can teach students and monitor their behavior as an alternative intervention.

Teaching the student to use behavioral skills for interaction with parents or for self-management purposes provides procedures that solve the initial problem. Improved self-esteem and social skills accrue as by-products of effective behavior. Constructs such as self-esteem are best taken care of by teaching students to emit behaviors that are positively reinforced in the target setting, not by trying to convince the student to have good self-esteem. No one knows how to do the latter, but we do know something about how to go about changing the environment that produced the poor result.

Assisting in the Inclusion Process

Deriving a good behavior analysis of the setting that the student is to enter can facilitate the process of mainstreaming a student from an intensive behavioral school or special education program to a normative school setting. Once the contingencies of the school and classroom are known, the psychologist, teachers, and teacher mentors can work together to prepare the student for the contingencies that exist in the setting to which they are to enter. The following inventory is used in CABAS schools and has been helpful.

Observing the Target Classroom and School

1. Identify the student's local school and make contact with the Principal and then the teacher. Set up an appointment to visit and do *a full-day observation*. If there is resistance and there is an alternative placement, determine if that alternative is more accepting. The goal is to develop a positive team relationship. We want to mainstream the student and as many of our teaching operations as possible, provided that the experience for the student results in rapid learning and a positive social experience.
2. Observe the teacher and the classroom throughout the day and determine the contingencies that are common for all students throughout the day. Find answers to the following:
   - What time does the student arrive and how does the student enter the classroom (e.g., is he/she delivered by parents, school bus, waits outside until bell rings)?
   - What are the implicit and explicit expectations of arriving and getting to work behaviors?
   - Obtain rate of teacher or monitor approval and disapproval responses to students during arriving and departing times.

Definitions and Data Collection

Approvals include vocal and nonvocal (smiles, pats, hugs) statements and delivery of prosthetic reinforcements—indeed any positive reinforcement operations. Obtain rate of disapprovals, which includes admonishments, criticisms, or any punishment or negative reinforcement operations that occur.

Determine the classroom rules, *both the explicit ones and the implicit ones*. Ask the teacher before school what his or her rules are for the classroom and the school-wide rules, and use those for the contingent noncontingent rate summaries that are to follow. The implicit rules are what you think really gets punished and reinforced by the teacher even though he or she may not be aware of her responses. These will be a best guess and should be discussed in your comments. Use the teacher's explicitly stated rules to guide the recording of contingent or noncontingent approvals/disapprovals.

When observing the teacher throughout the day, tally occurrences of all reinforcements of social behavior as a capital letter "A" and errors in reinforcement as a circled capital letter "A." If the teacher approves or provides a reinforcement operation that is consistent with the rules this would be an instance of contingent positive reinforcement (uncircled "A"). However, if the teacher reinforces a behavior that he or she should not have, consistent with *his/her rules,* that response is a circled "A" or an error in reinforcement (e.g., he or she calls on a student who blurts out rather than raising his hand is recorded as an error provided that the teacher has a hand-raising rule). When a teacher omits reinforcement operations of social behavior, you do not record an error. In the latter case, you record nothing.

Disapprovals (i.e., criticisms, admonishments, nagging (i.e., asking a student to do something more than once), physically punishing, frowning or glaring at a student as either a punishment or a negative reinforcement operation) for behaviors that are not consistent with the rules of the school or the classroom are uncircled capital letter "D." When the teacher provides a punishment or negative reinforcement operation that is not consistent with the rules circle the capital letter "D." *Allow 3 s between each approval and disapproval, before counting the behaviors as a new occurrence.* Thus, if the student is praised or criticized in long strings of statements, do not count a new one until there is a 3-s period of no speaking or glaring. Note prolonged disapprovals or praises in the comments.

Rate Calculations

For all rate calculations determine the elapsed time for each period. For example, if the first part of the day is ceremonial in nature (calendar and weather, prayers, opening ceremonies) label that as a ceremonial period in the subject matter space, and determine elapsed time (start at beginning and stop at end and the difference is the elapsed time). *Do this for each lesson time* (e.g., math, language arts, lunch, play, group free time). The form will allow you to do this and to record social teacher interaction in capital "D" and "A" (circled and uncircled and learn units or portions of learn units). On the form, each page is a period. You may also use additional pages for a single period as needed; however, label the periods so that the team can categorize the data accurately.

Learn Units

Record learn units and portions of learn units (errors) in two categories—vocal learn units and written learn units. Record vocal learn unit portions as you do for our regular TPRAs with some differences. Each time the teacher presents a vocal antecedent use a check (i.e., tick) mark—erroneous or ambiguous ones are circled an unambiguous ones are not circled. When the students are given an opportunity to respond, record their responses as correct or incorrect. If they do not respond and there was a clear opportunity record a minus for the student also. If the teacher provides no consequence, record a circled "C" as you would for our usual TPRA. Use the same symbols that we always use for teacher behavior circled and uncircled "R" and "C." Only true learn units (i.e., presentations with no errors and no omissions) constitute learn units. Count antecedents with no response opportunities as errors in addition to all standard errors (see chapter 9 for the instructions for doing the TPRA).

Conflicting teacher antecedents without a response time are recorded as a circled tick/check mark. Lectures or instructions without response opportunities are not recorded. The student must be given the possibility of responding such as a question in order for an antecedent to be recorded as a check mark. Record each correct as a plus and each incorrect as a minus. Record accurate and inaccurate reinforcements or corrections as circled or uncircled "R" or "C." At the same time *continue to record the approvals and disapprovals in the same period but in the social category*. Be sure to label each period or portion of the day by subject matter and categories such as free group play, outside play, or lunch time. If the person in charge of the class changes, note his or her name and role and treat his or her behavior with students as you would the class teacher's behavior. Include their behaviors in a separate tally total.

If tutoring occurs between all or some students, pick one tutor–tutee pair and count learn units or portions of learn units done for one pair, just as you would for a teacher and *multiply times* the number of student pairs in the class to obtain an *estimate* of tutoring learn units and correct/incorrect responses. These tutored learn units are added to the teachers total, just as the errors are added to the teacher's total. If the students record data use the students data and place in the written learn unit category.

Rate Determination

For each period of the day determine the rates for all categories. You will have rates of learn units (complete) and social interactions. You will also record rates of teacher ambiguous and nonambiguous antecedents and rates of correct and incorrect student responses—first for vocal and then for written responses. Written responses on the board with teacher vocal consequences are recorded as vocal learn units.

For the daily rates, sum the elapsed times and the rates across all periods. For the academic instruction time record the sum of the elapsed time. Elapsed time for each period is the time from when the teacher indicates the beginning of the period until the last consequence or the signaled end of the period such as when the teacher says, "put away your pencil box." You will determine *down time* as the difference between the elapsed time and the academic time. You will determine costs per learn unit by finding out the *prorated* tuition paid for each student multiplied by the number of students and divided by the number of complete learn units. Prorated tuition is the total money that the state and local community allocates for the year divided by the number of school days in the year.

The written learn units are a bit tricky. You will have to make a quick count of the student responses and teacher consequences or lack of consequences. Any papers returned to the students are marked as follows. If the student responses are marked as correct and the student observes them, they are recorded as correct student responses and full learn units including papers returned from the prior period or day. If the student written responses are incorrect they are not full learn units until the student corrects them. These would be error teacher presentations since they are not consequated to result in a learn unit. If the teacher works with the student individually and the student corrects her work, each correction is recorded as a full learn unit for the teacher. A sample summary form is shown in Table 1.

TABLE 1
Class Inventory of Instruction
(Summary across All Periods)

Student to be Mainstreamed_____

School_____Date_____

Observer_____

Teacher's Name_____

Number of Students in Class_____

Number Present_____

Principal or Head_____ _____School Phone_____

Number of Periods_____School Day Total Time_____

Day Begins_____Day Ends_____Sum of Elapsed Instruction Time_____

Rate of LUs_____Rate Correct Student_____Rate Incorrect_____

Rate Teacher Errors (incomplete or error presentations) _____

Rate Approval Contingent_____Rate Approval Noncontingent_____Rate Disapprovals
   Contigent_____

Rate Noncontingent Disapprovals_____

Total Rate Disapprovals_____

Tuition Costs per student_____total

Tuition Class_____(per student times number students)

Cost per Learn Unit (all learn units into total tution class)_____Costs per Correct Student
   Response (all correct LUs into total tuition for class)_____

List Subjects Taught:

List Other Periods That Are Not Instructional:

(continues)

TABLE 1 (*continued*)

Queries, Comments

1. Did the teacher have a classroom in which the student were quiet, busy and noisy, disruptive sometimes, or continuously disruptive?
2. Did the teacher get upset with disruption or was the teacher calm?
3. Did the teacher have well-planned instruction and curricula in advance of the class or did the teacher improvise? Did the activities the students in appear to be tied to a cogent set of curricular goals?
4. Did the teacher seem interested in having our student in the class or was the teacher reluctant? How about the Principal or Head?
5. What do you think we need to teach our student in order for our student to be successful in this class and this school?
6. Did you notice any students that you think might prove to be difficult for our student to interact with?
7. What time does the student arrive, and how does the student enter the classroom (e.g., is he/she delivered by parents or school bus or does he/she wait outside until bell rings)?
8. What are the implicit and explicit expectations of arriving and getting to work behaviors?

General Comments:

Once you have data on the classroom, you can provide instruction at the originating school to prepare the student for the frequency of learn units they will receive, the frequency of reinforcement or punishment they will encounter, how to recruit reinforcement, and how to use self-management or verbally mediated tactics to avoid difficulties. It is critical to prepare the students to perform academically well (e.g., at or above grade level academic skills). In CABAS schools we follow up these children with additional tutoring sessions. In addition we increase the time in the school contingent on the child's performance in the new setting.

ASSISTING RELATED SERVICES PERSONNEL

Psychologists can assist other related service personnel to apply positive behavioral procedures (e.g., social workers, speech therapists, occupational therapists, and physical therapists). They do this by (a) introducing programmed curricula for social repertoires in related service settings and by (b) teaching and supervising other personnel to use behavioral teaching operations. Furthermore, a psychologist can ensure the general application of social repertoire objectives, across the different teachers who work with a particular child. When a successful procedure and curriculum is implemented in one setting, a carefully scripted program is written by the psychologist and he or she sees to that other teachers are trained in this approach and that it is implemented systematically across school settings.

The written or scripted program should conform with (a) the guidelines concerning the programming of curricula described in Part III and (b) the components of pedagogy described in the chapters on teaching and supervision. Once the program is written, the psychologist reviews the program with those being trained and demonstrates the procedure with the target student for the benefit of the trainees. The accuracy with which the procedures are implemented and the trainees' accuracy in collecting the data are monitored and corrected using the TPRA observation procedure as described in the supervision chapter. As in the case of the supervisor, the psychologists' supervision is nonadversarial and is designed to: (a) provide feedback to the trainee, (b) monitor the adequacy of the program, and (c) maintain reliable data. Changes in the program are done in a manner that ensures that any changes that are made are done simultaneously with all of the professionals involved in the programming effort.

If a planned ignoring program is implemented, for example, the procedures used are consistently and reliably implemented across all professionals for the same target behaviors and only those behaviors and in the same manner. The goal is to apply a consistent set of contingencies and instructional procedures across the total school environment for the target student.

The psychologist must draw on tried and tested tactics and the most recent literature in applied behavior analysis or other research. In addition, the psychologists must have the necessary repertoires associated with the strategic science of instruction (e.g., contingency-shaped repertoires, a strong verbal repertoire about the sciences, and verbally mediated analytic repertoires). The psychologists must know how to shape the behavior of students, draw on the scientific literature, write appropriate scripted programs with adequate data collection procedures, adjust the program consistent with the findings, and see to it that the program is implemented and adopted accurately across all of the professionals involved.

In this role the psychologist is the catalyst for the systematic generalization of programs across professionals with reference to social repertoires. As such, the teaching and supervision operations and the curricular perspective described throughout this book are as critical to the practicing psychologist as they are to the supervisor and teacher.

ASSISTING TEACHERS TO BECOME STRATEGIC SCIENTISTS OF INSTRUCTION

The school psychologists can provide a critical role in providing classroom teachers with an ecobehavioral assessment of the existing contingencies in the classroom. This analysis can be useful to: (a) determine the existing contingencies in the classroom affecting appropriate/inappropriate social behavior and emotional affective responses and (b) analyze the instructional style of the teacher. This role is critical, particularly in classrooms not adhering to compre-hensive behavioral approaches. Classrooms that are not subjected to valid supervisor observations and the kinds of supervisory assistance and pervasive measurement found in CABAS schools are in desperate need of the expertise. In some cases, teachers who request assistance can be supplied with compon-ents of thoroughgoing behavioral instruction. In fact, a small cybernetic system like the CABAS model can be developed with the psychologist providing supervisory support similar to that provided in the teacher/supervisor relation-ship in CABAS schools.

How does one go about introducing the practices associated with the science in schools that are not designed to provide this kind of expertise? Actually there is a long tradition of behavioral psychologist assisting the teacher (Greenwood et al., 1991). Adequately trained psychologists can use that model to introduce sophisticated teaching operations in the classrooms. Typically, when the psychologist notifies teacher that he or she is available to assist with behavioral problems in the classroom with the teacher, new or inexperienced teachers will likely request assistance. Thus, the first step is to locate teachers who want assistance. The request for assistance may be a necessary setting for the success of the assistance.

The presenting problems are often identified as behavioral problems or problems of classroom management. Teachers typically characterize the prob-lem as bad behavior. However, as we have observed throughout this text, the real problem frequently is in the lack of expert instruction for the problem at hand. Thus, it is important to assist the teacher not only in dealing with the bad behavior via tested behavior analytic procedures; it is equally important to enlist the teachers cooperation in improving the design and practice of instruc-tion of all types in the classroom.

Clearly psychologists who serve in the spread of science of pedagogy to classrooms need to have within their repertoires the expertise that has been described in the chapters on teaching, supervision, and curriculum. Psycholo-gists who have the necessary pedagogical skills, or who arrange for experiences to acquire them, can replicate the procedures described in the chapter on supervision and in chapter 3. The steps for introducing instructional expertise to teachers can be done on a microlevel at the classroom level. These steps are adapted as shown in Table 2.

In this latter effort (i.e., the increase in teachers and classrooms with behavioral expertise), the necessity for instructing teachers or therapists in the precise and technically accurate use of verbal behavior about the science becomes critical. The verbal behavior about the science is, to a large part, the basis of any scientific community. Using the terminology accurately to describe the student and the environment is as important as is the repertoire of interventions represented by the vocabulary. Nuzzola-Gomez (2002) found that teaching teachers to use accurate scientific terminology using supervi-sor learn units acted to improve student outcomes. The accurate use

TABLE 2
Steps for Psychologists to Use to Introduce Behavioral Pedagogy to Classroom Teachers

1. Choose the teacher (one) based on the teacher's request and willingness to continue on a cybernetic basis with the psychologist as a consultant.

2. Eliminate bad behavior by training contingency-shaped classroom management skills to the teacher.

3. Target one or two academic areas to introduce the use of behavioral instruction.

   (a) Use model demonstration by the psychologist followed by TPRA observations conducted by the psychologist of the teacher as described in the chapter on supervision.

   (b) Develop continuous learn unit measures and visual displays of the target academic areas.

      (i) Review data and graphs daily/weekly with the target teacher; incorporate TPRA observations tied to learn unit measurement.

   (c) Introduce readings/quizzes based on the PSI teacher training modules described earlier (see supervision chapter). Enlist the teacher in setting his or her own assignments and target performance levels. Maintain a collegial and nonadversarial relationship at all times.

   (d) Log supervisor procedures done with the teacher as described in the supervision chapter and chapter 3.

4. As the teacher has success, expand the target academic areas.

5. Increase the individualization of instruction across the curricular areas as described in the chapter on the delivery of curriculum

   (a) When the teacher has reached a satisfactory level of expertise, move to another classroom (a second requesting teacher) and continue the contact with the first teacher.

   (b) Establish a small group of teachers involved. The small group meets and shares data on a regular basis. If they are teaching similar levels of students induce them to share curricular development.

   (c) As the teachers' expertise in teaching as behavior analysis grows, the psychologist will acquire colleagues and a critical mass of expertise. The support (e.g., reinforcement, assistance, and collegiality) of others who have a similar perspective and expertise is critical to the maintenance effective teaching operations and effective leadership by the psychologist.

of the vocabulary provides a necessary component for developing and maintaining technically and scientifically precise schooling programs.

Some of the strategies that have been found useful in CABAS, Direct Instruction, and Precision Teaching include the following. Devote a portion of meetings to sharing data. Each teacher or staff member can share a best and worst case with accompanying visual displays. Presentations of problem cases can be accompanied by the presenter's selection of alternative strategies and tactics *from the literature* that can be tried next. Colleagues can contribute still other ideas, but always with the *proviso that there be a data-based source for their suggestions.* This is done in order to avoid using hunch-based approaches that have no credible scientific or theoretical basis. Initially the psychologist will need to provide the tactics from the literature to the teacher.

Still another useful procedure is to have staff members report on research articles or books in the field. This serves two purposes: (a) it sharpens the presenter's synthesis of the material and (b) it disseminates the work to colleagues.

These meetings serve as forums to encourage research efforts on the part of teachers who, in turn, can reinforce the growth of expertise and collegiality. The shared goals of developing better teaching and supervisory expertise provides the environment for continued learning and maintains the focus of the school efforts on the joint concerns of the individual student and scientific expertise. Some have suggested that scientific and programmatic efforts that grow into schools or movements have to do with developing enough colleagues so that a *critical mass of expertise is developed*. When this is done the contingencies are in place for the school to be a center of inquiry for students, parents, teachers, and related personnel.

Many school professionals are hostile to the use of behavioral skills. It is not productive to attempt to force these individuals to engage in behavioral procedures. However, work with those individuals who are accepting and who ask for your assistance. New teachers are often eager for assistance and as the procedures work for them they will seek additional expertise. We have found over years of working in schools that those who are initially hostile become more accepting as the efforts of good behavioral programming become obvious. The best course of action is to provide assistance on a request basis.

PORTFOLIO AND INVENTORY ASSESSMENTS: ARCHIVAL RECORDS

Educators increasingly recognize the importance of maintaining archival records of students' achievements. As assessments become accurate records of the reliable objectives achieved by students, assessment files take the form of portfolios or inventories of achievement. As such, students' records are archival and cumulative lists of what they can do to date and the conditions under which they did them. Coordinating the students files such that they are consistent, reliable, up to date, and user-friendly falls to the psychologist in her revised role as *coordinator of measures of learning and instruction*.

The mastery of concepts, operations, and skills for academic and social repertoires are recorded in each student's file in a consistent manner across all of the students in the school. This means that the psychologist must see to it that the data from the classroom teachers, support personnel, and the parents are summarized in a reliable and consistent manner in each student's folders. These files need to convey to future teachers and employers what the student can do, not just overall ratings of arbitrary subjects.

When a *comprehensive inventory of the student is done on a regular basis*, a portfolio accumulates that identifies deficits and assets. Assets are those repertoires that the students have mastered *and the conditions under which the student can perform them*. The deficits actually pinpoint those repertoires that the student has not yet been taught. They need not, nor should they be, identified as student problems of personality, cognition, or affect. Rather the deficits represent repertoires that will require instruction. Those repertoires that were judged most important at the time are the ones that receive highest priority while others await the future. Once the target repertoires are identified for instruction, the pedagogy and curricular materials are matched with the target objective. Instruction continues with necessary modifications until mastery occurs. Unsuccessful instruction, as described in the early chapters of Part I, results in revisions in pedagogy or modifications in the curriculum (e.g., new subobjectives are identified or changes in the curriculum are made).

Selecting or Designing Inventories of Repertoires

A useful inventory will identify reliably (a) what the student *can do* and (b) what the student *cannot do*. It will be criterion referenced and individualized and will specify under what conditions the asset or deficit occurs. Academic skills are identified in standard ways. For example, a statement such as "the student has mastered (e.g., achieved criteria) four digit addition, division up to two decimal points, and reads fluently at the fourth grade level with appropriate descriptions of the components of reading" identifies the behavior and controlling contingencies. Statements like this are straightforward and are immediately useful for those teachers with the relevant expertise (see the chapter on teaching). The social repertoires are, however, the special curricular domains for which the psychologist has special expertise. Thus, they will be the subjects of discussion.

The examples and concepts used will be ones associated with preschool and the primary grades (e.g., preindependent reader stage). The inventory consists of three relevant domains for psychologists. They are: (a) school survival skills, (b) social repertoires, and (c) the emotional affective domain. Many state departments of education use these or similar domain/categories for curricula. Of course for more verbally advanced students academic literacy, self-management and the students' conditioned reinforcers and problem-solving repertoires are included in the full inventory.

School Survival Skills

Young children require certain repertoires that are associated with receiving the benefit of academic instruction as well as cooperating with their peers and teachers such that the classroom is not disruptive and the child stays out of trouble with those in authority. Traditionally, the survival or school self-sufficiency repertoires were identified only if the child created a problem. This experience suggests that it is important to *identify in advance what students need before the problem occurs*. Once the repertoires are identified, they can be subjected to task analyses, the identification or scripting of relevant curricula, and the implementation of adequate teaching operations.

Such repertoires for children in traditional school settings include for example: (a) sitting still and looking at the teacher when instruction is given (e.g., listener behavior under instructional control), (b) following the particular classroom social rules of teachers, (c) standing and staying in line quietly, (d) socially engaging with peers (e.g., conversing or playing) in those settings where it is appropriate, (e) socializing with peers and authorities in a manner that leads to positive reinforcement and little or no punishment, (f) following school-wide rules pertaining to using the bathroom and engaging in appropriate behavior in the restroom, cafeteria, hallways, and transportation settings (e.g., daily bus procedure, field trips).

The psychologist coordinates the collection of repertoires into inventory forms from teachers, supervisors, and support personnel (e.g., speech therapists, counselors). Repertoires that are deficient and seen as priorities are targets for instructional programs sometimes this results in simply seeing to it that the appropriate behaviors are subject to good reinforcement operations performed at high rates across different settings with concomitant planned ignoring strategies. At other times more elaborate programs, involving response cost and contingent observation, are coupled with token economics, peer intervention, or parental tutoring. The goal is for each student to achieve critical skills across the school settings that are habilitative to him or her in the school setting. They are no less in need of systematic instruction than are math, reading, or writing skills. The incidences of deficits that have eminent priority are targets for immediate instruction. This approach results in curricular programs to teach appropriate and habilitative behavior rather than simply targets for behavior decrease programs. The problem can always be reconceptualized in terms of what the student needs to learn to do in order to replace the inappropriate behavior with a behavior that results in reinforcement for habilitative responding. An example of these types of repertoires that is necessary for school is shown in table 3.

Social Skills

Social skills include repertoires that affect the student's well-being in school as well as beyond the confines of school. Can the student negotiate the world such that he obtains more reinforcing consequences than punishing consequences in social settings? Useful social repertoires are ones that minimize the frequency or intensity of punishment from others increase cooperative behavior for mutually beneficial ends, provide students with negotiation skills to use with those who are momentarily in charge, and engender repeated friendly overtures from others. They include repertoires described by laypersons as kindness, tact, and consideration. Although, crude and bully-like behaviors often evoke short-term benefits, their long-term effects are that the child is avoided, disliked, and potentially ostracized from society.

TABLE 3
Replacing Bad behavior with Habilitative Repertoires

Deficits or behavior of complaint

1. Touches peers in ways that peers avoid.
2. Assaults (e.g., kicks, hits, bites) peers.
3. Uses inappropriate behavior to obtain attention from others.
4. Threatens, cajoles, or harasses others (e.g., uses disapproval) to obtain cooperation.
5. Avoids other children while playing with other children.
6. Does not share.
7. In social settings, the student does not contribute.
8. Peers harass the student.
9. The student does not engage in social interaction.
10. The student functions as a listener rather than alternating speaker listener roles.
11. Does not greet, say good-bye, or engage in conversations with authorities such that authority likes the student.

Repertoire needed (see the tactics lists in chapters 5 and 6 for procedures to teach the student):
Use the existing tactics from applied behavior analysis to teach the student to:

1. Touch peers in a way that they do not avoid.
2. Obtain attention by engaging in positive statements to peers
3. Recruit attention by using verbal behavior or sharing.
4. Mand appropriately.
5. Condition other children as reinforcers by teaching the peers to reinforce the target student.
6. Teach sharing as a repertoire.
7. Teach conversational units via social scripts.
8. Teach student to reinforce others via tutoring assignments.
9. Use social skills and mand training to teach interaction.
10. Have the student deliver reinforcement to others via tutoring interactions.
11. Teach the student to reinforce authority figures.

A baseline assessment of social skills can serve as an inventory of repertoires that need be taught. Inventories are particularly useful for younger students. In their case, the repertoires are likely to be incipient; thus the appropriate repertoire can be taught quickly and subsequent problems avoided. Older students need to identify and work on their own deficits in ways described earlier.

The deficits are behaviors that are problematic for the student in the long term. One can decrease them by direct punishment or extinction procedures. *An alternate and comprehensive route* is one that teaches the desirable repertoire for which the noxious behavior is replaced. If the appropriate repertoire becomes a matter of instruction the inappropriate behavior may disappear entirely or need only a planned ignoring operation. If one simply punishes the behavior, it will not exist when the punishment is available but will recur in other settings because no alternative repertoire for obtaining reinforcement has been taught. In the example described earlier, had we not taught the boy or girl to reinforce certain teacher behavior (Polirstok & Greer, 1977), it would have been unlikely that he or she would have evoked more reinforcement from his or her teachers. We could have punished the response (e.g., removed points for disapproval of the teacher) without teaching the student how to approve. Had we not taught the student what to do, it is likely that the disapprovals would have ceased under the prosthetic contingencies, but once the short-term prosthetic contingencies were removed the student would not have had the repertoires to engage the natural contingencies in the classroom. If academic responding is not leading to reinforcements, it is likely that other and inappropriate responses will occur. Rather, by teaching our student a socially useful repertoire (e.g., reinforce teachers), teacher approval resulted and teacher disapprovals decreased. Of course, had the teacher used optimum teaching operations in the first place it is likely that the problem would not have occurred. Even had it not, the lack of a reinforcement operations repertoire for evoking social approval from authorities may still have needed to be trained for future settings.

The behavioral selectionist view is one that considers the long-term effects of repertoires as well as the short-term view. The stance is one that analyzes the existing contingencies and the repertoire that will be needed in the future and the effects on the environment as well as the behavior per se. The question is not what should we do to decrease the bad behavior but rather what deficit in the student's repertoire has served to evoke the aberrant behavior? The source may be in either the instructional history or even in genetic or traumatic sources (e.g., speech loss or no vocal apparatus). As in the case of an artist who after losing the use of his hands learned to paint with his feet, compensating repertoires can be taught that have the needed functional effects of the original behavior.

We all need a range of repertoires. When we have no behaviors that have long- and short-term benefits, we may develop behaviors that are useful in the short term but detrimental in the long term. Students who do not have positive repertoires that lead to peer reinforcement will emit noxious or clown responses that result in laughter or even disapproval from the teacher. No attention may be more disagreeable than disapproval; at least the data suggest this to be the case (Patterson, 1982). One may well ask to what end is the behavior emitted now? What behaviors can be substituted that are of long-term benefit that satisfies the function of the noxious behavior.

Emotional/Affective Repertoire

A common domain, which is treated as an important category in education, is one described traditionally as the emotional or affective domain. A problem results when the locus of a so-called emotional deficit is not identified in terms of stimulus control. A student is described as lacking in *affect*. In such cases, alternative behaviors can be identified and taught; yet, the category of the deficit still suggests that something inside—one's affect or emotion—is not quite intact. In fact, we know no such thing, and it is unfair, unjust, to treat the source of the problem as lying *within the child*. In behavior analysis we think that a more useful approach places the repertoire as responses that are controlled by the contingencies in the environment. From the view of behavioral selection, the source of the problem is the *range of conditioned reinforcers and punishers* that are part of the student community of reinforcers and punishers. In a manner of speaking, we may be said to have a community of reinforcers or punishers developed through our instructional history and our learning experiences.

Children who are afraid of pets, the dark, teachers, other children, crowds, heights, elevators, or any of numerous contingencies have developed conditioned punishers or respondently conditioned eliciting stimuli. *The controlling stimuli may have been the result of respondent or operant contingencies.* Those fear responses that have no survival value or that interfere with habilitation and education can be debilitating. Children who avoid going to school, interacting with individuals, going to the school bathroom, or even swallowing (Greer et al., 1991) have acquired conditioned aversive control that impedes their progress.

Another category has to do with the lack of conditioned reinforcers. Individuals who are said to read for enjoyment probably do so as a result of a conditioning history with reading experiences and reading instruction. The same may be said of performing or listening to music. Authority figures (e.g., teachers, police officers, supervisors, psychologists, medical personnel) also acquire reinforcing or punishing properties. Those who do not seek out others may have a history that did not provide reinforcement opportunities from others (Greer, 1980).

In still other cases, students do not acquire avoidance responses to some stimuli that are useful (that is, punitive control by stimuli in which the punitive control is to their long-term benefit). Examples include avoiding strangers, drugs, poisonous snakes, strange animals, street crossing that are dangerous, and peers who are likely to do them harm. In such cases avoidance is desired and habilitative instruction can be taught using pedagodical expertise.

A school designed and run in a manner consistent with the basic science and the applied science will condition new reinforcers as a function of the salient and frequent use of positive reinforcement, individualized instruction (e.g., the student is successful), and the planned management of back-up exchanges for tokens. Similarly, the avoidance of positive or type 1 punishment tactics and the minimization of contingencies that evoke escape and avoidance response serve to reduce the probability that stimuli will be conditioned to evoke undesirable stimulus control. Students who are reinforced frequently acquire new reinforcers as a natural course of events. Those attributes associated with school truancy (e.g., school is a conglomerate of stimuli that evokes avoidance and escape) are avoided. Students are functioning at ability levels that are appropriate for them; behavior is reinforced and stimuli are conditioned as positive reinforcers. Reading instruction is effective and verbal repertoires are taught and reinforced. In effect each student is emitting behavior that lead to reinforcement for him or her and the behaviors are those that are desired by the school and community.

NORM-REFERENCED AND PROJECTIVE TESTS

We have advocated that the measurement of real behavior or its products in real time under specification of the existing contingencies for the behavior is central to effective schooling. The pitfalls of explanatory fictions and the third or intervening variable have been shown to create problems that do not lead to useful solutions. The use of criterion-referenced objectives, programmed learn units, and inventories of repertoires as portfolio evaluation are instructionally valid and require no tautological constructs. What remains useful for the norm-referenced test?

Norm-referenced standardized tests are useful to ensure that criterion-referenced tests include adequate instruction in skills and knowledge. They serve to assess whether or not the criterion-referenced instruction is comprehensive and provide guidance in the development of curricula. In CABAS schools we have behaviorally defined state and national standards as our inventories of instruction and we use standardized tests of achievement to check the adequacy of these standards and our curricula.

Traditionally, norm-referenced tests are used for at least two purposes. First, because of the existing laws, specialized services (e.g., speech therapy, special education) are justified to a considerable degree by scores on projective tests or norm-referenced tests. The second purpose is to assess the performance of the school with regard to national norms. Newly created "syndromes" such as the so-called "attention deficit syndrome" or "minimal brain dysfunction" take on an existence and validity that exceed good science or maybe even good sense. Classifications lead to tests and tests serve to give the tautological explanations credence. However, without the test the special instructional services are difficult to obtain. When the tests are used appropriately they can lead to instruction that serves to eliminate the source of the problem; but when they are used inappropriately they become a lifetime assignment.

The multiple-choice nature of group norm-referenced tests is also a problem. There is reliable evidence (Chase, Johnson, & Sulzer-Azaroff, 1985) that multiple-choice responses and response production are different verbal response classes within instruction. Responding correctly to a multiple-choice question does not mean that the student can produce applications of academic literacy to solving problems or producing answers. Thus, if there is an *in situ* assessment of the student's repertoire in specified settings, the assessment is more valid than a subscore on a norm-referenced test. The original purpose of the norm-referenced test is to predict or estimate the existence of the true repertoire. If the true or *in situ* repertoire is reliably documented *in situ,* then that is the most valid measure. Unfortunately, this is not well understood. Individually run standardized tests overcome many of these problems.

Clearly, a revolution needs to take place in the types of measures that are used and believed in schools. As teachers, psychologists, and schools shift to comprehensive behavior–environment measures, the function and usefulness of standardized tests need to be reconceptualized, but not necessarily discarded.

Because of misuses of norm-referenced tests and the lack of knowledge about how they should be used many educators have developed an antimeasurement biases that extend to all types of measurement. The so-called "postmodern philosophy" is one example of the antimeasurement sentiment found in some educational circles. The antimeasurement sentiment that is found in some schools may be a partial result of the unfounded notion that norm-referenced tests tied to the normal cure are synonymous with educational measurement. The use of learn-unit- and criterion-based assessment can be reliably and pervasively used by teachers, as we have seen. They are direct measures of teaching and learning and can be used together with well selected standardized achievement tests and national and state standards to

Measurement is critical to effective teaching, but tests that serve only to select out are not useful tools for instruction. They are useful for many other purposes. If used carefully and judiciously (see chapter 2), they can serve as an auditing tool to determine *how the school as a whole is performing in terms of national norms.* But schools that approach teaching as behavior analysis are ones that perform from *four to seven times better than what resulted from prescientific approaches to teaching.*

The literature does not always suggest a strong correspondence between school success and postschool success (Mithaug, Martin, Agran, & Rusch, 1988). This is probably due in part to the inadequacy of standardized tests or even idemnotic measurement. A more adequate evaluation would be one that compared the reliable measurement of students' achievement of repertoires in school with repertoires that are useful in the world. We believe that the use of national and state standards together with good criterion-referenced measurement of instruction can be combined with appropriate use of standardized tests to improve the relationship between school instruction and lifetime outcomes. We will improve the relationship between schooling and students' future success through further behavior analyses. Such analyses will become increasingly feasible when schools measure and document repertoires in addition to the measurement of nomological constructs. School psychologists can lead in this effort as they learn to apply a comprehensive behavioral selectionist perspective on their role in schooling.

SUMMARY

There are several roles for the school psychologist who functions from the perspective of contemporary behavior analysis and behaviorism:

1. The psychologist can lead in an approach to the treatment of bad behavior that is revised around the practices and perspectives of the science such that the behavior is treated in a total environmental perspective.
2. The psychologist's role in the parent and school relationship draws on the operations and functions found in the literature on teaching parents to teach their children in collaboration with schooling efforts.
3. The psychologists can assume behavioral training roles with other therapists such that behavioral expertise is spread across the other therapists.
4. The psychologist can teach students to use behavioral skills to solve problems in the home and school
5. The psychologists can assist effectively in the mainstreaming and inclusion efforts using behavioral assessment of the mainstream setting,
6. The psychologist can provide the initiative and the leadership for revising classrooms around behavioral expertise.
7. The psychologist can assist in the appropriate use of norm-referenced tests combined with criterion-referenced classroom measurement guided by state and national standards instructionally to provide comprehensive, valid, and student-driven assessment.

References

Ayllon, T., & Roberts, M. D. (1974). Eliminating discipline by strengthening academic performance. *Journal of Applied Behavior Analysis, 7*, 71–76.

Carey, P. (1991). *Ilywacker.* New York: Harper Collins.

Carr, E. G., & Durand, V. M. (1985). Reducing behavior problems through functional communication training. *Journal of Applied Behavior Analysis, 18*, 111–126.

Chase, P. M., Johnson, K. R., & Sulzer-Azaroff, B. (1985). Verbal relations within instruction: Are there subclasses of intraverbal? *Journal of the Experimental Analysis of Behavior, 43*, 301–313.

Chu, H. (1998). *Functional relations between verbal behavior or social skills training, and aberrant behaviors of young autistic children.* Unpublished Ph.D. dissertation, Columbia University, New York.

Donley, C.R., & Greer, R.D. (1993). Setting events controlling social verbal exchanges between students with developmental disabilities. *Journal of Behavioral Education, 3*, 387–401.

Duvall, S. F., Delquardi, J. C., Elliot, M., & Hall, R. V. (1992). Parent tutoring procedures: Experimental analysis and validation of generalization in oral reading across passages, settings, and time. *Journal of Behavioral Education, 2*, 281–304.

Forehand, R. L., & McMahon, R. J. (1981). *Helping the noncompliant child: A clinician's guide to parent training.* New York: Guilford.

Greenwood, C. R., Carta, J. J., & Atwater, J. (1991). Ecobehavioral analysis in the classrooms. *Journal of Behavioral Education, 1*, 59–78.

Greer, R. D. (1980). *Design for music learning.* New York: Teachers College Press.

Greer, R. D., Becker, B.. Saxe, C. D., & Mirabella, R. F. (1985). Conditioning histories and setting stimuli controlling engagement in stereotypy or toy play. *Analysis and Intervention with Developmental Disabilities, 5*, 269–284.

Greer, R. D., Dorow, L. G., Williams, G., & Asnes, R. (1991). Peer-mediated procedures to induce swallowing and food acceptance with young children. *Journal of Applied Behavior Analysis, 24*, 783–79.

Iwata, B. A., Dorsey, M. F., Slifer, K. J., Bauman, K. E., & Richman, G. S. (1982). Toward a functional analysis of self-injury. *Analysis and Intervention in Developmental Disabilities, 2*, 3–20.

Kauffman, J. M. (1989). *Characteristics of behavior disorders of children and youth* (4th ed.). Columbus, OH: Merrill.

Kazdin, A. E. (1988). *Child psychotherapy: Developing and identifying effective treatment*. New York: Pergamon.

Kelly, T. M., & Greer, R. D. (May, 1992). *Functional relationships between learn units and maladaptive behavior*. Paper presented at the international conference of the Association for Behavior Analysis, San Francisco.

Lindsley, O. R. (1991). Precision Teaching's unique legacy from B. F. Skinner. *Journal of Behavioral Education, 1*, 253–266.

Martinez, R., & Greer, R. D. (1997). *Reducing aberrant behaviors of autistic students through efficient instruction: A case of matching in the single alternative environment*. Paper presented at the annual conference of the Association for Behavior Analysis, Chicago, IL.

Mithaug, D. E., Martin, J. E., Agran, M., & Rusch, F. R. (1988). *Why special education graduates fail: How to teach them to succeed*. Colorado Springs, CO: Ascend.

Nuzzola-Gomez, R. (2002). *Observed and direct effects of supervisor learn units on teachers scientific vocabulary and subsequent effects on student learning*. Unpublished doctoral dissertation, Columbia University, New York.

Nuzzola-Gomez, R., Leonard, M. A., Ortiz, E, Rivera, C. M., & Greer, R. D. (2002). Teaching children with autism to prefer books or toys over stereotypy or passivity. *Journal of Positive Behavior Interventions, 4(2)*, 80–87.

Patterson, G. R. (1982). *Coercive family process*. Eugene, OR: Castalia.

Paul, J. L., & Epanchin, B. C. (1991). *Educating emotionally disturbed children and youth: Theories and practices for teachers* (2nd ed.). Columbus, OH: Merrill.

Polirstok, S. R., & Greer, R. D. (1977). Remediation of a mutually aversive interaction between a problem student and four teachers by training the student in reinforcement technique. *Journal of Applied Behavior Analysis, 11*, 707–716.

Shapiro, E. S. (1987). *Behavioral assessment in school psychology*. Hillsdale, NJ: Erlbaum.

Sherman, T. M., & Cormier, W. H. (1974). An investigation of influence of student behavior on teacher behavior. *Journal of Applied Behavior Analysis, 7*, 11–21.

Skinner, B. F. (1956). *Science and human behavior*. New York: Basic Books.

Sulzer-Azaroff, B., Drabman, R. M., Greer, R. D., Hall, R. V., Iwata, B. A., & O'Leary, S. G. (Eds.) (1988). *Behavior analysis in education 1968–1987 from the journal of applied behavior analysis*. Lawrence, KS: Society for the Experimental Analysis of Behavior.

Sulzer-Azaroff, B., & Mayer, G. R. (1986). *Achieving educational excellence*. New York: Holt, Rinehart, & Winston.

Task Force on Comprehensive and Coordinated Psychological Services for Children: 0–10 (1994). *Comprehensive and coordinated psychological services for children: A call for service integration*. Washington, DC: American Psychological Association.

Trovato, J., & Bucher, B. (1980). Peer tutoring with or without home-based reinforcement for reading achievement. *Journal of Applied Behavior Analysis, 13*, 129–142.

Wahler, R. G., & Dumas, J. E. (1986). Maintenance factors in coercive mother–child interactions: The compliance and predictability hypotheses. *Journal of Applied Behavior Analysis, 19*, 13–22.

Yesseldyke, J., Dawson, P., Lehr, C., Reynolds, M., & Telzow, C. (1997). *School psychology: A blueprint for training and practice*. Bethesda, MD: National Association of School Psychologists.

Glossary

A

absolute units of measurement. The expression *absolute unit* refers to a level of scientific measurement. Variables may be categorized as (1) nominal, (2) ordinal, (3) interval, (4) ratio, or (5) absolute unit. nominal variable is one that is discrete by categories such as yes/no, black/white, male/female, and high locus of control/low locus of control. (2) Ordinal variables can be rank ordered, but the interval of difference is not the same between each rank (e.g., ratings or rankings of performance by judges). (3) Interval variables are measures that are scaled to a standard such as percentages correct, points on the normal curve, or standard scores on a test. (4) Ratio and absolute measures are measures that have a true zero. Ratio variables have the same incremental difference such as frequency or rate of responding or physical measures such as temperature or speed. Behavior analysis added still another level. These are responses and response classes that produce certain outcomes. While response classes have most of the characteristics of absolute units, they differ in some regards. That is, responses that produce the same outcome can belong to the same class but involve different levels of energy and different individual behaviors. Like absolute units, response in response classes is real behavior in real time; yet, they may not have the same physical energy measures (i.e., the calories produced by different absolute units may not be equivalent). The latter is not a problem for a science of behavior because the dimensions are not based on scales of energy; rather, the new dimensions reflect a science of the behavior of outcomes.

academic conversational units. An expression coined by the author to refer to verbal behavior episodes in academic settings in which both a speaker and a listener each receive three-term verbal contingencies. This term was replaced in the literature by the term *learn unit*. *Academic conversational units* or learn units are distinguished from social conversational units, in that the latter is under the control of the reinforcing effects of verbal exchanges per se, whereas in academic units, the reinforcement is under the control of the teacher. The *learn unit* is an *academic conversational unit*.

actual instruction time. The cumulative time in which teacher presentations, instructional trials, or learn units are actually done. This time includes only that time which is devoted to responding learn units or instructional presentations and excludes transition or downtime. See chapters 5 and 6 for a detailed description.

administrative repertoires. The outcomes and associated behaviors that are done to administer the schooling process. They include tasks concerned with the provision of materials, goods, and personnel services (e.g., payroll, recruitment, interviewing potential employees) to ensure the welfare of students and the preparation and maintenance of facilities to carry out the business of pedagogy. They are distinguished in the CABAS model of schooling from supervisory repertoires. See the detailed treatment in chapter 9. *See also* **supervision repertoires; supervisor rate; management by behavior analysis**.

allocated instructional time. The time allocated by the state or school for instruction. Most states specify that schools must provide a given number of hours of instruction daily; however, the time is simply allocated and does not represent actual engaged time. Thus, it includes transition or downtime, and it is the time used typically to estimate inter-learn unit latency. See chapter 9 for a detailed description.

alternating treatment designs. Tactics of research design that implement the Mill's canon of differences by rapidly alternating sessions between baseline and treatment conditions or two treatment conditions following a baseline phase. The procedures are similar, if not synonymous, with same of the procedures described under the rubric *multiple schedule design*. In designs that rely on the rapid alternation of conditions, there must be little or no carryover or spillover in the effects of one condition to the effects of another. Carryover effects will increase the probability that any functional relationships attributable to separate variables are obscured by the rapid alternation. If the possibility of carryover is obtainable or occurring, withdrawal or reversal procedures should be used. This design may be combined with a reversal design. *See also* **multielement**.

antecedent to behavior *or* **antecedent**. Refers to the S^d (stimulus discriminative) or *target* S^d in the three-term contingency (antecedent, behavior, consequence relationship). That is, the antecedent may be one targeted to gain stimulus discriminative status in the teaching process. Once taught, or learned, the antecedent functions as a stimulus discriminative. This function is acquired as a result of the selecting mechanism of the consequence (e.g., reinforcer or punisher). The reinforcing or punishing effects of the consequence pull the antecedent stimulus control along. The presentation of the target antecedent by a teacher should be unambiguous and done while the student is under instructional control. If there is also a teacher antecedent to the antecedent presented to the student, it should function as an existing stimulus discriminative to focus the student on the target antecedent. The teacher may also provide an antecedent (S^d) to the desired response for the student. In the latter case, the teacher behavior follows presentation of the target stimulus to the student and precedes the student response. Care should be taken to ensure that the target antecedent is one for which the student has had the necessary prerequisite experience. The salient or defining characteristic or characteristics of the controlling component of the target antecedent should be consistent across presentations, as should the teacher antecedent. See chapters 4–6 for a thorough treatment of the "antecedent" as a strategic source for locating potential learning difficulties. *See also* **discriminative stimulus; positive reinforcement principles and operations/tactics; negative reinforcement principle; negative reinforcement operations; operant behavior; learn units**.

applied behavior analysis. The study of socially significant behavior in natural or naturalistic conditions. The effects of certain interventions are said to be significant scientifically when the results of the intervention have an educationally or clinically significant effect on the behavior. Applied behavior analysis also eschews the practices of hypotheticodeductive science to solve real world problems and advocate the application of the experimental analysis of behavior in the natural environment to real world problems. Applied behavior analysts draw on the basic principles of the science and over 188 applied research tactics. We extended these scientific practices as described in chapters 5 and 6 in order to analyze and intervene in learning or teaching problems to ensure that the student learns.

appropriate contingency-shaped behavior. Behavior learned directly from the contingencies that act to the long-term benefit of the person who is behaving. In the example of the teacher, a teacher learns to ignore inappropriate student behavior and reinforces the student for behaving appropriately. Typically, the appropriate responses are learned with mediating operations from supervisors who arrange conditions that allow appropriate teaching behavior to come under design instruction to teach their students' contingency-shaped behavior.

a priori belief. One that cannot be attributed to authority, logic, or science. The philosophical doctrine that the knowledge of God is immediate and intuited (ontologism) and that all knowledge originates with this belief is an a priori belief. Such beliefs are nonreflective and nonanalytical and they accept no reflective, authoritative, or scientific argument as capable of correcting or modifying the belief.

archival records. Portfolio evaluation of students, teachers, and supervisors, for example. The science of behavior builds on archival record of the direct measurement of behavior rather than statistical summaries or measures of hypothetical constructs. Thus, when archival records are used to summarize the repertoires of students, a listing of three-term contingencies and settings acquired by a student becomes the portfolio assessment of the student's achievement. In CABAS schools, this takes the form of an inventory of repertoires that specify the antecedent–behavior–consequences repertoires of students under behaviorally defined categories such as (a) academic literacy repertoires, (b) self- or contingency management repertoires, (c) problem-solving repertoires, and (d) the community of reinforcers for students.

ascending trend. A trend of a linear visual display of data is said to be ascending when a straight edge placed across the display of data shows an upward line. The ascending trend may be slight, steep, or gradual. The trend may vary slightly or greatly around the slope (e.g., steepness of trend). *See also* **descending trend; variable trend; visual display**.

assaultive (or aggressive) behavior. Behavior of an individual that harms or threatens to harm another individual (i.e., has the potential to do tissue damage or evoke escape from the person targeted for the "assault"). It occurs in verbal or nonverbal forms. The response class is associated for some individuals with escape or avoidance. Thus, it is associated with punishment or negative reinforcement or stages of deprivation or satiation. We prefer the term assaultive because it has no reference to an intervening variable, while the term aggressive invokes an intervening variable (i.e., the assault is a manifestation of *aggression*). In the latter case, an assault is assumed to be an incidence in which one "exhibits" aggression.

Association for Behavior Analysis. An international association of individuals who engage in the experimental, applied, and theoretical analysis of behavior (e.g., behavioral selectionism, interbehaviorism). The organization publishes a scholarly journal called *The Behavior Analyst*, and the home office is located in the Psychology Department at The Western Michigan University in Kalamazoo, Michigan (www.behavior.org).

autism. A behavioral diagnosis involving a *nomological* network of behaviors including severely delayed language, self-injurious repertoires, assaultive or withdrawn repertoires, and stereotypy. It is a psychological construct for which there is no known genetic cause; indeed there may be multiple causes. Individuals who are said to be autistic are typically developmentally delayed. However, individuals so diagnosed vary widely.

autoclitics. A term coined by Skinner in 1957 to describe verbal behavior that functions to quantify, qualify, affirm, negate, specify, or in some way modify the effect of a speaker's behavior on a listener or audience. One may specify a mand as in "that cookie" or "do it now . . . please" or a tact as in "that is the correct answer," "in the box," "under the hat," "no, that is not it." Autoclitics, like other types of verbal behavior, are determined by their function, not their form. Thus, the form or topography cannot determine whether a unit of verbal behavior has a particular function. *See also* **mand; tact; intraverbal behavior; verbal behavior**.

automated programmed instruction. A programmed curriculum that replaces teacher antecedent and consequences and teacher-delivered shaping procedures with either a programmed text presentation or a mechanical or electronic delivery. It is said to engender learner-controlled responding in that the student's progress controls the automated teaching consequences. See chapter 8 for a thorough description.

B

bad behavior. Used in this book from time to time in lieu of maladaptive behavior. It is a term that is based on social consensus about appropriate and inappropriate behavior by the layperson but we use the term *bad behavior*, as behavior that leads to short-term reinforcement but long-term punishment, and thus the response is actually adaptive for the individual at the moment the behavior is emitted.

baseline. The phase in an experiment in which conditions are held constant and the experimental variable is not present or the experimental manipulation is not performed. It is the control condition against which a single difference or package of differences (i.e., the intervention) is logically tested. Baselines should be stable or exhibit a trend opposite to that of the expected effect of the experimental condition. Baselines are inserted between each experimental condition to test for within-subject variability. In teaching settings, the baseline becomes the current teaching operations and the intervention the systematic change in teaching operations. When baselines cannot be recovered, between-subject or intersubject designs are substituted. *See also* **between-subject research designs (intersubject designs); within-subject designs (intra-subject designs); Mill's canons of the scientific method**.

basic science of behavior. *See also* **experimental analysis of behavior**. The expression *basic science of behavior* or the bench science of behavior refers to the laboratory branch of the science (also called the experimental analysis of behavior) as distinguished from applied behavior analysis. Both follow species and human species are the subjects in the basic science, and the subject matter embraces the setting events (motivational and contextual), setting stimuli, antecedent, consequences, and adjuncts of the target operant behavior. The field originated with the publication of Skinner's book *The Behavior of Organisms* in 1938, and the original journal was *The Journal of the Experimental Analysis of Behavior*.

behavior acceleration program *or* **behavior increase program**. An expression that denotes that the program is devoted to increasing the occurrence of a behavior as opposed to decreasing or decelerating a behavior.

behavioral assets. Repertoires of behavior that the student possesses at the time of assessment. Typically, the expression is used when referring to an assessment of the presence or absence of certain repertoires as a precursor to determining what needs to be taught (e.g., deficits) versus that which the student has already learned (e.g., behavioral assets). Of course, the asset is not simply a behavior but three-term contingencies, their context, and a class of responses. See chapters 3 and 6–9. *See also* **inventory of student repertoires; repertoires of behavior; means objective (short-term objective)**.

behavioral community psychology. A field of applied behavior analysis devoted to the study of context and related three-term contingencies or their analogs in the community at large. Practitioners study tactics that reinforce or punish practices such as littering, using seat belts, conserving electricity, or engaging in safe sexual practices. See the *Journal of Applied Behavior Analysis* for examples of research in community psychology from the perspective of applied behavior analysis.

behavioral deficits. The lack of a repertoire as determined by a reliable observation. Behavioral deficits are the targets for instruction when education is individualized. *See also* **behavioral assets**.

behavioral medicine. A field of psychology devoted to the psychological aspects of medical treatment or medical conditions. Behavior analysis has been used in a range of applications including ones for the comatose, feeding disorders, medication compliance, preventive medicine, or dealing with seizures or pain or asthma, to name a few of many applications.

behavioral models of schooling. The behavioral models of schooling include Programmed Instruction, Precision Teaching, Direct Instruction, the Consulting Behavior Analyst Model, the Ecobehavioral Model, the Morningside Generative Model of Instruction, Personalized System of Instruction (PSI), and the Comprehensive Application of Behavior Analysis to Schooling (CABAS). See definitions of each model in this glossary.

behavioral selection. A term that is used synonymously with the terms radical behaviorism, thoroughgoing behaviorism, and reconstructed behaviorism. It is the philosophy of the science of behavior and is a component of behavior analysis or the science of the behavior of the individual. The position proclaims the value of the study of behavior per se, particularly with regard to the role of the selection of ontogenetic behavior by the consequences of that behavior. It eschews the necessity for pursuing intervening variables and substitutes the pursuit of inductively determined general principles concerning the operant and the instructional history and setting event factors associated with the operant. Both nonverbal and verbal behavior are the targets of study. Behaviors beneath and outside the skin are part of the purview of the science. The practices of the science are related to those of the natural sciences where functional relationships are seen as more critical than structural analyses. Behavioral selection is seen as an extension of Darwin's natural selection. In the latter conception, the environmental consequences select out the species or phylogenetic attributes, while in behavioral selectionism, the environmental consequences of behavior select out ontogenetic behavior (i.e., the repertoires acquired by the individual as result of the contingencies he experiences in his lifetime). In behavioral selection, the *Homo sapiens* shares the same deterministic basis, as do the follow species. Both are subject to the lawful effects of phylogenetic and ontogenetic variables. Thus, it is held that genetic and environmental events are the factors that contribute to behavioral repertoires. Intervening variables are not satisfactory explanations of behavior and are or can be replaced by principles of behavior consequence relationships. See also biologists who suggest cultural controls for behavior that supplement genetic controls.

behavior decrease programs *or* **behavior deceleration programs**. Instructional programs that have as their objective the reduction or elimination of an antecedent–behavior–consequence rate are referred to as behavior or decrease or behavior deceleration programs. There are ethical and professional guidelines associated with the application of such programs. The science and ethics of the treatment of bad behavior and how to deal with it are treated in chapter 10.

behaviorism. The philosophy of the various sciences of behavior.

behavior momentum. A tactic developed in applied behavior analysis to induce a behavior that occurs infrequently by interspersing behaviors easily performed. The momentum of known behaviors or ones in the student's repertoire creates new behavior as a result. Typically, the response desired has to belong to the same response class as the desired new behavior. The term is used somewhat differently in the basic science.

behaviorology. An alternative term used to describe the science of behavior—its applied, basic, and theoretical branches. The term was developed by a group of behavior analysts because they believed that the objectives of the science of behavior had become increasingly separate from the professional organization of psychology. Behavior analysts are associated with education, psychology, therapy, speech therapy, and social work, to name a few roles in which the science of behavior is directly or indirectly related to professional activities. Behavior analysts study behavior and environmental relations particularly with regard to selections by consequence and an enlarged analysis of the operant unit. The associated scientific society is The Society for Behaviorology. For a description of the activities and origin of the term, see the journal *Behaviorology*.

behavior therapy. The application of principles of operant and respondent conditioning to the treatment of referring conditions such as anxiety, neuroses, and lack of assertiveness, among numerous other conditions. Some procedures are based on operant analyses while others are based on respondent analyses.

belief. *See also* **prediction**. Beliefs have to do with the expected relationship between events. If one switches on a light switch, lights will, it is believed, result. If a teacher performs certain reinforcement operations, it is believed that reinforcement affects will follow. Our culture has developed certain problem-solving repertoires to develop beliefs such as seeking authoritative sources, using methods of logic, or using scientific methods. *Beliefs* that do not result from the aforementioned operations are intuitive or a priori. See the discussion of belief systems in chapters 6–8. *See also* **a priori belief; prediction; method of authority; method of tenacity/logic; method of science**.

between-subject research designs (intersubject designs). An *intersubject design* is one in which the control, baseline, procedure, or the experimental procedure are assigned to different individuals to test the independent variable. In single-subject designs, one subject (individual, behavior, setting) is in baseline while another is in treatment to implement the method of difference. In group designs, the groups are equivocated by population criteria, and one receives the control condition and another the experimental condition. The logic of the design suggests that the subjects are comparable; therefore, differences in performance are attributable to treatment versus control variables. The procedure involves the logical notion that in some cases within subject comparisons may be substitutes with between-subject comparisons. In some cases baselines are not recoverable, and between-subject designs may be necessary. *See also* **method of disagreement (method of difference); within-subject designs (intersubject designs)**.

biospecification of behaviors. A term used by Davidson in 1978 to describe the lack of correspondence in psychological research studies between behavioral reports, measures of introceptive systems, and measures of proprioceptive systems. Davidson suggests that differences or lack of correspondence between measures of these human behavior systems is explainable in terms of the evolution of the species. That is, each has evolved such that each may be susceptible to independent effects by the same antecedent and consequent relationship. For example, physiological stress responses may occur while an individual reports no stress. *See also* **correspondence teaching**.

breaking down goals. *See also* **task analysis**. The process of breaking down goals refers to dividing means or short-term goals into even shorter-term goals. The process of programming involves dividing unit or terminal goals into short-term or means objectives. If the subobjectives are set appropriately for the individual student, he or she will achieve each subobjective rapidly. If the objectives are too difficult for a given student, it is wise to break down the instructional material to goals that are realizable for the student. See chapter 8 for a detailed treatment of programming. The expression *breaking down goals* is an informal term that relates to the principles of successive approximation and tactics associated with task analysis and the design of frame steps or scripted learn units. *See also* **successive approximation principles and operations**.

C

canonical correlations. Statistical procedures used when multiple correlations are performed on the same variables in order to mathematically "weigh out" the likelihood of obtaining erroneous correlations that can occur as a result of simply chance relationships due simply to the number of calculations performed on the same data.

certified teachers. Teachers who are certified are ones who have met the academic and practice requirements of a state such that they are recognized by the state as certified teachers. Typically, this involves the completion of a Baccalaureate (four-year degree) or a Master's degree in which certain courses and practice requirements were completed.

chaining. The process of teaching operations whereby a series of learned operants is brought under sequential control as in the process of chaining together steps in a task analysis such that they flow from one to another. The operations involve applications of tactics to engender successive approximation whereby reinforcement results at the end of the chain.

changing criterion design. The changing criterion design proposes that the logic of a functional relationship is shown by repeated phases in which the subject attains the target criterion. It has utility in school settings wherein the attainment of new objectives may be said to affirm that the pedagogy being used is functionally related to the achievement of criterion. More precisely speaking, the obtained relationship is more in the nature of concomitant variance in that the teaching interventions vary consistently and in the same direction as the student's achievement of objectives. As such, it is an evaluative and correlation design. When more rigorous analyses are called for, designs that adhere to the Mill's canon of differences should be used (e.g., reversal, withdrawal, multiple treatment, or multiple baseline designs).

coercive traps. Relationships between two or more individuals that are characterized as negatively reinforced reciprocal behaviors. Disapprovals by one party result in disapprovals by another, and the relationship is said to be caught in a coercive trap.

cognitive problem-solving routines. Used herein to refer to the solving of problems based on the application of recognized methods of determining courses of action associated both with particular subject matter and functional repertoires that go across different subject matter (e.g., methods of authority, science, logic). Courses of action are often based on verbal or mathematical abstractions for classifying stimuli. These verbal mediations (or other symbolic systems such as math or music) allow the solving of problems without trial and error. The relation between the verbal mediation, the verbal category, and the phenomenon so characterized is a repertoire of behavior and, if so conceptualized and operationalized, it may be taught, thereby placing an important desired attribute of education as an observable and replicable educational objective.

compliance (as in student compliance). The following of instructions by a student or the behavior of the student that is consistent with rules of a classroom or the directions of a teacher or parent. It is conceptualized also as listener behavior wherein the student as a listener complies with the verbal behavior of a speaker (usually the teacher or parent).

Comprehensive Application of Behavior Analysis to Schooling (CABAS). A behavioral model of schooling that draws on (a) the other behavioral models of schooling, (b) the tactics and strategies form the applied and experimental branches of behavior analysis, (c) the epistemology of behavioral selectionism, (d) research on the model and its components, and (e) demonstration applications to several schools. These components are applied to all of the individuals involved in the school community—(a) students, (b) parents, (c) teachers, (d) supervisors, (e) the university training program—on a system-wide basis. This system-wide analysis is used to determine or evoke cybernetic relationships between all of the parties such that the effects on the students' learning are the controlling variable for the relationships between roles (see also www.cabas.com).

computerized data management. The use of a computer and computer program (i.e., spreadsheet software) to summarize, categorize, or organize data for purposes of managing the data. The device and its usage replace hand calculations and pencil and paper operations.

computerized spreadsheet. A software program for a computer that is used to organize and summarize data. For example, a computerized spreadsheet may be used to assist in data management of a data-intensive school.

concepts. Behaviors that result from discriminations between and within categories. They are classificatory in nature and involve classification according to salient or relevant variables of stimuli and not the irrelevant attributes.

consequence of behavior. The result, effect, or event that follows a behavior. Consequences that increase the probability for future occurrences of behavior are reinforcers, while consequences that decrease behavior are punishers. Consequences may be the natural effect of the behavior or they may be unrelated to the behavior. A natural consequence to doing a math problem is the answer, while the reception of a token is an unnatural or prosthetic consequence. The lawful relationship of consequences to behavior is sometimes referred to as "the law of effect" and is attributed to E. L. Thorndike. Skinner in 1938 showed that the consequence is the major component for the selection of operant behaviors for the organism. *See also* **positive reinforcement principles and operations/tactics; negative punishment or type II punishment**

operations; behavioral selection; teacher performance rate/accuracy (TPRA) observation; behaviorism.

consulting behavior analyst. The expression *consulting behavior analyst* was, I trust, originated by this author to describe the particular contribution of educational applications of applied behavior analysis to a science of pedagogy. Tactics and, by inference, strategies of applied behavior analysis grew out of the work of behavior analysts at the University of Kansas, The University of Washington, and the University of Illinois. (Vance Hall, Sidney Bijou, Bill Hopkins, Donald Baer, Betty Hart, Todd Risely, Montrose Wolf, Judith LeBlanc, Joseph Delquadri, Charles Greenwood, Eugene Ramp, and Tom Lovitt are just a few of the individuals who made seminal contributions to the application of behavior analysis to schooling.) Many of the contributions have concerned the social behavior of students and the management of classrooms by teachers. A collection of articles selected from The *Journal of Applied Behavior Analysis* (edited by Sulzer-Azaroff and an ad hoc board of editors, 1987) is representative. The Ecobehavioral Model is an outgrowth of the Consulting Behavior Analyst Model. The movement developed certain types of measurements (i.e., interval observation) and prototypical single-case experimental designs (i.e., reversal, multiple baseline designs) and numerous tactics that apply the strategies of the science to educational and pedagogical problems. Two prominent and seminal contributors, Bijou and Beer, have also contributed to an enlarged behavioral perspective on developmental psychology. This movement has also introduced concepts of interbehaviorism to the behavioral schooling movement.

context. The context in which something is learned or taught is a critical component of instruction. In behaviorology and behavioral selectionism, context is operationalized as the setting events and setting stimuli, both motivational and nonmotivational, that affect three-term contingencies. *See also* **motivation and the science of behavior; establishing operations; setting events; setting stimuli**.

contingencies of reinforcement and punishment (three-term contingences). The antecedent–behavior–consequence relationship that characterizes the operant. The occurrence of operant behaviors is a function of the contingencies of reinforcement and punishment. That is, behaviors decrease or increase as the result of the selection by consequences of the behavior or response classes. Punishers function to decrease or decelerate while reinforcements, reinforcers, positive or negative, function to increase occurrences.

contingency-shaped behavior. Operant behavior shaped directly by its consequences. It is differentiated from verbally mediated or rule-governed behavior wherein the verbal behavior of a speaker shapes the behavior over the direct consequences or serves in lieu of the direct consequences. Learning to do something as a result of vocal or textural verbal behavior rather than experiencing the natural contingencies is an example of verbally mediated behavior. Learning to do the same task as a result of experiencing the direct consequences is an incidence of contingency-shaped behavior.

contingency-shaped teaching behavior. Teacher operations that occur as a result of direct contact with the contingencies in the classroom. Supervisors work with teachers in the classroom to ensure that the contingency-shaped operations that are used are controlled by the practices of effective teaching from the science rather than ineffective practices. The concept is discussed at length in chapter 9 and is introduced in chapters 3, 5, and 6. *See also* **appropriate contingency-shaped behavior; inappropriate contingency-shaped behavior**.

contingent. A consistent relationship between an antecedent, behavior, and consequence. If the relationship between all three is consistent according to predetermined criteria, the relationship is said to be contingent. *See also* **contingency-shaped teaching behavior; contingent reinforcement; noncontingent reinforcement; contingent correction; contingent reprimand**.

contingent correction. A teacher or automated correction of a student response that was done consistent with a curricular program or an instructional objective. For example, after a student incorrectly performed a mathematical operation, the teacher demonstrates the correct procedure and the student also does the operation again but correctly. That is, the student was incorrect and the teacher's correction operation was appropriate. If, on the other hand, the student is correct and the teacher erroneously has the student correct the response, the teacher has presented a noncontingent correction. *See also* **teacher performance rate/accuracy (TPRA) observation; teacher-contingent consequation**.

contingent educational activities. *See also* **Premack principle and operation**. Several studies have documented the effects of the use of educational activities as a reinforcer for instruction. That is, if another activity (e.g., reading) is, at the moment, slightly more preferred than the current instructional task (e.g., addition), the reading activity has a high probability of reinforcing the current task (e.g., addition responses). *See also* **Premack principle and operation; establishing operations; motivation and the science of behavior; response delay**.

contingent praise or approval. When a teacher praises or approves the behavior of a student when that behavior is part of the student's social or academic objectives, the teacher is providing contingent praise or approval. It is an accurate teacher response for the setting and response of the student, when the praise functions as a positive reinforcer for the student's behavior contingent praise is a frequent response in classrooms that are based on the science of behavior and pedagogy. *See also* **teacher performance rate/accuracy (TPRA) observation**.

contingent reprimand. A teacher-delivered operation (e.g., reprimand, overcorrection, disapproval) that is correct according to the response of the student and a preplanned program that prescribes the use of reprimands. The use of reprimands as acceptable teaching operations is rare. Contingent refers to the fact that the reprimand consequation was the appropriate one. *See also* **teacher-contingent consequation; teacher performance rate/accuracy (TPRA) observation**.

continuous interval recording. *See also* **interval recording**.

control. *See also* **controlling variables of behavior**. The term *control* is synonymous with determinism in science. Science is the process of locating controls or determiners of certain phenomena. The uncovering of new controls leads to new questions about controls. When teaching someone to read or perform some other operant or class of operants, a teacher seeks to bring the student under the control of these now, for the student, stimuli. The student's environment is, so to speak, enlarged by the development of new controls (e.g., new languages, new music, new scientific findings). The world is seen and responded to differently as a result of these newly taught controls. Control is used by laypersons to refer to the aversive manipulation of people by individuals (e.g., Machiavellian dictators). Such is not the meaning of the word as it is used in the science of behavior or the strategic science of instruction. In a manner of speaking, the more that we know about the controls of behavior, from a scientific perspective, the more one can use such controls towards habilitative purposes. *See also* **determinism; behavioral selection; method of science**.

control conditions in experiments. The control condition in single-case experiments is typically the baseline. In group designs, the control condition is exactly like the experimental condition but minus the independent variable. *See also* **independent variable; functional relationship; baseline**.

controlled presentation recording. The terms *controlled presentation recording* or *discrete trial recording* are used interchangeably to denote that responding rates are controlled by presentations, typically by a teacher or trainer. When a learn unit, discrete trial, or program frame is controlled directly by the teacher, in part, the term is apt. Typically, such procedures provide an intraresponse period or window of time for the student to respond (i.e., 5 s). Responses are recorded as incorrect, correct, or no response. Learn units that are more directly controlled by students (e.g., doing a series of math problems, multiple-choice items, or timings or completing automated instruction frames) are learn units that are only indirectly controlled by teachers. Learn units that are scripted and therefore presented under the partial direct control of teachers are controlled presentations. *See also* **discrete trials; learn units**.

controlling variables of behavior. The known and hypothesized phenomena that determine the emission of behaviors. These include the antecedent and postcedent of behavior as well as the setting events, setting stimuli, instructional or reinforcement history, and, as well, phylogenetic factors. Instruction or education is the process of developing new controlling variables for individuals. Individuals are said to learn independence but, in fact, what occurs is a now set of controlling relationships between the individual's behavior and the individual's environment. The individual never becomes independent of his or her environment on the phylogenetic and ontogenetic factors. The science of behavior and, in turn, the science of pedagogy are concerned with the study of the controlling variables of behavior and the development of practices that load to the prediction of behaviors in a more precise fashion. *See also* **determinism; behaviorology; behavioral selectionism; strategic science of instruction**.

correct and incorrect rates of responding by students and teachers. Just as number per standard unit of time is the primary datum of the science of behavior, its application to pedagogy suggests that rate of responding is the primary datum of schooling. Response rate measures include both rate correct and rate incorrect since the two are independent. That is, unlike percentages, the correct–incorrect components are not ipsatic (i.e., the mirror image of each other). Correct and incorrect rates are applicable to student repertoires and teacher repertoires (e.g., teacher performance rate/accuracy observation). Correct and incorrect may be visually displayed together. Alternately, the rate of learn units received and the rate of correct responses to learn units may be displayed with the differences between the two data points representing the rate incorrect. A standard semilogarithmic visual display for plotting rate is available from Precision Teaching. *See also* **rate of responding; Precision Teaching**.

correction errors. When a teacher, tutor, or teaching device fails to perform a correction or performs a correction procedure incorrectly per the learn unit, it is termed a *correction error*. In some cases, this means that the student did not correct him- or herself or the correction response was reinforced when it should not have been. See chapters 5 and 6 for discussions of the issues associated with identifying correction errors. *See also* **corrections as prompts**.

corrections. Consequences of postcedents presented by teachers or teaching devices that serve to remediate incorrect responses and prompt subsequent correct responses to learn units. They substitute for instructional programs that do not provide errorless learning. When done correctly, they serve as prompts for incorrect responses, or, in rare cases, they punish incorrect responding. Decisions about the designed function of corrections are based on the criteria described in chapter 6. *See also* **contingent correction; correction errors; teacher performance rate/accuracy (TPRA) observation; punishment; differential reinforcement**.

corrections as prompts. Corrections to incorrect responses are boat presented in ways that have them serve as prompts for correct responses in subsequent learn units. There are exceptions in which the correction functions better as a punishment (see chapter 6). Corrections are useful typically if the student emits the correction response in addition or after the teacher performs the correction. See chapter 6 for a description of when correction responses of students should and should not be followed by reinforcement operations.

correct teacher operations (accurate teacher operations). Those that are consistent with the known principles of behavior and those teaching or pedagogical tactics known to be associated with evoking, maintaining, and strengthening operants in relevant settings. This includes but is not limited to instructional control of the student prior to presentation, the accurate delivery of all components of a learn unit (antecedent, behavior, consequences) as prescribed for a student and program, accurate and contingent reinforcement and punishment, accurate data collection, and appropriate shaping procedures as needed. *See also* **incorrect teacher operation (teacher errors); teacher performance rate/accuracy (TPRA) observation; successive approximation principles and operations; unambiguous exemplar; teacher-contingent consequation**.

correspondence teaching. Teaching correspondence between what is said and what is done. Research literature in applied behavior analysis has grown around the correspondence between verbal and nonverbal behavior. The issue is related also to the independence of behaviors that has been a common finding in the study of operant behavior. For example, verbal behavior about nonverbal behavior may be under the control of intraverbal contingencies, whereas the nonverbal behavior may be contingency shaped via nonverbal contingencies. The notion that both relate to a central "understanding" is an a priori notion and is, in fact, an explanatory fiction. Thus, the lack of correspondence between doing and saying is not perplexing from a behavior analytic view. If both are needed, one can expect that both need to be taught. *See also* **biospecification of behaviors**.

cost analysis by learn unit. Procedures are described in chapter 5 for determining the financial costs of learn units. The numbers of learn units presented and correct responses to learn units, respectively, are divided by the tuition for a student or all students for a given term. The result is the costs per learn unit.

counterbalanced designs. Ones in which the independent variables are systematically rotated in an attempt to control for order effects on the introduction of variables or to "weigh out" the effects of extraneous variables. Certain statistical tests are used in conjunction with the counterbalancing scheme such as a Latin-square analysis of variance. Counterbalancing is a logic of experimental manipulation that attempts to ameliorate the effects of extraneous variables on the objective of the experiment by systematically rotating certain variables across the presentations of the independent variable and control condition. *See also* **Mill's canons of the scientific method**.

creative solutions. Solutions to problems that are brought about by the combining of learned repertoires in ways that have not been observed previously by the person solving the problem. Unique solutions or art works that occur without a learned history or combining of learned repertoires are instances of accidents, not creation.

criteria. Used in the context of learning and instruction to refer to more than one criterion-referenced objective. For example, the student achieved 10 criteria in verbal behavior or the student achieved the fourth grade and fifth grade criteria for reading fluency.

criterion-referenced objectives of instruction (e.g., learning objectives). Instructional objectives that specify performance requirements that must be met. These requirements or criteria remain standard for all students. Typically, the criteria include the components of the learn unit (e.g., antecedent behavior, consequences for the student and teacher behaviors) and specify the mastery (e.g., 60 wpm correct and 0 incorrect on two consecutive sessions or 90% correct on two consecutive sessions). The teacher or teaching device that records the behavior of the student in terms of the criterion must be calibrated to a constant standard. A record of those criterion objectives that a student has achieved is the most valid measure of learning provided that the measurement is reliable. See the discussion of criterion referenced objectives in chapter 2. The terminal criterion-referenced objectives remain the same for all students; however, subobjectives or means objectives may differ for individual students. In all cases, however, the criteria are specified. *See also* **criteria; learn units; terminal objectives (long-term objectives); means objective (short-term objective); breaking down goals**.

criterion-referenced quizzes. Ones that have a predetermined criterion. The criteria are typically 100, 90, or 80%. The phrase *achievement of criterion* is an expression that characterizes performance on a quiz as having met the preset criterion. Criterion-referenced quizzes are standard features of behavioral approaches to instruction. This feature was probably an outgrowth of the practices from experimental psychology called "trials to criterion." The object is the achievement of mastery by all students or teachers. Typically, students or teachers who do not achieve criterion on a quiz the first time study again and are availed tutoring opportunities to teach those items not mastered. An alternative test (written or oral) is given to the student after the additional study. The procedure is repeated until the student achieves the criterion on an examination.

critical mass of expertise. The presence of a minimum number of individuals with a given expertise necessary to maintain the necessary contingencies that occasion a community that continues to carry out the practices.

curricular analysis. The logical or empirical analysis of curricula with regard to goals and subgoals. Chapters 7 and 8 describe the behavioral selectionistic perspective on the analysis and construction of curricula.

curricular goals. The long-term or terminal objectives or the short-term or means objectives associated with a particular program of study. They include the behaviors, contingencies, and setting events prescribed for a given course of study.

curriculum. The terminal and means objectives taught and the sequence and materials used to teach the objectives. Curricula (or curriculums, both are correct) have structural and functional components. The subject of curriculum is treated in depth in chapters 9 and 10. *See also* **functional curricula; structural curricula; programming**.

curriculum validity. A term used in this book to refer to the relationship between learn units and curricular objectives. Those learn units that are valid are tied to curricular objectives. It is also used to refer to whether or not test items for standardized tests are valid measures of what was taught in class. For an item to be valid, it must have been associated with classroom learn units and objectives.

cybernetic system. A cybernetic system is a complex system of interrelated and interdependent components in which changes in one part of the system affect the other parts. In this book, schools are seen as cybernetic systems in which all of the roles of the consumers and professionals are interrelated in terms of the science of behavior. The management and analysis of all the components are tied to an analysis of the interrelated consequences of the behaviors of all parties involved.

cyclical changes or variability. Cyclical change is a form of variability that recurs concomitantly with time cycles (i.e., the circadian cycle, days of the week). For example, decreases in learn units following holidays are a common source of cyclical variability in CABAS schools. Any form of variability that is associated with time periods alone is cyclical in that it recurs correlated with time or time latency. Of course, the time is not necessarily the functional source of the variability. *See also* **method of concomitant variance**.

D

data management. An expression that refers to the operations and devices to manage data that have been collected as, for example, the management of data in a CABAS school. It also refers to the way in which the data are organized and schedules of use and display.

deductive scientific practices. This expression refers to the scientific practices used by those who apply hypothetico-deductive strategies to implement the Mill's canons of science. The practices are tied to group design logic and the use of related statistical analyses and logic. Both inductive logic and deductive logics are used, but the primary approach is one in which hypotheses are formulated and experiments done on a deductive basis. Principles or constructs are deduced based on confirmation of the hypothesis. *See also* **inductive scientific practices; hypotheticodeductive statistical procedures; inferential statistical tests; theory; group designs**.

delayed multiple baseline design. A boilerplate single-case experimental design with all of the characteristics of a multiple baseline except some subjects are introduced to the baseline conditions in a delayed fashion compared to other subjects. *See also* **multiple baseline design**.

demonstration study. A descriptive study that demonstrates a relationship between certain variables and that relationship is summarized in forma of concomitant variance between certain variables. Numerous studies conducted on stimulus equivalence are demonstration studies. *See also* **method of concomitant variance**.

dependent variables. Those behaviors that are measured in an experiment. They must be free to vary only as a function of the experimental variables (independent variable) used in the experiment. The degree of variability in the repeated measures of the dependent variable attests to the tightness of the experimental control. In the science of behavior, the dependent variables are absolute units or idemnotic measures (e.g , real behavior in real time), whereas those who use deductive scientific practices may use ratings or scaled measures. *See also* **independent variable; single-case experimental design; group designs; functional relationship**.

deprivation of stimulus in a chain of operations. Also known as interrupted chain. A procedure in which a familiar object or device used in a familiar chain of responding is hidden in order to occasion the emission of behavior, usually verbal. For example, a child assembling a favorite puzzle is deprived of one piece to occasion or evoke the omission of a request for the piece in the presence of a listener who in all probability can or will deliver the piece. It is one of several establishing operations used to increase the reinforcement effect of the consequence of using verbal behavior, therefore increasing the probability of the emission of the verbal behavior (e.g., in this ease, a mand).

descending trend. A trend of a linear visual display of data is said to be descending when a straight edge placed across the display of data shows a downward direction. The trend may be slight, steep, slight with or without variability, or steep with or without variability. *See also* **ascending trend; variable trend**. Descriptions of trends are global rather than precise, but they are critical in ascertaining whether or not instructional changes are warranted. One acquires expertise in reading graphs with repeated experience in making data-based decisions. The expression "the supreme court" method was used by Johnston and Pennypacker in 1981 to describe advanced expertise. That is, one cannot always describe the process without prior experience with positive and negative exemplars of particular trends.

describe function. Curriculum and pedagogical procedures that have as their goal that the student describe an operation, concept, or event are described in this book as objectives that have a describe function. "Describe" repertoires are associated with functional objectives. They can be applied also to numerous structural curricula (i.e., social studies, English composition, biology, literature). The contemporary behavioral perspective on curriculum suggests that the development of describe objectives and curricula are needed and their omission explains some of the problems that students have in that they have not been taught describe responding and the related contingencies as a direct objectives.

descriptive research. Structural in nature. That is, it describes the phenomena in question and summarizes the data typically in forma of central tendency and variance. It does not result in the affirmation of a functional relationship between variables. *See also* **method of concomitant variance**.

determinism. The belief that events occur as the result of other events. For example, behaviors are the result of ontogenetic and phylogenetic histories of the individual and the species of the individual. The scientific method is used to uncover the functional or correlation relationship between variables. Variability in responding that is not attributable to known variables is held on faith to be traceable at some future date to variables that will increasingly come under the purview of empirical scientific procedures. The Heisinger uncertainty principle is, for example, seen to be a problem in measurement, not a denial of determinism. Levels of "known" relationships are continuously changing as a result of scientific efforts, particularly in the development of new instruments for measuring phenomena.

differential reinforcement (DRO, DRL, DRH, DRI, DRA). *Differential reinforcement* refers to the practice of reinforcing only certain behaviors or rates of behavior and not reinforcing other behavior. It is basic to the process of teaching discriminations. The contrast between reinforced behavior and nonreinforced behavior is consistently invoked. DRO refers to the reinforcement of behaviors other than a target behavior that you seek to decrease. DRL refers to differentially reinforcing successively lower rates of behavior with the goal of eventually eliminating behavior. DRH refers to differentially reinforcing successively higher rates of responding to increase behavior. DRI refers to differentially reinforcing behaviors that are incompatible with a behavior that you wish to reduce. DRA refers to differentially reinforcing an alternative behavior that has the same effect as the behavior that you wish to reduce. With the exception of DRH, all of these operations are cosigned to reduce behavior as negative punishment (Type II) operations whereby the reinforcement is unavailable when the target behavior is present and contrasted with reinforcement occurring at high rates when the behavior is not present.

direct instruction. The expression *direct instruction* (no capitalization) indicates that something is directly taught. Rather than teaching someone to type indirectly by teaching a so-called generic class of fine-motor responding, an individual is taught directly how to type. Rather than have a student learn an objective as a result of rediscovering a concept or principle, the operation or concept is taught directly. The educational research literature has repeatedly affirmed the validity of direct over indirect instruction. Individuals are taught discrimination or chains of behaviors across multiple exemplars or settings such that the salient discriminations are learned directly. The behavioral model termed Direct Instruction (DI), like the other behavioral models of schooling, emphasizes direct instruction procedures, but DI is a complex of curriculum and pedagogical precepts that extends beyond the simple notion of direct instruction.

Direct Instruction. One of the behavioral models of schooling. Direct Instruction wedded behavior analytic pedagogy to logical analyses of the curriculum. The logically analyzed curriculum is arranged into scripted teacher presentations. Both antecedent and consequent teacher behaviors are scripted or written for the teacher. Students respond frequently and in groups (although it may be done individually). The result is that individual students do receive high numbers of complete learn unit (*see* **learn units**). The sequence of these presentations or learn units or programs is based on principles of behavior, logical analysis, and "run-throughs" with students. The development of scripted and programmed instruction has a similar rationale. Direct instruction developed and extended the general case instructions into Direct Instruction curricula. The curricular materials of DI (e.g., Distar reading, Distar math, Reading Mastery) are well tested. DI also has developed teacher training courses and a national consulting network to teach accurate and skilled presentations. Teachers use attention signals or teacher prompts in a consistent manner. These procedures are somewhat similar to the "chorale responding" characteristics of the "scientific management" approach to education of the late 19th century. However, this is only a topographical similarity. Engelmann and Carnine's 1981 treatise on Direct Instruction is an important and unique source of curricular analysis. The chapter on learning theory is, however, not an accurate treatment of the science of behavior or behavior selection.

direct replication experiments. *See also* **systematic replication experiments**. An experiment that is a *direct replication* is one that repeats all of the components of the original experiment with the exception that a different subject may participate. The same subject may experience all of the components of the experiment again also. Typically, the same practices used to implement the method of difference in the first experiment are applied in subsequent experiments to test the replicability of the phenomenon in question. *See also* **method of disagreement (method of difference)**; **systematic replication experiments**.

discrete trials. Discrete presentations of a stimulus and response opportunity to a learner by a teacher or teaching device. When the trials involve all of the components of a learn unit, they are discrete learn units. They are discrete because they involve controlled presentations rather than learner-controlled or free operant responding.

discriminative stimulus. The discriminative or discriminated stimulus refers to the antecedent in the three-term contingency. In an operant the discriminative stimulus *evokes* the response that is in turn followed by a reinforcing consequence. Much of instruction is devoted to teaching discriminations and the discriminative stimulus (S^d) is taught by the consequence following the S^d and the response. *See also* **operant behavior**.

duration recording. Used when the length or duration of engagement in behavior or response class is of primary interest. For example, the stimulus control of an episodic reinforcer (i.e., television show, music recording, or the distribution of time in ongoing activities) is one such interest, particularly when the relative reinforcement value is the dimension of behavior that is of interest. Duration is a special case of rate. That is, an operant measure may require key pressing or toggle switch movement to maintain the presence of a stimulus. Alternatively, a controlled switching apparatus that requires a touching response or blockage of a photoelectric cell may be used as a direct transducer of attention to stimuli.

E

eclectic instruction. The term *eclectic*, in the context of teaching methods, refers to teacher practices or variations in practices that are not tied to a consistent epistemo-logical, scientific, or pragmatic set of organizing principles. It often refers to teacher operations based on hunches, whims, or prescientific or antiscientific assumptions. Some educators and psychologists take the view that eclecticism is a wise approach to educational practice. Reform is one of the constants in the history of education. Indeed, education and to a certain extent psychology seems to be driven by fad and fashion. Perhaps there is good reason for educators to eschew a consistent philosophical position. However, the development of a sciences of behavior and individualized instruction represent over 50 years of programmatic research coupled with the devel-opment of a philosophy that embraces the complexity of human behavior and schooling. Applied behavior analysis and teaching as behavior analysis are not fads; rather they are a collection of research findings and an epistemological view that provides measurably effective outcomes. Moreover, a few schools have embraced this approach to education for over 25 years

ecobehavioral analysis. An approach to the analysis of schooling procedures in the natural or existing school setting that uses an expanded observational procedure to identify setting events, antecedent, behavior, and consequence relationships that are occurring. It is an applied behavior analysis of the ecological variables in a classroom. Ecobehavioral analysis is, perhaps, an evolution from the Consulting Behavior Analyst Model.

echoic verbal behavior. Point-to-point correspondence between a vocal antecedent by a speaker and the subsequent echoing of the vocal antecedent. For example, a teacher says, "dog," followed by a student saying "dog." The student's response is an echoic. Skinner distinguished imitation from echoic behavior in that in imitation the imitator can see the response that is imitated, while in echoic behavior the person echoing cannot see the behaviors involved in the echoic response.

effective teaching practices or operations. Ones that bring students' target behaviors under the control of those antecedent, consequent, and setting stimuli specified in a curriculum. They are distinguished from ineffective teaching operations that do not result in the student mastering the behavior in the target setting under the target antecedent and consequent conditions. Effective teaching operations result in more correct learn units, fewer incorrect learn units, and more criterion-referenced objectives than those that are not effective. Effective operations are ones that are measurably superior to ineffective practices and they are functionally related to student learning. *See also* **pedagogy**.

elicited behavior. Elicited behaviors are respondents. They are elicited by unconditioned or conditioned respondent antecedent stimuli. Respondents differ from operant behaviors in that operant behaviors are evoked. Respondent repertoires are phylogenic in origin. *See also* **evoked behavior**.

emit. The verb associated with operant behaviors. Operant behaviors are emitted, whereas respondent behaviors are elicited. The organism emits the operants due to consequences that follow. The term implies that the actual first instance of the behavior may or may not be traceable—it simply occurs. On a theoretical note, infants emit thousands of behaviors prior to and immediately after birth. Certain of these behaviors are selected out to a result of reinforcers. For example, the range of vocal sounds that are emitted by the infant are shaped by the environment into words or communicative behaviors.

emotional/affective repertoires. From the perspective of behavioral selection, emotional and effective repertoires are seen as a function of conditioned reinforcers and punishers that control students' approach, engagement, or avoidance responses. Some of the stimuli that evoke or elicit the emotional responses are conditioned respondent stimuli, and others are conditioned operant stimuli. Students for whom textural stimuli reinforce observing or textually responding have come under the conditioned reinforcement control of printed text. Students who do not read in their free time are said not to have acquired print stimuli as a conditioned reinforcer. Positive affect is directly teachable as conditioned reinforcement control. Negative affect is counterconditioned by procedures that engender positive reinforcement control, where formerly the same stimuli engendered avoidance.

environments by people. We use the expression "environments by people" to clarify to the reader that (1) people or their behaviors constitute the environment vis-à-vis behaviors as much or more so than traditional notions of the environment and (2) the contemporary world is one in which environments by people are critical controlling parameters of our daily life. Environments by people have replaced many of the natural environments that controlled our predecessors.

epistemology. The branch of philosophy that concerns how we come to know what we know and, thus, what it is that we know. Issues such as is something known when it is observed or not or whether the occurrence must be observed by an independent observer are relevant. Beliefs or predictions about causes of action, for example, are epistemological. Is the word of someone else, a scientific result, or a cultural belief adequate for setting a course of action? The methods of authority, tenacity, a priori, or science refer to different sources of knowing. The arguments about the validity of these sources of knowledge and the relative validity vis-à-vis kinds of knowing are concerns of epistemology.

errorless learning and corrections. In a program in which students are taught such that they emits no errors in the course of learning a new repertoire or operant is referred to as an errorless learning program. Occurrences of errorless learning are held in high esteem in the science, but occur infrequently in school or other applied settings. This is probably due to (a) inexpert application of what is already known, (b) incomplete knowledge at the basic science level, or (c) the lack of availability of carefully tested instructional programs. One applied tactic that seeks to implement errorless learning is the constant or progressive time delay operation. Still other stimulus fading and prompting procedures can be used in scripted or programmed instruction along with general case operations to strive for errorless learning. Few programs that are used outside of the laboratory have the necessary research background to result in errorless learning. As a result, corrections are done in lieu of programming expertise; in such cases, efforts are extended to have the correction serve as a prompt for subsequent learn units.

essay learn units. Essay learn units are created or occur when the student observes and responds to feedback regarding the written responses the reader of the student's essay. See chapters 2, 5, 6, 9, and 10 for detailed discussions of essay learn units.

essential stimulus control. The term used by Engelmann and Carnine in 1982 to describe the control of a target stimulus in general case instruction. *See also* **general case instruction**.

establishing operations. Events or interventions that affect the reinforcing or punishing effects of the consequences of behavior. Tactics that implement establishing operations are manipulations of setting events or setting stimuli that in turn affect reinforcement/punishment. In behavior analysis establishing operations are the motivational events that affect the three-term contingency. They result from deprivation, aversive stimuli, or satiation events that act *momentarily* to enhance or detract from reinforcement or punishment effects. Naturally occurring satiation and deprivation experiences affect the reinforcement/punishment effects of postcedent stimuli and in turn antecedent stimuli. In order for the procedure to be termed an establishing operation, change in the effect of behavior consequences must occur and the frequency of the behavior must correspond with the change in the consequence effect. The phenomenon is probably related to the effects found with response deprivation wherein the reduction of response opportunities below operant baseline rates acts to increase the frequency and reinforcement value of the response that was inhibited. The phenomenon is also probably related also to portions of the Premack principle. The Premack principle (i.e., the well-known component) states that a more preferred activity will function to reinforce a less preferred one. The Premack principle also states that a formerly preferred activity can be reinforced by a formerly less preferred activity as satiation and deprivation of each of the activities are manipulated. Unlike conditioned reinforcement/punishment, changes in consequent effects are momentary.

event recording. A term used in applied behavior analysis to describe the practice of counting occurrences of discrete behavior. Discrete behaviors have easily discernible beginnings and cessations. The procedure is distinguishable from interval recording procedures or duration recording procedures. The latter are procedures used to record response classes that occur either too quickly to reliably count or are prolonged or have indiscrete articulations. Event procedures are typically associated with free operant behaviors. Counts of events are typically reported as number per standard unit of time or per preset constant number of opportunities. Sometimes the event counts are converted to percentages but percentages are problematical. *See also* **rate of responding; absolute units of measurement**.

evoked behavior. Those engendered by motivational conditions (setting events and stimuli) and operant stimulus discriminatives. The discriminative antecedent stimuli are selected out by the consequences of the behavior. Operants are omitted behavior and they are evoked, whereas respondents are elicited. Antecedents in the three-term contingency relationships evoke behavior, while antecedents elicit respondents such as knee jerks. Our repertoires of elicited behavior are genetically determined, while evoked behavior is determined by our instructional history or ontogeny. *See also* **elicited behavior; operant behavior; three-term contingency; learn units**.

exemplar identification. Student responses that involve identifying a correct answer from a group of answers. The responses take the form of circling, drawing a line between, underlining, selecting, or pointing.

experimental analysis of behavior. One of the three prominent activities engaged in by behavior analysts. These include theoretical analyses, applied behavior analyses, and experimental analyses. Experimental analyses are those associated with studies done with the human or fellow species in laboratory conditions; applied analyses refer to studies done in settings that are naturalistic, such as schools. *Both endeavors use the experimental method*. Thus, the term does not refer to the use or lack of use of the experimental method of science, but to the setting in which behavior consequence relationships are studied. Perhaps a better term might be laboratory behavior analysis.

experimental design. The manipulation of the independent variable under controlled conditions such that changes in the dependent variable can be attributed to the independent variable. *See also* **experimental method**.

experimental method. Any set of scientific practices that implements the Mill's canon of difference or the joint method. Conditions are constant between control and experimental conditions with the exception of a controlled independent variable that is present in the experimental phases and not present in the control conditions. Alternately, a control group or control individual not receiving the independent variable is substituted for the addition and withdrawal of the independent variable with the same group or individual. Thus, there are both within-subject or group experimental procedures (e.g., repeated-measures group designs or reversal individual designs) and between-group or individual procedures (e.g., Solomon five-group design, multiple baseline across individuals, the same individual across different settings, or across different individuals).

experimental test. The use of the strategies of experimentation (method of disagreement, joint method) and the related tactics (reversal design, probe, multiple baseline) to determine functional relationships between given interventions and changes in a target behavior (i.e., dependent variable). It is the primary way that teachers who are strategic scientists isolate variables that lead to effective instruction. If certain questions about the difficulty a student in experiencing suggest that the student is lacking a prerequisite repertoire, a teacher performs an experiment to determine if this is, indeed, the variable.

explanatory fictions. A term used to refer to psychological terms that draw on intervening variables to explain behavior. These include, but are not limited to, IQ, locus of control, personality, attribution theory, and diagnostic labels not grounded in environmental or genetic sources.

extrastimulus prompts. Prompts that are extraneous to the target stimulus. The target stimulus is unchanged but is highlighted, underlined, or in some way marked with an extrastimulus device designed to serve as a cue or prompt to evoke discriminative responding. It is faded or vanished in order to progressively shift the stimulus control from the prompt to the target stimuli that is to evoke a discriminative response with no extrastimulus prompt in evidence.

F

fading or vanishing tactics. Teaching or automated instruction procedures that are designed to develop the control of a target stimulus by shifting stimulus control from a functioning stimulus discriminative to one targeted for stimulus control. *See also* **extrastimulus prompts; intrastimulus prompts**.

faultless presentations. Presentations designed such that they cannot be confused with another stimulus. Shape, hue, texture, or dimension components of the target stimuli, for example, are perfect and nonconfusing.

feedback. A term associated with systems theory. It refers to consequences that purportedly provide "information" to the organism. The science of behavior suggests, however, that the consequences of behavior reinforce, punish, or have no effect. Consequences such as measures or graphic displays may function as conditioned reinforcers or punishers. In effect, the term feedback is a less precise description of postcedent effects on behavior. Feedback may act to increase or decrease behavior or it may have no effect.

fluency. The ease or facility that a student demonstrates relative to operant responses they have mastered. *Fluency* refers to rate and accuracy of the performance. The related literature in other areas of psychology is research on automaticity. Norms for fluency can be established by determining the rate of responding (mean or range) of a given task performed by an expert. Once a student has mastered an operation or an intraverbal or rote skill (e.g., multiplication), it is important to develop the fluency of correct responding that is or will be needed by the student. Sometimes, a student's difficulty with a complex operation may be traceable to the lack of fluency with a component repertoire.

Precision Teaching theory maintains that fluency predicts maintenance, the learning of more complex repertoires, and reductions in distraction. Fluency may represent a measure of the shift from verbally controlled to event governed responding. *See also* **criterion-referenced objectives of instruction; mastery learning; Precision Teaching; number per standard unit of time; rate of responding**.

follow-up data. The collection of data for periods of time following the completion of an applied experiment or an instructional procedure. Typically, it is done to test for the maintenance of experimental effects following the official termination of the experiment.

frame of programmed instruction. A frame of instruction in automated programmed instruction refers to an antecedent textural presentation, a textural response from a student, and a presentation of the correct response that then allows the student to proceed. Or, if the students was incorrect, he or she corrects an incorrect response before proceeding. It is an automated learn unit. See chapter 10 for a detailed discussion. *See also* **learn units**.

"free operant learn" units. Learn units whose rate or occurrence is learner controlled more so than those that are not free operant. Some types include automated instruction frames, student-controlled practice, or tests with flash cards (e.g., 1-min timings) or captured or incidental learn units. *More precisely the learn units are learner controlled rather than teacher or automatically controlled*. Free operant responding requires careful control of all conditions such that a range of responses can be emitted.

functional curricula. Curricula cosigned to teach individuals certain functional repertoires. Those who teach individuals with developmental disabilities to distinguish between a useful or functional instructional goal and one that is not sometimes use the term. However, the expression is just as applicable for all types of students. Instruction in traditional content areas frequently does not specify functional objectives. Building on the distinction between the functional focuses of verbal behavior versus the structural approach of linguistics, a behavior selection analysis of curricula, points to a reconstruction of curricula as both functional and structural. See chapter 7 for a thorough discussion. *See also* **verbal behavior**.

functional relationship. One in which an experiment implements the logic of the Mill's canon of difference such that the experiment demonstrates that changes in the dependent variable are a function of the independent variable. That is, when the independent variable is present, the dependent variable (e.g., behavior) is significantly different than when the independent variable is not present. The effect on the dependent variable may be sufficient alone for the change or it may even be a necessary condition for the change in behavior. Additional research typically reveals additional relationships such as catalytic or additive relationships involving several independent variables. Functional relationships are only determined in true experiments (e.g., at least at the level of method of difference). *See also* **independent variable; dependent variables; Mill's canons of the scientific method; single-case experimental design; between-subject research designs (intersubject designs); within-subject designs (intrasubject designs); group designs**. Note that a functional relationship can loosely be described as a "causal" relationship between the independent and dependent variables. However, the word "cause" is avoided since the true cause, and all that notion entails, is difficult to demonstrate in many cases. That is, the functional relationship always occurs within the context of certain events and settings.

G

general case instruction (multiple exemplar instruction). Curricular programming sequences that teach essential stimulus control and the general case. Categories, concepts, or operations are relevant. The essential or salient attributes of a concept are identified. Learn unit presentations rotate presentations of positive exemplars (e.g., correct example) and negative exemplars (e.g., negative exemplars or foils). The number of learn units needed to achieve criterion decreases depending on the fineness of the initial discrimination between positive and negative exemplars that an individual student can differentiate. To increase the control of the essential component of concept or the generalized control of the stimulus aggregate, irrelevant aspects of positive exemplars are rotated across exemplary presentations. See chapters 5–8 for examples and applications.

general case teaching operations. General case teaching operations or programming refers to the presentation of positive exemplars with a range of negative exemplars. Each time that the positive exemplar is presented, irrelevant attributes to the salient aspect of the stimulus are varied. Negative exemplars are rotated with positive exemplars. Contrasting presentations begin with the finest discrimination for which the student is capable, given his instructional history. The range of positive exemplars with irrelevant attributes and the range of negative exemplars determine the range of the control of the salient aspect. *See also* **stimulus generalization**.

generalization. *See* **stimulus generalization principle and operations**.

generalized reinforcers. Reinforcers that affect a wide variety of behaviors; that is, they are not specified by the particular antecedent–behavior relationship. Tokens, praise, grades, checkmarks, and visual graphs of behavior are all examples of generalized reinforcers. In same cases, they function as prosthetic reinforcers. Generalized reinforcers are conditioned reinforcers. The acquisition of a community of generalized reinforcers is one of the primary objectives of education since it is the reinforcers that select out and maintain behavior. *See also* **natural conditioned reinforcer; prosthetic reinforcers**.

graphic display. The visual display of data.

group contingencies. A group contingency specifies certain antecedent behavior relations that must be performed by a group or member of a group. For example, if each member of a group meets an individualized criterion or a common criterion, a reinforcing operation is performed for the entire group. An example is the good behavior game. There are various permutations.

group designs. Designs that interrelate inferential statistics and the logic of experimentation. The basic experimental control group design is an example. Typically, pretests are followed by interventions and subsequently by posttests. Measures occur before and after interventions, not during the interventions. The means of the groups are compared statistically before and after the treatment. If the means were not statistically different before the intervention but were different after the intervention and the experimental design, assignment to groups, and derivation of the sample were done properly, the posttest differences are deduced to be the result of the intervention. *See also* **inferential statistical tests; Mill's canons of the scientific method**.

group instructional control. An expression that pertains to the management of students in a classroom with regard to the teacher's skill at keeping all of the students under instructional control. It is a critical initial repertoire for a teacher and involves the teacher seeing to it that students are reinforced effectively and therefore responding effectively when they are or are not receiving teacher-controlled instruction. Teachers initiate instructional control of groups by acquiring reinforcement operations for students following classroom rules. However, following classroom rules is but the means to the real purpose of group instructional control; that is, the student is continuously responding to valid learn units. A teacher with advanced skills for maintaining group instructional control will have classrooms in which the latency between learn units, correct responses, and criterion objectives is small. As a by-product of this latter effect, students will appear task engaged and will follow the social rules of the classroom. See chapters 6, 7, and 9. *See also* **bad behavior; individual instructional control**.

H

habilitation *or* **habilitative repertoires**. Repertoires of behavior that lead to more long-term reinforcers for the individual and fewer punishers. Hawkins in 1977 provided this definition. See chapters 7 and 8 for elaboration.

history of instruction. An individual's history of instruction consists of his or her operant and elicited stimulus control repertoires. In the case of operants, the antecedent–behavior–consequence relationships and associated setting events experienced in the lifetime of the individual constitute their instructional history. The instructional history of an individual also includes the conditioned eliciting stimuli. The expression reinforcement history is used as a synonym for the history of instruction. The history includes not only the acquired discriminative stimuli and behaviors, but also the individual's community of conditioned reinforcers that, though subject to rank order fluctuation, also serve an equally important role. Whenever histories of instruction are more precisely known, as in the case of data-based portfolios or inventories of the repertoires of students, the search for the locus of a current instructional problem is simplified. Otherwise, experimental probes or test sessions are necessary to identify potentially missing instructional experiences. The more scientifically precise the description of histories (i.e., the setting events and consequences of the behavior are known also), the more probable that problems associated with instructional history can be rapidly and efficiently located. The concept of instructional history plays an important role in the analysis of learn unit responding and the use of effective tactics with appropriate curricula.

hypothetical constructs (intervening variables, mentalistic constructs). Explanations of the relationship between environment and behavior that posit unobservable sources for the relationship. The intervening variable, cognitive or mental, is seen as the source for the behavior. Such constructs are derived from a nomological network of behaviors that are said to point to the existence of the construct. For example, a series of test responses using a test that purports to measure one's personality type suggests that the person who has taken the test scores high on the inferiority complex scale. This is then taken to be the source of an individual behavior in other experiments or life settings. Because such constructs are unobservable and not parsimonious explanations, those who adhere to behavior selection as mentalistic constructs believe that they are not useful. Those who study cognition from a behavior–environment perspective posit various theories as to the existence of intervening variables to explain behavior. The notion of the intervening variable is closely associated with Hull's book on the principles of behavior. However, Hull was seeking principles of intervening variables. The result of the book was a series of hypothetical mathematical and logical questions. The experiments were not done. Such theory is in opposition to the epistemology and purview of the science of behavior. *See also* **behavioral selection; hypotheticodeductive statistical procedures; projective measurement; nomologically derived constructs (nomological network)**.

hypothetico-deductive statistical procedures. *See also* **statistical significance level; inferential statistical tests; split-plot designs**. Used to describe a set of deductive scientific practices in which theories are developed and then tested by experiments that implement the logic of the scientific method via the use of the logic of group statistical procedures. A testable hypothesis is subjected to a procedure whereby one group (or within-subjects counterbalancing action) receives a control condition and another receives the experimental conditions. Of course, there may be multiple conditions and multiple groups. Pretests and posttests surround the treatment or control conditions. Measurements of the groups are summarized into means, standard deviations, and least squares estimates. Statistical tests are selected consistent with the requirements of the design. These tests are then used to determine whether the differences between group means are statistically significant (.05, .01, .10, .20 significance levels, for examples). These differences are then inferred to confirm or deny the hypothesis. Typically, although not always, the measures are said to be measures of nomological constructs that are often intervening variable–constructs.

I

idemnotic measurement. Data that consist of absolute units that represent the displacement of space in time.

identification function. In the analysis of curriculum, the expression *identification function* refers to the functional effects and controls of learning as distinct from the structural effects. Mathematical computations that are used to solve a particular problem for a student when the student needs to know the solution are examples of a functional effect for the learner. Curricula that have as their effect the functional consequence of identification have an identification function. Identifying environmental events in a scientific manner are typical functions of certain scientific activities.

imitation. The duplication of another person's behavior. It is one of the primary ways that behaviors are acquired or done. For example, much of the communicative behavior of individuals is initiated by imitation; however, unless those behaviors themselves become operants through direct reinforcement, they remain imitative rather than functional. Planned presentations or models of behavior are used to evoke imitative responses as an initial way to induce a behavior that the teacher wishes to bring under the control of target consequences. However, *echoic verbal behavior is not a form of imitation* since the movements that result in speech are not observable by the individual echoing' the behavior. *See also* **echoic verbal behavior; modeling; observational learning**.

inappropriate contingency-shaped behavior. Contingency-shaped behavior is behavior shaped directly by the contingencies. The expression *inappropriate contingency-shaped behavior* is used to describe teacher behavior that is not consistent with sound pedagogical practices, but was, nevertheless, learned by teacher/student interactions Teachers who learn to disapprove student responding for the momentary reinforcement effect have learned inappropriate contingency-shaped behaviors. As in the case of a student's bad behavior, the inappropriate behavior of the teacher leads to short-term reinforcement but long-term punishment.

incentive systems. Sets of procedures invoked to function as positive reinforcement for target accomplishments. While they may seek to reinforce, they do not necessarily have a reinforcement effect. For example, removing the capital gains tax might be done to provide an incentive for investment. It may or may not function to reinforce investment. Only a functional analysis (e.g., an experiment) could determine whether, in fact, the incentive actually reinforced investment for particular individuals.

incidental trials. Responses to naturally occurring or naturally simulated antecedent stimuli and setting events that are consequated by the delivery of natural consequences. They typically involve communicative behavior and are conceptualized as mands and tacts in verbal behavior. They may be conceptualized also as captured learn units. See chapter 2.

incorrect teacher operation (teacher errors). Those operations, performed by a teacher, that are inconsistent with the principles of the science of behavior and the existing known correct behaviors of teaching. Such errors include, but are not limited to, omission of reinforcement, noncontingent reinforcement, use of positive punishment procedures (e.g., disapproval) unless planned, ambiguous antecedent presentations, ambiguous stimulus or response prompts, failure to fade prompts, failure to provide the appropriate intraresponse time, use of corrections that functions as punishment, reinforcing or not reinforcing a correction response, failure to obtain a correction, presenting an antecedent when the student is not under instructional control, recording the data incorrectly, and omitting the recording of data. *See also* **teacher performance rate/accuracy (TPRA) observation**.

independence. A repertoire whereby the natural or conditioned contingencies control effective behavior without the use of prosthetic contingencies or a prosthetic environment.

independence of behaviors. *See also* **biospecification of behaviors**. Behaviors are often independent of each other. Verbal behavior is frequently independent of nonverbal behavior (i.e., lack of one-to-one correspondence), and extroceptive responses may be independent of introceptive responses. The notion that behaviors are independent is a characteristic approach to the study of behavior that is in opposition to those who see intervening variables as catchments or way stations that group behaviors into constructs reflected by notions such as personality or cognitive stages. When students respond correctly to multiple-choice questions that "tap" the intervening variable in question, the student is believed to "understand." Thus, where the student does not write or speak authoritatively about the concept in question, there is something wrong with the understanding mechanism. However, from the perspective of behavioral selection, each repertoire (e.g., multiple-choice or description) may indeed be under the control of a different set of contingencies. Grouping the responses or emitting one-to-one correspondence is a repertoire that one would expect needs to be learned, not an automatic assumption. The issue of the independence of behaviors is treated at length in chapter 7. *See also* **correspondence teaching**.

independent variable. The variable that differentiates the treatment or intervention phase from a baseline phase. In well-designed experiments, there is only a single variable that differentiates the baseline and treatment phases. In group designs, the independent variable is the single variable that differentiates the control group from the experimental group. However, in statistically derived research designs that use groups, more than one independent variable may be tested in a single experiment.

individual instruction control. An expression that refers to the maintenance of the student's attention to instructional details during the presentation of instruction by a teacher. It is similar to **group instructional control**, but involves the instructional control over an individual student rather than a group of students. Students who are under instructional control respond reliably to the "attention signal" of teachers, previously learned textural and nontextural stimuli, and newly targeted antecedent stimuli. The teacher can specify what it is that the student is to do, and the teacher can expect that the student will do what was asked. The student's repertoire relative to instructional control can be described also as listener or reader responses. Instructional control is necessary for the process of teaching. However, instructional control is not to be confused with disallowing students to disagree with opinion-related issues provided that disagreement is done with respect for others (e.g., not interfering with the reception of learn units by others). *See also* **group instructional control**.

individualized instruction. An expression that refers to the provision of pedagogy and curricula that is designed to meet the instructional needs of the individual student. The concept embraces the notion of individualization within classrooms in which each student is at a different level and each student is receiving optimum instruction. The tactics and operations for doing so are drawn from the science of pedagogy and the perspective on curriculum and the behavior of students based on behavioral selection Chapter 6 includes a detailed explanation of the process of arranging for individualized instruction for all children. *See also* **Personalized System of Instruction (PSI); Programmed Instruction; Comprehensive Application of Behavior Analysis to Schooling (CABAS); modes of curriculum**.

inductive scientific practices. Associated with the natural sciences and behavior analysis. The results of experiments lead to the formulation of principles. While both inductive and deductive logics are used, the primary approach is inductive. For example, varying the ratio of responses to reinforcement in a series of experiments led to the formulation of the principle of ratio reinforcement. A deductive approach would have posited a hypothesis and then done the experiment to test the hypothesis. *See also* **single-case experimental design; Mill's canons of the scientific method; deductive scientific practices**.

inferential statistical tests. Tests that infer the presence or absence of differences between groups based on laws of probability (ANOVA, *t* test, ANACOVA). The analysis of variance or covariance or *t* tests are types of inferential statistics commonly associated with group designs and hypotheticodeductive scientific practices. *See also* **hypotheticodeductive statistical procedures; statistical significance level; experimental design; Mill's canons of the scientific method; deductive scientific practices**.

in-service training. An informal term for continuing education that is associated with the work place. Traditional approaches to in-service training for teachers have consisted of workshop presentations. The evidence, however, suggests that workshops have little actual impact on teaching In CABAS schools, the training is done by (a) PSI operations devoted to the three repertoires of teaching and by (b) classroom demonstrations for teachers by supervisors and the use of teacher performance rate/accuracy (TPRA) observations. This approach to in-service training is effective and has been shown to affect actual teaching practices. The manner in which teachers are trained is detailed in chapter 9.

***in situ* training (training in the natural setting).** The process whereby an individual is trained in the natural or applied setting to perform certain tasks. In this book, it is used to describe the training of teachers or teacher assistants by supervisors or master teachers in the classroom setting. The training includes systematic observation, specific feedback, verbal and nonverbal prompts, and demonstration. See chapter 9 for a detailed treatment. *See also* **supervision repertoires; modules of instruction for teachers; teacher performance rate/accuracy (TPRA) observation**.

instructional design. A term that refers to any or all instructional activities that result from planned procedures. Instructional design includes the curriculum, pedagogy, and related measurement specified by a plan or design. Instructional design also refers to the process of designing instruction. Learning that occurs by unplanned experiences is the opposite of learning that occurs by instructional design. Incidental instruction, however, can be a component of instructional design in those cases where the adventitious use of incidental settings were planned in advance or captured *in situ*. *See also*

scripted curriculum; automated programmed instruction; learn units; captured learn units.

instructional errors. Errors made by teachers or teaching devices in the delivery of teaching operations. They include, for example, omission of reinforcement or correction, errors in reinforcement or correction, ambiguous antecedent presentations, failure to gain instructional control, omission of data collection, and errors in data collection.

instructional variables. Those environmental events that either function to teach or have the potential to teach. They include the antecedents–behaviors–consequences of the learn unit as well as learn unit histories associated with each component of the student's three-term contingency. Similarly, they embrace the instructional history associated with all components of the learn unit. They include the setting events and setting stimuli that influence or potentially influence the learn unit. They do not include intervening variables such as personality, locus of control, learning disability, emotional disturbance, and intelligence quotient groupings, for example. These latter are not related to instruction and are usefully replaced by analyses of the contingencies, setting events, and instructional histories for each student. See the discussion of verbally mediated strategies in chapter 6. A related set of instructional variables concerns curricula. See chapters 7 and 8 for a discussion of curricular variables.

interobserver agreement. An index of agreement between a data observer and an independent observer. The term "reliability" is often used to refer to interobserver agreement. However, reliability refers to issues other than interobserver agreement alone, such as the reliability of treatment or the reliability of effects.

interspersal of known items. The interspersal of known items, to which the student readily emits the correct response and receives reinforcement, is a tactic that has been shown to increase the acquisition of new discriminative stimulus control. Presumably, it functions like a reinforcement schedule, wherein the effect of the occurrence of successive incorrect responses does not punish student responding. The effect may be related also to behavioral momentum.

interval recording. A procedure developed for applied settings when the frequency of the behavior is not humanely possible to count without a special automated device. It is a less preferable measure but sometimes necessary, as in the case of a high-frequency behavior or one with indiscrete boundaries such as self-injurious behavior or stereotypy. In such cases, the most reliable estimate is obtained by using very short intervals (2 to 5 s) in which the presence of the behavior is counted if it occurs throughout the entire interval (e.g., whole interval procedure) or occurs at all (e.g., partial interval procedure). It is also a more reliable procedure when the observation intervals are consecutive. Sometimes short observation intervals are rotated with short record intervals (e.g., interval rotation procedure), and these produce less reliable estimates. Variations include interval rotations across groups where members of the group are rotated for observation purposes. It is a sampling procedure that estimates the actual rate of responding or duration of responding.

intraresponse time. The instructional time prescribed for a student response or, in some cases, the time from the presentation of an antecedent response until the student responds. *See also* **learn units**.

intrastimulus prompts. Prompts that are inserted within the stimulus itself to evoke discriminative responses from students. Changing the appearance of the stimulus is typical. As correct responses increase, the intrastimulus prompt is vanished or faded until the unprompted stimulus occasions a correct response from the student.

intraverbal behavior. Verbal behavior under the control of other verbal behavior. The use of strings of words as in a memorized poem or the creation of a poem are examples (i.e., eeny, meeny, miny, mo) where the verbal behavior is controlled by a history of antecedent consequent verbal arrangements. Intraverbal responses also occur between individuals in which case the response of the listener does not have one to one correspondence to that of the speaker. *See also* **verbal behavior**.

inventory of student repertoires. An expression that originated in CABAS schools. It refers to a criterion-referenced assessment of student performance across all of the educational domains or repertoires for which the school is responsible. It provides a continuous record of what the student has learned and what the student has not learned. The assessment is done across several instructional sessions or days until a reliable picture of the student's performance emerges. The repertoire specifies ante-cedent–behavior–consequence conditions and the setting stimuli or events. The achievement of a criterion for a particular repertoire (e.g., uses vocal speech to specify reinforcers) can be observed without instruction, or the criterion may be the result of a student achieving several criteria on curricula directly related to the repertoire. Some educators have begun to use the expression portfolio evaluation to refer to an archival record of what a student has learned. The inventory serves a similar function, as discussed at length in chapter 10.

irrelevant stimulus attributes. The nonsalient attributes of a stimulus presentation targeted to gain stimulus control for a student. If color hue is the target stimulus, control, shape, size, and dimension are irrelevant to the essential stimulus control sought. A good program will shape stimulus control such that the student is not misled by irrelevant attributes. If shape is the desired or essential attribute of the stimulus or stimulus congregate, color, size, and dimension will not control responding. Vanishing procedures, intrastimulus or extrastimulus prompts, and general case instruction oper-ations are tactics used to ensure that the irrelevant attributes of stimuli do not produce erroneous responding. Frequently, student's difficulties with concepts, operations, or simple discriminations are traceable to the control gained by irrelevant stimulus attri-butes in the student's instructional history. See chapters 5, 6, and 9.

J

joint method. The joint method draws on more than one of the Mill's canons of the scientific method. If the method of difference affirms a functional relationship, the method of concomitant variance may also be in effect.

K

Keller plan. *See* **Personalized System of Instruction (PSI)**.

L

law of parsimony. A tenet of science that suggests that if a simple or less abstract explanation is plausible in contrast to a more hypothetical one (e.g., an intervening variable), the scientist should choose the simple, direct explanation simply because it is the more parsimonious one. *See also* **Mill's canons of the scientific method**.

learning theory. A term used in certain branches of psychology to describe the study of learning whereby learning is viewed as an intervening variable (e.g., cognition, understanding). The term was not seen the purview of the experimental analysis of behavior, applied behavior analysis, or behavioral selection. Skinner' a article "Are Theories of Learning Necessary?" established that the objective of the science of behavior was not the study of the intervening variable. This is not to suggest that behavioral selection is not concerned with theory, but only that theory that is related to the epistemological tenets of radical behaviors or behavioral selection. *See also* **behavioral selectionism; radical behaviorism; theory**.

learn unit interresponse time *or* **inter-learn unit response time (ILRT)**. The time, typically, between the presentation or reception of learn units. This result gives the mean number of minutes or fractions of minutes between learn units. The lower the interresponse time, the more instruction is occurring.

learn unit prerequisites. The repertoires needed by the student to respond adequately to new learn units that are presented faultlessly. If the student does not respond correctly, and the presentation is errorless, one source to pursue is the instructional history. The instructional history should include the prerequisite stimulus control. Other prerequisites include the setting events or context in which the learn unit is presented.

learn units. Consist of a three-term contingency for the student and two or more three-term contingencies for the teacher in which the response of the student occasions a reinforcement or correction operation from a teacher or teaching device. Learn units occur in scripted, automated, discrete, dispersed, and massed forms. Learn unit rates serve, together with criterion-referenced objectives, as the basic measure of the teaching process and as the nucleus for analyzing instructional variables. Typically the mean numbers of learn units to criterion are measures of a teacher's instructional skills. A teacher who can achieve instructional objectives with the fewest number of learn units is the most skilled teacher. Learn units can also be used to determine the cost and benefits of education. See chapter 2 for an in-depth discussion of the learn unit.

levels of teacher skills or expertise. In the CABAS model (see chapters 5, 6, and 9), the three components of teaching expertise are themselves composed of various levels of skill or expertise. See chapter 9 for a detailed explanation.

linguistics. The study of language and language usage in terms of the practices of a particular linguistic community or all linguistic communities. It is the study of the structure of language or languages and typically uses structural or descriptive research tools. For an excellent summary of linguistics and its relationship with verbal behavior, see Catania's 1992 *Learning*. Verbal behavior, on the other hand, is the study of the function of communicative behavior in the extended operant. It is not necessarily incompatible or competitive with linguistics, but is complementary in many instances. *See also* **verbal behavior**.

listener behavior. In verbal behavior, there are two large divisions. Listener behavior refers to behavior that is under the control or designed to be under the control of a speaker. When a student responds consistently and accurately to the verbal behavior of a teacher, the student has a listener behavior repertoire. Compliance to verbal instructions is a case of listener behavior. *See also* **verbal behavior**. An extension of listener behavior to textural presentations is reader behavior.

locus of learning problems. When a student is having difficulty learning, the sources of the problem are instructional variables. These variables (i.e., learn unit components, instructional history, setting events) are the locus of the problem. Strategies for seeking the source of the problem are described in chapters 4–6 and 9. They involve the verbally mediated strategies of one who has expertise in the science of pedagogy. The locus of the school problem is not located in the student or the parent, but rather in the instruction provided by teachers for which supervisors are also accountable. Accountability for this source of instruction is limited only by the limits of the scientific expertise presently available.

logs of supervisor behavior (counts of tasks of supervisors). Supervisors log a count of their completion of tasks that are consistent with their job description and two general objectives. Tasks meet result in learning or its probability or the provision of services and goods to meet the schooling needs of students and by extension teachers. The tasks are summarized and visually displayed in terms of numbers per hour (rate). The completion of certain tasks has been functionally related to teacher and student performance (see chapter 9). Supervisors typically have a checklist with standard tasks and a space for initiated tasks on which they record task completion and allocated work time as well as overtime. In order to count a task, there must be a permanent record of the performance of the task. It is a critical component of the management of administration and supervision by behavior analysis. See chapter 9 for a detailed treatment. *See also* **supervision repertoires; administrative repertoires; management by behavior analysis; rate of responding; correct and incorrect rates of responding by students and teachers**.

loose stimulus control. The practice of varying the presentation of antecedent stimuli and setting events in order to engender stimulus generalization. Such practices are avoided until the student is under unambiguous antecedent control. The desired as opposed to undesired "loose stimulus" control is probably and ideally best arranged through systematic programming. *See also* **stimulus generalization principle and operations; faultless presentations; fading or vanishing tactics; successive approximation principles and operations**.

M

mainstreaming. A term used in special education to refer to the introduction of children who have "handicapping conditions" into mainstream classrooms and schools. It refers to the movement to integrate classrooms such that children with and without handicapping conditions are taught together. The term *inclusion* refers to a form of mainstreaming whereby even the most severely handicapped individual is taught alongside those without handicaps.

maladaptive or nonadaptive behavior. *See also* **bad behavior**. Behavior that is not in the best long-term interest of the student. Assaultive behavior, self-injurious behaviors, or noncompliant behaviors of students in school settings are examples. These behaviors lead to short-term or idiosyncratic reinforcement for the student, but in the long term lead to punitive or coercive effects. The result in reinforcement that is not habilitative either for the individual student or for his or her community. Typically, the objective of good instruction is to develop for the student a habilitative repertoire that replaces the maladaptive one. The concept is treated at length in chapter 10. *See also* **bad behavior; coercive traps**.

management by behavior analysis. The application of behavior analysis to the management of personnel. It includes the components of management by objectives, but requires extensive and ongoing measurement. In schools that use behavior analysis comprehensively (e.g., CABAS schools), the supervisory and administrative responsibilities of the school personnel are carried out via the use of management by behavior analysis. See chapter 9 for a detailed discussion. Also see the *Journal of Organizational Behavior Analysis*.

management by objectives. *See* **management by behavior analysis**.

mand. A category of verbal behavior. The mand may also take different forms (e.g., vocal or gestural, many or a few words or gestures). Regardless of its form, the mand functions to specify its own reinforcer. A speaker emits a mand form and a listener delivers that which we manded on a predictable basis. Mands are occasioned by deprivation conditions or motivational setting events that occasion the mand when a listener is present. In the "pure" mand, there are no verbal antecedents. Rather, the mand is under the antecedent control of nonverbal discriminative stimuli (e.g., a particular listener) and setting events (e.g., the listener has access to the reinforcer and the speaker is under same deprivation of the reinforcer item). The function determines whether an incident of verbal behavior is a mand. The word or gesture "bread" is a mand only if all of the above conditions are in effect. The word in this case specifies its reinforcer. The word could also function as a "tact" in which case the speaker is making verbal contact with the nonverbal environment under the control of generalized reinforcement such as "Yes, that is bread." *See also* **verbal behavior; tact; autoclitics**.

massed trials. The practice of repeating trials one after the other rather than dispersing them. Teacher-controlled learn units that are presented in rapidly consecutive fashion are referred to as massed learn unit presentations. *See also* **discrete trials; learn unit**.

mastery learning. Instructional criteria that demonstrate that the individual has mastered the learning objective. The learned antecedent–behavior–consequence has been learned to the point that it is in the individual's repertoire. For example, a student can be said to have mastered the operations involved in double-digit addition. *See also* **repertoires of behavior; Personalized System of Instruction (PSI); fluency**.

means objective (short-term objective). The term *means objective* is synonymous with the term *short-term objective*. It refers to an objective that is a subobjective that successively approximates a terminal or long-term objective. When a student achieves an objective in a task-analyzed sequence under partial gestural prompts, he or she is said to have achieved a short-term goal. When a student reads on grade level at 20 words per minute correctly at leas than 1 word per minute incorrectly, he or she may be arid to have met a means objective that approximates the terminal objective of reading 60 words per minute and zero incorrect words.

mental retardation. A term used for several years by psychologists to refer to subnormal performance on standardized tests of intelligence. The categories have included individuals who were once said to be severely retarded, profoundly retarded, or mildly retarded depending on the fashion of the times. Sidney Bijou describes a more appropriate designation (e.g., developmentally disabled) in his chapter in the book *Design for Educational Excellence: The Legacy of B.F. Skinner*, published by S. West in 1992. A current term that is under consideration is intellectually disabled

metacognition. The tacting of certain operations by the person performing those operations. One is said to carry out cognitive functional while analyzing those functions at the same time. The concept could be loosely translated by behavioral selectionists as forms of verbally mediated self-observations.

method of agreement. One of the five canons of the scientific method described by John Stuart Mill. It states that if in all situations in which a given phenomenon occurs, and there is only one variable in common, that variable is probably related in same manner to the occurrence of the phenomenon. For example, the occurrence of a virus in conjunction with certain symptoms is found to be in effect in all instances of the symptoms and that virus is the only variable found to be common. The virus and the symptoms are probably related in some manner. It is the weakest form of scientific explanation, because it does not specify the degree of concomitant variance or the functional relationship between the variables found to agree.

method of a priori. A belief or course of action predicated or derived on a priori assumptions. Courses of action that do not involve the use of logic, authoritative sources, or scientific approaches are predicated on a priori assumptions. Like the other methods, a priori-based action may or may not be identified by the person or persons using them. Much of formal schooling involves the replacement of courses of action based on a priori beliefs with ones drawing on the methods of authority, logic, and science using one or more disciplines (e.g., history, behavior analysis, paleontology, architecture, law, medicine).

method of authority. *See also* **Pierce's four methods of fixing belief**. A belief or course of action derived, predicated, or rationalized on some authoritative source. The rationale for a law or the interpretation of a law is often based on method of authority. In essence, it is the "supreme court method." The interpretation of history or the writing of history is based on the adequacy or primacy of primary sources. The method of authority is the basis for practices in history, philosophy, law, or religion in which a course of action or interpretation is based on locating or authenticating an authoritative source. Certain problems may be described as problems for which the method of authority is the best, or only, course of action to follow to solve a given problem. How one determines the best method and the practices associated with the method is one of the problem-solving methods students need to learn.

method of concomitant variance. One of Mill's five canons of the scientific method. It states that when two variables vary in a systematic and consistent manner across subjects, settings, or behaviors, the two are said to be related. For example, the occurrence of praise increases with the occurrence of correct responses in a systematic manner. As teacher praise increases, student correct responses increase. This relationship suggests that the two variables are strongly related. One of three possibilities exists: (a) teacher praise increases student correct responses, (b) student correct responses increase teacher praise, or (c) a common variable or common set of variables affects both in a functional manner. Statistical indices of concomitant variance are percentage agreement or correlation.

method of disagreement (or method of difference). One of Mill's five canons of the scientific method. It states that if all conditions are the same except for the systematic and controlled alternation of one variable, any differences that occur on a reliable basis are a function of the presence of the single variable. It is the basis of the experimental method. Practices of science that use the experimental method employ various tactics or research designs to realize the requirement or to approximate the requirements of the canon of disagreement/difference.

method of residuals. One of five canons of the scientific method formulated by John Stuart Mill. It states that in an experiment or application of any or all of the other canons, there remains some variability not attributable to those variables tested directly, then subsequent scientific tests will uncover other factors to which the variability may be traced. In the science of behavior, the source of variability becomes a stimulus for future experimental research. In the logic of statistics, the residual is a mathematical difference not attributable to the variables that were directly tested and is a residual sum of squares.

method of tenacity/logic. *See also* **Pierce's four methods of fixing belief**. A belief or cause of action derived, rationalized, or predicted on rationality. In those cases where the tools or repertoires of logic are used, then the method of tenacity describes the congregate of operations performed to set or rationalize a course of action. Like the other methods (i.e., methods of a priori, authority, and science), tenacity is often combined with other methods. The method of tenacity plays a prominent role in interpretations associated with methods of authority and science. As an instructional objective, the method of tenacity can serve as one of several problem-solving repertoires to be taught students such that they can use them functionally across schoo. subjects.

method of science. *See also* **Mill's canons of the scientific method**. The method of science applies several tests to determining the validity or predictable success of course of action based on certain criteria. First, the phenomenon must be empirical or observable and it must be seeable or observed by independent observers or their extensions. Second, the countability of the phenomenon (e.g., measurement) must be validly and reliably tied to aspects of the phenomenon. Third, relationships between phenomena must be determined by various procedures generically described as methods of concomitant variance, agreement, disagreement, residuals, and the joint method.

mind/body dualism. The historic and prescientific assumption that the mind is separate from the body—in effect, the mind transcends the body. Sometimes a tripartite separation is suggested as in mind, body, and spirit. In behavioral selection, the notion of "mind" is but behavior beneath the skin. While mind or behavior beneath the skin is recognized, it is seen as behavior per se.

Mill's canons of the scientific method *or* **Mill's canons of science**. The Mill's canons of the scientific method consist of (a) the method of agreement, (b) the method of difference or disagreement, (c) the method of concomitant variance, (d) the joint method, and (e) the method of residuals. They are a summary of the basic logic of science and as such are related to tactics and strategic of scientific research. Each is described separately in the glossary and at length in chapters 7 and 8.

modeling. The process of presenting a target three-term contingency relationship to another as an initial step in the teaching process. The teacher presents the model or behavior example and the student is to imitate the teacher. Textual models or examples may be used as antecedents for textual imitations also. *See also* **imitation**.

models (teacher, student, schooling models). The term *model* is used, first, to describe the principle of behavior that characterizes learning by imitation of a mode. and second to describe the person or exemplar presented. The effect may be direct as in direct observation or indirect as in vicarious modeling. Much learning occurs through intentional and unintentional effects of modeling. The effects of modeling are attributable to reinforcement operations and effects. This book employs a third use of the term *model* in which the epistemological and operational aspects of a school are tied to a consistent paradigm or model. The behavioral models of schooling are examples of particular approaches to schooling. *See also* **peer contingencies; positive reinforcement principles and operations/tactics; response delay**.

modes of curriculum. Used in this book to categorize four aspects of pedagogical and curricular activities. The concept is related to Whitehead's 1929 book *Aims of Education* in which he uses the term modes of curriculum, but it is recouched in terms that have more current usage. The four modes of curriculum are: (a) academic literacy, (b) self-management, (c) problem-solving repertoires, and (d) community of reinforcers. While they are interrelated in a comprehensive approach to curriculum, at least from the perspective of behavior selection, each requires careful attention in terms of curricular analysis as well as strategies and tactics of pedagogy. These components of curriculum are described in detail in chapters 7 and 8.

modules of instruction for teachers. The units of instruction delivered to teachers in PSI fashion. Each module includes objectives associated with the three repertoires of teaching. Teachers pass components of modules by meeting classroom performance criteria that are individually set and by passing written or oral examinations covering levels of expertise about components of the science of behavior and pedagogy. Incentives are used frequently to increase the probability that the completion of modules is reinforced. *See also* **Personalized System of Instruction (PSI)** and the detailed treatment of the use of PSI for teacher training in chapter 9.

modus operandi. *Modus operandi* is a Latin term that refers to standard operating procedures. One who does something in modus operandus fashion follows the typical or standard procedures to accomplish the task in question.

motivation and the science of behavior. Variables or conditions that affect the reinforcement value or punishment value of consequences to behavior. The frequency of engagement or consumption of events, objects, or activities or their deprivation affect the reinforcement or punishment effects of consequences. Deprivation or satiation, momentary or prolonged, affects the reinforcement value of food, activities, or events relevant to particular responses or response class. The deprivation of sleep makes sleep opportunities more reinforcing and at the same time affects the reinforcement value of activities requiring concentration. A sleep-deprived child is less likely to engage in behaviors that would typically have high preference for that child. Periods of prolonged physical exertion make setting more reinforcing and vice versa. Teaching operations that function to make consequences more reinforcing are called establishing operations. Analyses of potential variables that affect the reinforcement value of stimuli suggest strategies and, eventually, tactics to use motivational conditions to desired instructional ends or to avoid detrimental effects from certain instructional operations. The analysis of motivational variables is critical to the use of reinforcement operations that have reinforcement effects. While one cannot, according to the old admonition, make a horse drink, one can provide motivational setting activities (e.g., exercise) that will in fact lead the horse to drink. The use of briefly deprived access to activities can be designed such that those activities function to reinforce other activities. Consumption of something not reinforcing to a child can under certain conditions be manipulated such that the consumption functions as a reinforcer.

motivational context. Those aspects of the setting events that determine the reinforcement value of reinforcement operations or natural consequences. They involve deprivation or satiation occurrences or intentional operations (establishing operations) that affect the reinforcing or punishing effects of the consequences of behavior. Instruction given in a context in which movement has been momentarily deprived increases the reinforcement value of movement as a consequence of a behavior and, therefore, the previous antecedent control. The context also incorporates setting stimuli or events that are not motivational; they may be conditional stimuli but are not associated with deprivation or satiation conditions or historic. For example, a set of verbal instructions may serve to affect the three-term contingency relationship.

multielement design. The term *multiple schedule* is also used to denote the comparison of two or more schedules in which the session using each of the schedules are rapidly rotated. In the multielement design two independent variables or the baseline and experimental variable are rotated from session to session as in the multiple schedule design. This rapidly alternating design is referred to also as an alternating treatment design. The procedures differ from reversal or withdrawal tactics in that treatments are rotated from session to session rather than waiting for stability to occur. The believability of effects is enhanced when the two treatments are preceded and followed by baselines and when each treatment is subjected to a prolonged repeated treatment. In the latter case, elements of the reversal or withdrawal designs are incorporated also in the overall strategies to implement the strategy of the canon of method of disagreement. The design was described by Schwartz and Baar in 1991 as a multielement design.

multiple antecedent control. Used herein as an expression that characterizes learn units in which the antecedent component involves more than one antecedent. Responses that are to be under the discriminative control of both the teacher's directions and the textural antecedents are one example. The teacher presents several colored objects and says, point to red. Both the hue of the object to be pointed to and the teacher's vocal directions (e.g., conditioned stimuli) were the antecedent stimuli. Instruction under the control of several antecedents, as in stimulus equivalence, is another case of responding under multiple antecedent controls. The term is used here for target discriminative stimuli. However, it could be used to describe the fact that multiple antecedent events are associated with all operant responding in that instructional histories, setting events, and discriminative stimuli are all multiple antecedent events associated with the response and consequence.

multiple baseline design. A research design from applied behavior analysis in which the logic of the method of difference is approximated by replication of treatment effects across individuals, behaviors, or settings without reversals to baseline conditions. In the multiple baseline, baseline and treatment effects are compared between subjects (i.e., behaviors, individuals, setting) such that one or more target subjects remains in baseline while one or more target subjects are receiving the treatment variable. A functional relationship is affirmed when the treatment onset, introduced in staggered fashion, produces (a) an effect on the dependent variable and (b) baseline performance of another subject, setting, or behavior occurring simultaneously does not change. The control setting is another subject, behavior, or setting; thus the design is an intersubject or between-subjects design. The design is useful when for a variety of reasons the individuals may not be returned to baseline condition. The logic of the design is the same as the logic for control and experimental group comparison designs.

multiple schedule design. A tactic to implement the method of difference, whereby one schedule is rotated with another. To avoid spillover or compound schedule effects, a separate discriminative stimulus is associated with each schedule. It may be the generic source for other designs such as the alternating treatment or multielement design. It is an intrasubject or within-subject design, and if there are no carryover effects and differences in behavior are reliable, the design is strong. *See also* **alternating treatment designs; multitreatment designs; multielement design**.

multitreatment designs. The comparison of two or more (rare) treatments simultaneously or separately. The term *multitreatment design* was used by Kazdin in 1978 to describe variations on reversal or withdrawal designs in which more than one treatment is compared. In the latter case, baseline phases (i.e., repeated sessions leading to stability) are compared with two or more treatments. In each comparison, stability is achieved for the treatment variables with baselines occurring between each treatment phase (e.g., ABACACA or ABACAB).

mutually reinforcing relationships. Reciprocal behaviors between two or more individuals in which the behaviors of each are positively reinforcing to each other in an episodic manner. Approving behaviors evoke approving behaviors from each of the parties involved. It is the opposite of coercive relationships. *See also* **coercive traps**.

N

natural conditioned reinforcer. Natural conditioned reinforcers refer to conditioned reinforcers that are acquired by most of us in daily interactions with environments. Praise is usually acquired as a reinforcer for most of us as a result of conditions that naturally occur. Food is paired with verbal and nonverbal indications of approval and approval becomes a generalized reinforcer as a natural course. Music, art, appreciation of nature, and reinforcement value for our peers are all potential conditioned reinforcers that are acquired as a matter of course. Events may occur, however, that interfere with the acquisition of such reinforcers. In such eases, care must be taken in the instructional process to condition such stimuli as conditioned reinforcers by the types of teaching operations used.

negative punishment or type II punishment operations. (1) *Principle of behavior*. The removal of a reinforcer for a behavior immediately after the occurrence of an operant behavior will, if done on a consistent and continuing basis, results in the reduction and extinction of that behavior, providing no alternative sources of reinforcement occur. Typically, the response or response class increases in frequency and intensity/duration in the early stages of the application. Following this burst of behavior, the response declines swiftly. The behavior may reoccur occasionally, but if the negative punishment operation is applied, extinction will result. (2) Negative punishment operations include planned ignoring, token removal, or time out from reinforcement (e.g., contingent observation). Negative punishment operations may also be seen to embrace the operations associated with DRO, DRI, DRL, or DRA (*see also* **differential reinforcement**) where the presence of high rates of reinforcement for behaviors in the classroom create a setting whereby momentary removal of reinforcement has the result of creating negative punishment.

negative reinforcement operations. The process whereby the removal or cessation of a stimulus is occasioned by or contingent on the emission of a response or response class acts to increase the behavior. Observational studies suggest that negative reinforcement operations are more common than positive reinforcement operations by teachers who are not trained in the use of positive reinforcement operations. Typica. teaching operations that function as negative reinforcement are disapproval or nagging until a response occurs, dispensing demerits until a behavior occurs, or cajoling or threatening students until behavior occurs. In order to be effective, the student must not be able to escape by any route other than the teacher-desired behavior. The use of negative reinforcement frequently leads to escape or avoidance responding. Typically, negative reinforcement operations are avoided or compensated for by those practicing the science of pedagogy.

negative reinforcement principle. Events, activities, or environmental occurrences (beneath or outside of the skin) that act to increase a behavior following the removal of the events after the behavior is emitted. The behavior of escape from the stimulus must increase the probability or frequency of responding. The cessation of the stimulus by a response or response class strengthens the behavior consequence relationship. Negative reinforcers are classified as conditioned, unconditioned, prosthetic, or natural. They are affected by and affect antecedents, instructional histories, and setting events or setting stimuli. Stimuli that occasion the acquisition or maintenance of behavior in this manner are avoided or to be used minimally in the teaching process since their usage occasions escape and avoidance responses from students.

nomologically derived constructs (nomological network). Psychological constructs such as personality, attribution theory, learning disability, minimal brain damage, or anxiety disorders are created by inferring the construct said to explain the behavior as a manifestation of the construct. This grouping of behaviors is the nomological rationale for the construct. The occurrence of several types of behaviors or measures on different tests is logically related to posit the construct. A person who reports that they have no control over their life and behave in ways that are said to relate to their reports are said to have an external locus of control. The "attribution of control" is posited as an intervening variable between the environment and behavior.

noncontingent praise or approval (noncontingent social reinforcement). The delivery of approval, praise, or other forms of social reinforcement for a response that is inconsistent with the pedagogical and curricular objectives for a particular student is a noncontingent praise operation. A teacher who recognizes a student who has blurted out a response, in a classroom that specifies hand raising for speaking out, has provided noncontingent praise. The praise operation will likely reinforce a response that is contrary to the target response. *See also* **teacher performance rate/accuracy (TPRA) observation; teacher-contingent consequation.**

noncontingent reinforcement. Reinforcement that does not correlate with the occurrence of a target response or correlates with the occurrence of a response other than the target response is noncontingent. Reinforcing a behavior that is in contrast with following classroom rules or reinforcing a student's incorrect response are incidences of noncontingent reinforcement. Teacher errors include noncontingent responding. *See also* **teacher performance rate/accuracy (TPRA) observation.** Noncontingent reinforcement is also a tactic used to provide students frequent and noncontingent reinforcement and functions under certain conditions to reduce the occurrence of bad behaviors. The effect over an entire day is questionable, since the bad behavior is not replaced with appropriate responding and the frequent use of the noncontingent reinforcer will likely lead to satiation of the reinforcer used. See the matching law.

nonexemplars, negative exemplars, or foils. In discriminative stimulus instruction, nonexemplar or negative exemplars are examples that are used to contrast the positive exemplars in order to develop discriminative responding. They are also referred to as S delta. *See also* **positive exemplar; general case instruction (multiple exemplar instruction).**

nonverbal antecedent. Nonverbal antecedents to responses typically used to describe verbal behavior that is evoked by nonverbal antecedents. Pure mands, for example, are evoked by the deprivation of something and the stimuli that occasion its possible delivery as a result of verbal behavior. A thirsty child after a period of water deprivation says to her parent, "Water, please." No verbal behavior preceded the response. Certain instances of verbal behavior that is under the control of nonverbal antecedents are said to be spontaneous.

normal schools. Teacher-training institutions prior to the development of 4-year teacher training programs. The publishers of the school texts that the novice teachers were to use with their future students frequently sponsored normal schools.

normative education. A term used in this text to distinguish scientific approaches to education from the eclectic approaches that are common to most schools. Thus, schools that use normative education approaches have no consistent epistemological view other than the prevailing opinion in education that teaching is still an art.

number per standard unit of time. Rate (number per unit of time) refers to the frequency of behavior in a given time period. Some behaviors are expressed in number per minute, and others in number per hour or even number per week. It is determined by dividing the elapsed time into the frequency of behavior. A student who completes 10 correct responses in 10 min and 5 incorrect responses per minute has a number per minute correct of 1.00 and an incorrect number per minute of 0.5. Rate refers to a specific learn unit type. Thus, rate objectives for multiplication are different than rates of fluent oral reading. *See also* **rate of responding; fluency.**

O

observable behavior. Behavior that can be observed beneath or outside of the skin by the behaving one or another observing person. It is objective behavior when another observer can observe the behavior independently or when the observer has been trained to a publicly based calibration.

observational learning. Observational learning occurs when the observer responds to the contingencies as a function of observing the effects of contingencies on the behavior of another. For example, controlled experiments have shown that female guppies will mate with male guppies of light color hue as a result of watching other female guppies mate with lighter hued males. This is striking because natural selection controls female guppies mating with brightly hues males. Thus, the unusual mating with lightly hued males is a function of observational learning. *See also* **imitation.**

ontogeny. A term that encompasses the developmental, experiential, and learning history of the individual (e.g., infancy, childhood, adolescence, adulthood). It also embraces the learning of particular repertoires associated with the physical development of stages. It is distinguished from phylogeny, which is the evolutionary history of the species. Behavioral selection and behavior analysis are concerned with the ontogenetic development of various species vis-à-vis the determination of environment/behavior relationships that hold widely. The fact that contemporary behaviorism concerns itself with the ontogenetic factors does not mean that the science or epistemology holds that the environment holds greater sway over individuals than their genetic factors. Rather, behavior selectionism grants the importance of genetic factors or really the inseparatability of the two (e.g., phylogeny and ontogeny). But the field of behavior analysis is devoted to the study of ontogenetic variables particularly.

operant behavior. Acquired behavior, unlike respondent behavior, which is fixed by phylogenetic factors. Operant behaviors are evoked by learned antecedent–behavior–consequences; they can come under the control of numerous antecedents and consequences. Operants are behaviors selected by consequences, and even the antecedent behavior relationship in the operant is selected or affected by the consequence.

operant chamber. An experimental chamber that is designed to maximize the emission and controlled study of operants and their determinants. The subject is subjected to certain establishing operations, controlled instructional histories, and even genetic histories prior to their placement in the operant chamber. Lighting, ventilation, temperature, and manipulanda are designed to occasion the responding that is under study. See *The Behavior of Organisms* (Skinner, 1938).

operations. The behaviors and the contingencies that are associated with producing or attempting to produce a given effect. For example, teaching operations (e.g., reinforcement operations) are those designed and performed to result in particular effects on the behavior of a student and the associated stimulus conditions and setting events. *See also* **contingencies of reinforcement and punishment (three-term contingences); verbally mediated repertoire.**

ordinal units of measurement. Rank orderings. Phenomena or behaviors can be ranked from low to high but the differences between the rankings are not equal intervals. One is different from two or three and so on, but the degree of difference is unspecified. Letter grades that are not based on ratio or percentage criteria are examples of ratings. Questionnaires that, for example, ask the respondent to rank teachers according to scales from 1 to 5, 1 to 3, or 1 to 7 by certain qualities are ordinal.

organizational behavior management (organizational behavior analysis). The application of behavior analysis to the management of personnel and the analysis or organizations in the business or industrial world. In the development of the CABAS behavioral system of education, organizational behavior analysis is used in the administration and supervision of all aspects of schooling.

P

parent education. A term used for instruction to parents that is designed to teach the basic components of the science of pedagogy to parents. See chapter 8 for an elaboration.

parenting groups. Small groups of parents who meet together to receive instruction in parenting and to share their data on instructional efforts with their own children with other parents and the parent education. They serve as a support groups for using the science of pedagogy systematically by providing a critical mass of individuals with similar expertise. See chapter 8 for an elaboration.

parent performance rate/accuracy observations. The application of teacher performance rate/accuracy procedures to training and monitoring parents to use the science of pedagogy with their own children. The observation procedures are described in chapter 9.

participatory management. An expression used in models of management that promote the full participation of personnel at all levels in decisions about management. In the CABAS model, participatory management is tied on the degree of pedagogical expertise demonstrated by instructional staff. Those professionals with the greatest expertise are accorded greater roles in the management of the instruction. It is also incorporated in the principle that teachers and supervisors participate in setting their goals, their reinforcement, and the criteria for achievement of objectives and reinforcement.

pedagogy. The study of, or application of, teaching operations that occasion the acquisition of new discriminations or conditioned reinforcers that result from the teaching operations. Such operations may result from direct teacher-controlled learn units or from indirect ones involving automated instruction or peer tutoring. This definition is, of course, a definition of pedagogy based on the sciences of behavior and pedagogy and is more specific than the traditional or prescientific definition. In a scientific sense, true pedagogical operations are ones that are functionally related to the student acquisition of new discriminations, operations or conditioned reinforcers. That is, the student would not have learned that which is being taught without the particular pedagogical operations. See a standard dictionary for generic definitions of pedagogy.

peer contingencies. The three-term contingency relationship between peers. Typically, systematic tutoring involves teaching tutors to use the three-term contingency accurately in the process of teaching a peer. The source of certain repertoires or behaviors can also be traced experimentally to peer contingencies as in the case of students who acquire behaviors from their peers as a result of peer approval and peer modeling. Numerous tactics are associated with peer contingencies such as peer-mediated time out, peer-conditioned reinforcement, peer modeling, peer-establishing operations, peer tutoring to name a few.

Personalized System of Instruction (PSI). The personalized system of instruction, developed by Fred S. Keller in 1968, is a behavioral model or system of instruction. Initially developed for and tested in college courses, the model has also been applied to high school, elementary, and music classrooms. The model is also used as a component of CABAS for the organization and management of individualized instruction for students, teachers, supervisors, and parents. The model has six components. These include the following: (a) students move through the material at their own pace. (b) The subject matter is divided into subunits and objectives identified for each unit of study (e.g., short-term and long-term objectives for specific repertoires and subject matter). (c) Students must achieve preset criteria for examinations given for each unit. (d) Students are not penalized for not achieving criterion, but are tutored (or study on their own recognizance) and given other examination opportunities until they do achieve criteria. (e) Lectures, if given at all, are for motivational purposes, not for the transmission of information. Information or subject material is transmitted in written form between the teacher and students. (f) Proctors or tutors (peer, older, and more experienced students; graduate students; or the professor) provide individualization of instruction. The professor or classroom teacher functions as a designer of instruction and the measurement of instructional outcomes. There are numerous studies testing the PSI model—analyses of the total package and analyses of the contributions of component parts. See chapters 6–8 and 10 for an extensive treatment of PSI.

Personalized System of Instruction for teacher training. In the CABAS model of schooling, the in-service training of teachers is done according to the precepts of the personalized system of schooling or the Keller plan. The subject matters of the science of behavior, the science of pedagogy, and curricular analysis are divided into units that are presented to teachers in the form of modules. Each module contains three components of teaching (see teacher training modules in chapter 9). The particular subject matter and the sequence are based on the hierarchy of the sciences of behavior, the existing expertise of the teacher, the composition of students in the teacher's classroom, and the suggestions of the teacher. Teachers move through the modules at their own pace with encouragement and tutorial assistance from more advanced professionals. They read material and take quizzes or vocally demonstrate their mastery of the material to the supervisor or tutoring teacher. Quizzes are, of course, criterion referenced. Performance that does not lead to criterion initially results in restudy or tutorial assistance, which continues until the material is mastered. Each module has performance criteria for the classroom, and supervisors or tutoring teachers work with the trainee until the repertoire is mastered, as demonstrated by reliable observations or learn unit data. Much of the communication between trainer and trainee occurs in written form and lectures are not used. Tutoring sessions are based on the trainee's request. In some CABAS programs, salary, rank, and promotion are tied to progress through the modules to serve as incentives and potential reinforcement in addition to the learning of new repertoires. The components of this system and a description of how one develops an in-service PSI system is found in chapter 9.

phylogeny. A biological term that refers to the developmental evolution of species. Debate over the relative contributions of phylogenetic or ontogenetic factors is a common argument in many introductory psychology texts. While the science of behavior concerns itself with the role of the environment over the individual's behavior (the very reason for its close relationship with a science of pedagogy), the science grants the importance of phylogenetic variables and their necessary interaction with ontogenetic variables. Interestingly, both biology and behavior analysis concern themselves with selection by consequences. For biology, the consequences are ones that affect the evolution of the species, while for behaviorology, the consequences are ones that affect the evolution of behavior within the individual's life. *See also* **ontogeny.**

Pierce's four methods of fixing belief. Charles Pierce, an American pragmatist (see chapter 7), categorized the manner in which beliefs are fixed. These epistemological categories, or ways of fixing beliefs, included the methods of a priori, authority, tenacity, and science. A course of action is traceable to the application of one or more of the above. In this book, the methods served both as a heuristic device and a potential useful repertoire to teach. The choice of the repertoires and their use functionally to solve problems is associated with subject matter instruction as an operationalized approach to teaching problem solving in the chapters devoted to curriculum. By using each of these as verbally mediated strategies for solving problems, one can avoid the pitfalls of relying on no operational constructs such as cognition. See chapter 9 for an extensive treatment. *See also* **method of a priori; method of authority; method of tenacity/ logic; method of science.**

planned ignoring as a correction. In some curricular programs (e.g., teaching sit still or eye contact), the provision of a correction in which the student performs or is guided through the correct performance is not possible or desirable (see chapter 6). In such cases, the consequence for an incorrect is planned ignoring; that is, the student is clearly ignored or briefly timed out from the teacher' a attention and reinforcement. This is followed by the presentation of the antecedent for a new learn unit when the student is under instructional control.

portfolio evaluation. *See also* **inventory of student repertoires.**

positive exemplar. A faultless presentation of the stimulus targeted to gain stimulus control. It is the essential stimulus attribute or attributes being taught. Teaching a child to discriminate the shape of a square in an exemplar instruction results in representations of the positive exemplar (square) in rotation with negative exemplars or ''not examples'' such as circles, rectangles, or diamonds. Negative exemplars are also sometimes referred to as foils; they set the stage for discriminative responding and are rotated. *See also* **general case instruction (multiple exemplar instruction); nonexemplars, negative exemplars, or foils**.

positive and negative correlations. Correlations are descriptive statistical indices used to ascertain the degree of concomitant variance or agreement between two or more variables. A perfect correlation is expressed as 1.00. If the variables vary in the same direction perfectly, the correlation would be $+1.00$ or a perfect positive correlation. If they vary perfectly in inverse fashion (e.g., one increases in direct proportion to a decrease in the other), they are perfectly negatively correlated (e.g., -1.00). For example, if a negative correlation of $-.80$ is found between learn units and bad behavior, then the bad behavior decreases as the learn units increase or vice versa.

positive reinforcement principles and operations/tactics. (1) *Principle of behavior.* Events, activities, or environmental changes (beneath or outside the skin) that follow behaviors or classes of responses function as positive reinforcement under certain conditions. The consequences to behavior must increase the probability or frequency of responding. The consequences select out or strengthen the frequency of preceding behaviors or response classes. Positive reinforcers are classified as conditioned, unconditioned, prosthetic, or natural. They are affected by and affect antecedents, instructional history, and setting events or setting stimuli. (2) *Operation.* The positive reinforcement operation is a process whereby a consequence is delivered after a response or response class. The operation is designed to produce the effect of positive reinforcement. Consequences that function as positive reinforcement vary across individuals, vary for individuals across time, and are affected by the latency between the response and the delivery of the consequence. As a result, reinforcement operations do not necessarily result in consequences that function as reinforcement at the time of delivery. Reinforcement operations incorporate tactics such as approval, token economies, self-monitoring, checkmarks on papers, or contingent access to objects or activities.

postcedent. A term coined by Ernest Vargas to be consistent with the use of the term antecedent stimulus that occurs prior to the behavior. Thus, a postcedent is an event that follows a behavior. *See also* **consequence of behavior; three-term contingency.**

praise, ignoring, approval, and disapproval. Behavior emitted by parents and teachers that function as reinforcers or punishers of behavior. The appropriate use of these operations and the avoidance of them in certain instances is a key repertoire in teaching and parenting. *See also* **contingencies of reinforcement and punishment (three-term contingences).**

Precision Teaching. Precision Teaching, founded by Ogden Lindsley with significant contributions by Pennypacker and others, is an application of the science of behavior to teaching. Its contributions to the science of pedagogy are numerous, but were particularly critical with regard to measurement. Precision Teaching puts the measurement practices from the science of behavior into teachers' hands. It emphasizes rate of responding and the standardization of visual displays. Teachers and students use short timing periods (e.g., 1 min) to evaluate the student's mastery and "fluency" with tool or component skills of more complex repertoires. During the timings, the student controls learn units (i.e., learner-controlled responding). Precision Teaching also translates the basic epistemological bases of the absence of behavior into several succinct and user-friendly statements. The *Journal of Precision Teaching* is composed of numerous school-based experiments by teachers. Teachers serve as scientist contributors to the science and technology, like PSI, DI, Programmed Instruction, and CABAS. The technologies and strategies are devoted to academic instruction. Advocates for the use of the standard celeration chart point out the usefulness of a standardized visual display. Critics point out that the differences between lines are of a rank order nature; the behavior of interest is not the center of the display, preinstructional performance is not represented, and the measurement does not include the teaching process per se. The components of Precision Teaching are particular useful to teachers who are not trained extensively in teaching as behavior analysis, because it provides a user-friendly approach to measurement.

prediction. The objective of much scientific activity. Research into the functional relationship between events leads to increased accuracy in predicting the relationship between events. Thus, there is predictability concerning the relationship between certain teacher behavior and student behaviors. The degree of predictability allows the practitioner to plan certain interventions in advance that function to effectively result in the intent of the plans. That is, if a teacher carries out a particular operation that has a history of predictable results, certain effects on students are likely to occur.

preferred activities, events, or objects. Those that students engage in or consume on a free operant basis. They frequently, but not always, predict reinforcers for those individuals, but are subject to relativity. *See also* **Premack principle and operation.**

Premack principle and operation. (1) *Premack principle.* Responses or response classes that are occasioned or maintained as the result of the occurrence of momentarily more preferred activities following momentarily less preferred activities. What were less preferred activities can also come to reinforce more preferred activities under certain conditions. That is, the deprivation of a less preferred activity and the satiation of a more preferred activity can reverse the relationship. However, across expanses of time, a hierarchical relationship can be found for the ranking of reinforcer behavior relationships. This is referred to as a hierarchy of reinforcement value. Reinforcement value or the reinforcement effect of consequences is relative to their use (e.g., deprivation and satiation). Setting events or motivational variables act to affect more preferred and less preferred relationships. (2) *Premack operations.* The use of operations designed to bring about or increase the occurrence of low-frequency responding by the contingent delivery of access to operant lovers of high-frequency responding. Using the opportunity to read or watch television (short periods of each) as a consequence for performing mathematical operations is one such example. Delivery of sips of chocolate milk following consumption of spinach, for example, is a common example of the application of what is referred to as the **Premack principle.** The principle is related to motivational tactics such as those described under the rubric of establishing operations. Studies have demonstrated that listening to low-preference music by students (e.g., classical music) can function to reinforce even lesser preferred activities (e.g., math responses). The activity need be only slightly more preferred to function as reinforcement given certain motivational settings.

prerequisite repertoires. Whether or not a student has the necessary stimulus control (antecedent–response–consequence together) in his or her repertoire to emit correct responses to new instructional programs. The term *instructional history* is used synonymously. One of the basic strategies of the science of pedagogy is that a student's difficulty with a discrimination, concept, or operation can in some cases be traced to the absence of a prerequisite repertoire. Once that has been done, the repertoire may be taught and the student returned more successfully to the new target curriculum. See chapters 4–6 for an extensive treatment.

prescientific characterizations. Both laypersons and educators characterize the teaching process, and the performance of students, typically in prescientific terms. *Prescientific,* in this case, refers to characterizations of behavior that predate the science of behavior and the epistemology of behavioral selection. Terms such as personality, locus of control, or emotional stability are often treated as if they have scientific precision (e.g., the ability to predict). However, such terms can be replaced with layperson terms and they have no predictive power. Scientific characterizations, on the other hand, precisely describe behavior, provide sources for tracing the problems scientifically, and lead to successful solutions or avenues of research. See chapters 1, 2, 6, and 11 for clarification.

principles of behavior. The principles of the science of behavior refer to statements that summarize and characterize environment/behavior relationships for a large number of species. They are statements of the substantive findings of the science at any given time. Like all scientific ''laws,'' they are subject to change as further research leads to new perspectives on the subject matter of the science. *See also* **positive reinforcement principles and operations/tactics.**

private instruction. A term for instruction done on a one-to-one basis—one teacher and one student. The value of such instruction is associated with the probability that such instruction will be individualized in terms of pedagogy and, to a lesser extent, curriculum. However, the degree of individualization actually is determined by the sophistication of the teacher in individualizing instruction.

probes. A session of trials designed to test antecedent, consequent, or setting ovens control of behavior in a setting unlike the one being used currently for instruction or an experimental phase. Probes are also pretraining sessions that test the occurrence of behaviors that are targeted for instruction. Generalization probes test for generalized stimulus control and maintenance. Probes also test for the maintenance of previously taught objectives. They are useful to analyze the locus of instructional problems in setting events, instructional histories, or varied antecedent presentations. See chapter 6 for a detailed discussion of probe operations as a key strategy in the search for the locus of instructional difficulties.

prognosis. The outcome of a course of activities. Traditionally, it is used in the field of medicine to refer to the outcome of a disease or treatment. Herein, it is used to encompass the outcome of curricular and pedagogical treatment.

programmatic research efforts. Research programs are thematic research efforts involving numerous experiments often over long periods of time. Greer in 1999 provided a summary of a research program devoted to the analysis of the total CABAS package and to the contribution of various components. Programmatic research may also be associated with issues such as conditional discrimination, the matching law, reinforcement conditioning, stimulus equivalence, or the establishing operation. Thus, the program may concern applied issues such as effective schooling or the basic principles in applied or laboratory settings. Although the programs are directed toward particular problems, the particular direction taken in the research is driven often by the findings of ongoing research.

programmed instruction. Programmed instruction is a term used generally throughout this text to describe the sequencing of curricular presentations and operations designed to teach new behavior/environment relations. It is used generically to describe intentional teacher presentations of curriculum, whether they be scripted, automated, or improvised. Historically, the term referred to learned controlled responding to carefully designed instructional frames or learn units. The term automated instruction is used herein as a more specific description of the latter. See chapter 8 for an extensive discussion.

programmed textbooks. Programmed textbooks were initially developed to provide a cost effective simulation of teaching machines. *See also* **teaching machine; programmed instruction.** The extensive history is described by Vargas and Vargas in 1991. In 1967, Stephens in his review of educational research from the 1960s provided a summary of the comparative studies associated with programmed textbooks. An exemplary prototype of a programmed textbook is Holland and Skinner's (1961) *Analysis of Human Behavior.*

programming. A generic term that refers to the arrangement of instructional steps or learn unite designed to lead the student to perform a target objective or repertoire in certain contests. Programming is used in scripted or automated presentations of learn units. See chapters 6 and 10 for detailed discussions.

projective measurement. Measures that are said to be degrees or manifestations of hypothetical constructs such as personality, intelligence, aptitude, or deviance. Such measures are said to predict degrees of hypothetical constructs that are not directly observable such as locus of control, introversion, giftedness, learning disability, or emotional disturbance. *See also* **vaganotic measurement; scaling for measurement; rating scales; explanatory fictions.**

prompting or cueing tactics. Teaching operations that are designed to evoke attention to a stimulus or stimulus prompt (i.e., extrastimulus or intrastimulus prompts) or to evoke a correct response (i.e., response prompt). They are operations engendered to shape stimulus control and correct responding.

prompts. Designed to serve as stimulus discriminatives to evoke (a) attention to a target stimulus or (b) a correct response to a learn unit presentation. These teacher operations or automated instruction operations are designed to decrease errors. They are faded or vanished consistently with the accrual of stimulus control by the target antecedent resulting in a new and unprompted three-term contingency for the student.

prosthetic reinforcers. Ones that reinforce a behavior but they are not the natural consequences of the behavior. Thus, incentives that serve to reinforce responding are examples of prosthetic reinforcers. Other examples include food/liquid reinforcement, tokens, praise, measurement, or other consequences that are not the natural result of a behavior. Prosthetic reinforcers are necessary when the natural consequences of responding are not reinforcing. One step toward natural control of natural reinforcers typically calls for the use of prosthetic reinforcers initially. Prosthetic reinforcers may be either conditioned or unconditioned reinforcers. The development of a target three-term contingency may include the three-term contingency under prosthetic reinforcement control. Of course, this later stage is an intermediate objective leading to the acquisition of the three-term contingency under natural reinforcement control. *See also* **criterion-referenced objectives of instruction; learn units; natural conditioned reinforcer; inventory of student repertoires; generalized reinforcers.**

PSI modules for parents. Similar to teacher PSI modules with the exclusion of the verbally mediated component of each module. The latter is not typically included in the parent modules. The criterion for each parent module is also less stringent than those set for teachers and more closely approximates modules for teacher assistants. See chapter 8 for an elaboration.

punishment. The behavioral principle of punishment refers to consequences that immediately follow a behavior which act to decrease the probability of that behavior occurring in the future. Stimuli that are added to the environment following the behavior, such as a verbal reprimand, are referred to as positive punishment or type I punishment. When the removal of stimuli, such as the removal of a reinforcer for a specific behavior, acts to decrease the behavior the effect is referred to as negative punishment or type II punishment. *See also* **negative punishment or type II punishment operations.**

R

radical behaviorism. *See also* **behavioral selection; thoroughgoing behaviorism.** *Radical behaviorism* is the term used by Skinner (1974) to describe and differentiate the philosophy of the science of behavior from the science per se. It is the epistemology of the scientific practices of the experimental and applied branches of the science. Issues associated with radical behaviorism are discussed in journals such as *Behaviorism, The Behavior Analyst,* and *Behaviorology.* Jack Michael used the term thoroughgoing behaviorism to describe the same perspective. Vargas suggested the term behavioral selection.

rapid presentation of learn units and differential reinforcement. The use of the rapid presentation of learn units is supported by the literature. Rapid presentations are important to correct incorrect responses for which no reinforcement or a correction results with responses that are reinforced, thereby facilitating the known effects of differential reinforcement for teaching discriminative stimuli in concepts or operations.

rate (number per unit of time). Rate is a primary form of summarizing the frequency of behavior in time and number per minute is a frequent rate measure. Rates of responding also include other time standards such as the hour, day, week, month, or year. The standard celeration chart (available from Precision Teaching) is a standard visual display for reporting rate. Both correct and incorrect responding are reported since the two may be independent and both must be known. Teaching performance rate/accuracy observations (see chapter 7) also allow the determination of accurate and inaccurate rates of teacher behavior. *See also* **rate of responding.**

rate of responding. The primary datum of the science of behavior and is still the preferred measure. Rate refers to the frequency of behavior in a standard unit of time (e.g., number per minute, number per hour, number per day, number per week). The most valid measure of responding in school settings includes both the rate of correct responding and the rate of incorrect responding. The use of rate does not require constant time periods or constant opportunities to respond. Rate is affected neither by the vagaries of percentage nor by the fluctuation of time. The Precision Teaching model of schooling uses a standard celeration chart (semilogarithmic graph) to represent rate across time periods such that the celeration of learning is presented in a standardized format.

rate of response. *See* **rate of responding.**

rating scales. A procedure associated with hypothetico-deductive scientific practices wherein a behavior is rated rather than counted or transduced. A conspicuous use of rating scales is found in college course ratings of professors. In such cases, the students rate certain characteristics of the professor and course on scales from 1 to 3, 5, or 7. The actual rating is an inference rather than a count of a behavior or behavior outcome. Rating scales are ordinal levels of measurement. *See also* **vaganotic measurement; projective measurement; hypothetico-deductive statistical procedures; absolute units of measurement; scaling for measurement; ordinal units of measurement.**

ratio units of measurement. Measures that achieve equal intervals between counts. Unlike ordinal rankings the difference between 10 and 11 and 10 and 20 is the same. They do not necessarily represent direct transduction of deflection of space in time. An intelligence test score is an example of ratio measurement. They are measures that are scaled and are of equal intervals (e.g., the difference between each unit is equal). Unlike absolute units, ratio unite do not have an absolute zero.

recycle. *Recycle* is a term that originated in courses and in-service modules to refer to the tutoring and self-study operations required to assist students to master an examination or performance objective in those instances in which they do not achieve criterion the first time they are assessed. Thus, the process of teaching, or having students teach themselves through self-study until they learn something to which they previously responded incorrectly, is referred to by the term *recycle*.

reinforcement of correction responses. The reinforcement of the corrected responses of students to correction operations (i.e., the student repeats the corrected response) is done when any kind of responding is a weak for a specific student. In instruction in which reinforcement and the successive approximation of instructional objectives one typically avoids reinforcing correction responses by the student. However, if the effects of the contrast of differential reinforcement are in effect (that is, in most adequately presented instruction), correction responses are not reinforced in order to avoid the elimination of the differential reinforcement effect for the student. *See also* **learn units; corrections; schedules of reinforcement.**

reinforcer and reinforcing operations. *See* **positive reinforcement and negative reinforcement principles and operations.**

reliability. The stability of behavior under constant conditions. It is also used in reference to the reliability of effects in which the effects of an independent variable is consistent. In addition, reliability is used in conjunction with the reliability of treatment. In the latter case, it refers to consistency in the application of the independent variable in an experiment or the accurate implementation of a tactic from research in an applied setting. Often reliability is used to refer to interobserver agreement, although this is thought to be an imprecise usage by some behavior analysts. In test construction, the term refers to whether repeated tests of a subject's performance on a test are consistent when there is no environmental intervention. In the latter case the reliability of a test refers to an index of agreement, usually expressed in terms of correlations. *See also* **interobserver agreement.**

repertoires of behavior. The term *repertoire* refers to a class or category of antecedent–behavior–consequence relationships. The range of behaviors associated with reading, writing, speaking, dancing, or managing one's time are respective repertoires. The repertoires of individuals are behaviors that were learned in the individual's history. Given the learned setting event and antecedent, the student will perform the behavior. Repertoires, unlike memory, exist in the learned environment behavior relationship. A concert pianist has certain pieces in his or her repertoire; that is, he or she can play them on demand. The same is true of the behavioral scientist view of the repertoires of individuals. That is, if the situation demands it, so to speak, the individual can perform the necessary repertoire. The repertoires that are needed by students can be assessed in the naturalistic setting in order to identify those repertoires that the student does or does not have. In this, the assessment is referred to as an inventory of repertoires. It serves as the means of determining each student's assets and deficits per specific repertoires. Kindergartners may be observed under specific contingencies of reinforcement and setting events associated with behaving in school settings to determine whether or not they have sufficient "school survival" repertoires. That is, can they follow classroom rules, raise their hand for attention, walk in line without talking, and follow teacher directions under prosthetic or natural reinforcement conditions? The cluster of behaviors associated with "school survival" category is described as a repertoire in that, given a classroom setting, the student performs the behaviors that are consistent with the rules of behavior of the classroom and school. The behaviors are said to be in the student's repertoire when they perform them under the specified conditions. See the description of "repertoires" in chapter 9. *See also* **inventory of student repertoires; portfolio evaluation.**

replication of experimental procedures. When experiments are duplicated with other subjects or within subjects, or between settings, the experiment is termed a replication. It is only through replication that one can test the generality and validity of research findings. A direct replication is one that reproduces the same conditions with a new subject or setting, while a systematic replication varies from the original procedures. Replications may be done by the original experimenter and research group or by other scientists. *See also* **validity of research findings.**

reprimands (e.g., disapproval). Disapproving comments or statements usually used to decrease occurrences of the behavior that were the target of the reprimand. The use of reprimands is discouraged for several reasons: (1) they have the potential to evoke and reinforce disapprovals and can occasion coercive traps between students and teachers or parents and children. (2) They function to detract from conditioning the subject matter taught as a conditioned reinforcer for the student's behavior. They serve to occasion the student to avoid contact with that which is taught and the source of the reprimand. (3) They do not function to teach new behavior or maintain useful behavior. (4) They are unnecessary in settings in which the science of pedagogy is applied in adequate fashions.

research posters. Abbreviated summaries of research designed to be presented to groups of people who peruse the poster for only a few moments. Poster sessions at scientific conventions allow the presentation of numerous research papers in abbreviated form. The audience circulates among many posters reading the abstracts and viewing visual displays. The author is available for discussion and more detailed written papers are available.

respondent behavior. Behaviors that are unlearned. They are present at birth or develop according to the evolution of the species. Certain stimuli (i.e., loud noises) elicit respondents even though the individual has no instructional history with the stimuli. Antecedent stimuli elicit respondents. Conditioned antecedent stimuli that are paired with unconditioned antecedent stimuli (i.e., loud noise with light) can be conditioned such that the conditioned stimuli (i.e., the light) alone will elicit the respondent. Escape or fear responses to certain stimuli are frequently traceable to respondent conditioning experiences. An applied operation that is designed to eliminate a conditioned respondent relationship is termed desensitization.

response. A specific incident of a class of responses. For example, the *mand operant* refers to a class of verbal behavior that specifies reinforcers, and a *mand response* is a particular incidence of the class (e.g., "cookie, please"). *See also* **rate of responding; number per unit of time; operant behavior.**

response classes. A class of responses is defined by its common postcedent effect. Numerous responses beneath and outside the skin are involved, for example, in reading, writing, or speaking. One can define the various topographies of behavior and they may differ greatly, but when they have the same consequent effect, they belong to the same class of responses. For example, one may mand for an item using signs or vocal verbal behavior; the mand may involve numerous words (e.g., "I beseech thee kind lady for a glass of water") or a single word (e.g., "water"). *See also* **response.**

response delay. A teaching tactic in which the response of the student is delayed in order to increase the probability of a correct response. It is done for students who respond to learn unit presentations prematurely and who seem to be responding before they give careful attention to the stimulus. This procedure may be used to create instructional control for some students. It is used alone or in conjunction with stimulus delay. A protracted time is inserted between the stimulus presentation and the intraresponse period; that is, the student is not allowed to respond for a period of time in order to interrupt the immediacy of responding. Students who respond before attending to the relevant stimuli or who respond "without reflection" are candidates for this tactic. Related procedures are constant time delay and progressive time delay.

response generalization. Variations in responses or a response class associated with a previously conditioned antecedent–response–consequence relationship. Variations of verbal behavior, vocal or textural, that are evoked by a previously conditioned stimulus are examples of response generalization.

response latency. A measure of operant behavior that is used when the time between responses or *latency* is the dimension of the behavior that is of interest. Latency between presentations of learn units is an example. See chapter 2 for a discussion of latency and learn units.

response prompts. Tactics designed to evoke a correct response or its successive approximation. They are distinguished from stimulus prompts in that the latter evoke attention to the stimulus or an attribute of a stimulus. Response prompts are done immediately prior to or during the intraresponse period by the teacher or teaching device. Examples include constant time delay prompts or progressive time delay prompts. In a constant time delay prompt, the correct response is given to the student with a zero-second delay initially; then a constant, i.e., 1–s, time delay is used. The effect is to prompt a response and to fade the prompt. Fading prompts is a critical component of using response prompts.

responsibility. In behavior selection and the context of this text, we use the term *responsibility* to refer to the collection of research-based tactics that should occasion the use of the science of behavior to seek solutions to instructional, managerial, and curriculum issues by those trained to do so. It is a repertoire of verbally governed behavior occasioned by an instructional history and a set of designed or adventitious contingencies of reinforcement and punishment.

reversal design and withdrawal design. The reversal design is fundamental design logic for implementing the method of difference. In this design, baseline procedures are reversed (e.g., baseline disapproval becomes treatment approval) for the experimental phase. Once repeated baseline measures show stability, the reversal or experimental condition is implemented until another steady-state occurs followed by a return to baseline conditions. The number of times that this cycle is replicated increases the validity of the design. Each replication that results in the same effect increases the believability of a functional relationship for the variable reversed. The withdrawal design functions similarly and they are often used interchangeably even though such use is imprecise. In the withdrawal design, a stable baseline is followed by a change in one variable as the experimental condition. Once stability is achieved under the experimental condition, the baseline condition is reinstated. Each occurrence of this cycle, A or baseline followed by B or treatment followed by baseline, increases the validity of the design. These within-subject or intersubject designs are the strongest tests of the method of disagreement. Between-subject designs, such as the multiple-baseline design or the experimental control group design, substitute between subject procedures for within-subject procedures. Thus, they are not as strong as reversals or withdrawals or counterbalanced within-subject group designs. In practice, intersubject and intrasubject designs are often mixed, thereby implementing the joint method. The validity of findings is enhanced when researchers incorporate features of both inter- and intrasubject logics. For example, one may use a design logic that combines a withdrawal–within-subjects feature with a multiple-baseline–across-subjects feature There are numerous design logics that can be used to implement the strategies of Mill's canons of science.

rotated-interval according. *See* **interval recording.**

rule-governed behavior. The term *rule-governed behavior* was the term used by Skinner in 1957 to describe behavior controlled by verbal rather than nonverbal contingencies. Skinner was increasingly disenchanted with the term because of its unrelated use in cognitive research, wherein the term *rule-governed* was used to invoke intervening variables as the "source" of some types of behavior. A cognitive interpretation would posit that an individual may behave in a certain way because of a kind of cognitive rule, while a behavior selection view would point to the relationship between observed verbal stimuli and observed behavior that corresponds to the stimulus. In both interpretations individuals engage in complex behavior such as problem solving. Ernest Vargas in 1991 suggested the term verbally governed as a replacement for rule governed behavior because the former more accurately describe the relationship that was consistent with the way Skinner described the particular functional relationship. *See also* **verbally mediated repertoire.**

S

sameness generalization. The capacity of an attribute (or attributes) of a stimulus (or stimulus congregate) to evoke the same response under various settings or across time. Examples of sameness of generalized stimulus control include identifying animals as mammals, female gender of *Homo sapiens,* sonata allegro musical form across two musical style periods, or "squareness" under deformation of shape, size, texture, and dimension and colors under deformation of form, shape, size, and texture. One of the first repertoires that is typically taught to very young children is matching "same with same." It appears to be a prerequisite repertoire for most instruction devoted to discrimination or differentiation.

scaling for measurement. Measures that scale behavior include intelligence tests, achievement tests, and other tests that impose a standard scale on behavior. It is also a way to create a class of behaviors when there may or may not be a natural fracture that groups those behaviors into a class. Differences between points on the scale are said to be represent equal differences; thus, they meet the criteria of consistent interval measurement. However, the measure need not have an absolute or real zero. A conspicuous example is the process of converting numbers of correct responses into percentages whereby 90% could be 9, 90, or 900 correct responses. The thing measured may be a construct such as personality whereby a personality scale is constructed based on a nomological network said to demonstrate the existence of the nonobserved hypothetical construct. The use of such psychological constructs is traceable to "Hullian" behaviorism. Hullian behaviorism is a particular methodological behaviorism that promotes the usefulness of psychological constructs as intervening variables between the organism's behavior and its environment. In order to measure these constructs methodological behaviorists need to scale certain behaviors. Physiological psychologists also scale stimuli, as in the case of substituting measure of cents for Hertz; however, in this latter case the stimuli are scaled rather than the behavior.

schedules of reinforcement. Reinforcement effects are determined in part by their schedule of delivery. Ratio schedules of reinforcement refer to the number of responses that occur prior to reinforcement, while interval schedules refer to the reinforcement of a response after periods of time. Ratio or interval schedules may be fixed or intermittent. Fixed schedules require a constant time or frequency requirement (e.g., each response after 5 s or FI 5 s or every fifth response FR 5). Intermittent schedules require variation in the delivery of reinforcement (e.g., variable 5-s intervals result in reinforcement or VI 5 seconds or variable emissions of five responses result in reinforcement VR 5). Schedules of reinforcement may be combined with one or more behaviors; schedules for two or a sequence of behaviors are referred to as compound schedules.

scientific characterization. The expression *scientific characterization* is used in this book to describe the manner in which the vocabulary of a science characterizes or describes pedagogical operations or the responses of individuals to their environment. The principles, operations, and method of science developed for the science of behavior and its extension to a science of pedagogy consists of a verbal community. The vocabulary of this community allows a writer or speaker to describe setting events, behaviors, and consequences that are the purview of teaching or research in precise term. Others who know the vocabulary can be apprised about what has been observed even when they were not present. The vocabulary allows one to characterize teaching and learning such that successful practices can be adopted and unsuccessful practices avoided. Scientific terms summarize, prescribe, and depict phenomena (in this case teaching and learning) in ways that lead to reliable implementation of tested interventions. In more advanced sciences the phenomena of the science are often characterized mathematically.

scripted curriculum. One that specifies the behaviors of the teacher vis-à-vis learn unit presentations as well as the objectives of the student. The curriculum is programmed in the generic sense as described in chapter 8. *See also* **programmed instruction; programming; automated programmed instruction.**

selection by consequence. Either the selection of the species by environmental consequence or the selection of responses or repertoires as a result of the consequences of behavior (e.g., reinforcement or punishment). *See also* **behavioral selection.** For example, one may characterize certain behaviors as choice; however, a selection by consequence interpretation of choice would invoke the "choosing" or selecting effect of the consequence of the individual rather than the individual as an organism that is behaving independent of its environment. However, some methodological behaviorists misinterpret the role of the selection of behavior by consequences to be one-sided. The choice that the organism or the environment makes, according to behavior selection, is a function of the immediate environment and the individual's history of instruction/experience with that environment. In effect, both the organism and its environment are "operating" on each other. However, the organism and its controlling environment are conglomerations of phylogenic and ontogenic contingencies. A more cognitive interpretation would suggest that the choice is cognitively controlled.

self-discipline. Self-discipline as viewed from a behavioral selection perspective refers to repertoires that come under the control of rule-governed or verbally mediated contingencies. Verbal contingencies displace "short-term reinforcers" that interfere with the attainment of "long-term reinforcers." Abstract generalized reinforcers, such as self-monitoring or tacts of one's own progress, also play a critical role along with verbal behavior and in fact verbal behavior is an abstract but observable control. Self-discipline is taught by progressively extending the ratio of accomplishment to reinforcement, teaching the components of self-monitoring (e.g., time and response self-recording) as reinforcers, and teaching self-editing functions in which one functions as a reader to reinforce one's own writing or problem-solving repertoires. See chapters 6–8 for detailed explanations.

self-editing. Skinner in 1957 used the term "self-editing" to describe the process whereby one functions as a critical reader of one's own written or spoken verbal behavior. The process is exemplified as a repertoire of curriculum in chapters 6–8. *See also* **verbal behavior; writer as own reader.**

self-injurious behavior. The expression self-injurious behavior is often abbreviated as SIB. It refers to behavior performed by an individual that results in injury to the individual performing the behavior. Such behaviors or response classes are done on a repeated basis. That is, when someone accidentally injures him- or herself, it is not an incidence of SIB. However, repeated and frequent occurrences of the response class by a single individual are said to be self-injurious. Such behaviors include, but are not limited to, eye gouging, head banging, scratching, body dropping, self-hitting, or overconsumption (e.g., aerophagia, bulimia). *See also* **behavioral deficits; reinforcement of correction responses; maladaptive or nonadaptive behavior; bad behavior.**

self-management *or* self-regulation. The coordinated use of self-monitoring and self-reinforcement to realize short-term and long-term habilitative repertoires targeted in advance by the individual who uses them. One might, for example, set a goal of running a marathon. The individual arranges a monitoring, shaping, and reinforcement system to lead to the goal. During the process of striving for the goal, measurement procedure will continue and reinforcement operations will be modified until success is achieved. The repertoire is, at least initially, one that is verbally governed.

self-measurement *or* self-monitoring. The counting of behaviors, response classes, or their products by the individual engaging in the behaviors. All of these are forms of verbal behavior and verbal stimuli that control nonverbal as well as other verbal behavior. It is seen as an important component of self-management, self-discipline, and self-determination. Self-monitoring is a critical component of organizational man-agement and the use of the strategies of the science of behavior (e.g., verbally mediated teaching repertoires) by teachers who function as strategic scientists of instruction. The relationship between self-monitoring and the development of expert independence by students is described in chapters 6–9. Cognitive behaviorists, those who adhere to a kind of methodological behaviorism, see *self-management* as cause for invoking interven-ing variable controls of behavior (e.g., cognition). They see the incidence of self-management as evidence of cognitive control, whereas behavior selection sees *self-management* as a form of behavior under the control initially of verbal behavior, and as such its occurrence does not call for an intervening variable explanation since verbal behavior and its control are simply incidences of behavior and environmental contin-gencies of which verbal stimuli are a part. The view that the process is observable does not discount the occurrence of neurological physiological mechanism beneath the skin; however, such behaviors beneath the skin are in the purview of other natural sciences. *See also* **self-reinforcement; self-management.**

self-motivation. From a behavioral selectionist perspective, *self-motivation* refers to the planned arrangement of deprivation or satiation conditions by an individual that function to increase the reinforcement value of certain activities and decrease the reinforcement value of other activities in the interest of achieving a goal or goals. If one arranges situations that increase the probability of certain events and minimizes the probability of other events, one can affect the reinforcers and hence in the behaviors in which one engages.

self-reinforcement. The arrangement of context, antecedent–behavior–consequence relationship for one's own goals. Often the process depends on the development of newly conditioned reinforcers for the individual. The development of self-reinforcement repertoires is discussed at length in chapters 6–8.

sequencing learn units. Learn units should be arranged in sequences that are char-acterized as good programmed instruction. Such good programming uses successive approximation to progressively increase the student's progress toward short-term or long-term goals. One may also sequence learn units across different instructional programs or within the same program. *See also* **massed trials.** See Chapter 2 for a detailed treatment.

setting events. Events that affect the antecedent–behavior–consequence relation-ship. Often, they are motivational in nature in that they affect the reinforcement or punishment value of the consequence in a subsequent operant or three-term contin-gency. Disputes between individuals, sleep deprivation, prior activity, and deprivation of activity, to name a few examples, affect the control of the consequence in the three-term contingency. Setting events are not discriminative stimuli, although discriminative or eliciting stimuli are associated with them, but rather they are events. Planned interventions that use or ameliorate the effect of setting events are tactics that are used as establishing operations. Setting events and setting stimuli together constitute the context in which instruction takes place. *See also* **setting stimuli; motivation and the science of behavior; context; establishing operations.**

setting stimuli. Those stimuli that affect the nucleus operant. Certain verbal instructions, for example, change antecedent–behavior–consequence relationships. The emergences of new and untaught stimulus equivalents that result from stimulus equivalence training procedures are still another example. Setting stimuli are critical factors in the instructional process and are associated with setting events. *See also* **motivation and the science of behavior; discriminative stimulus; setting events.**

simultaneous treatment designs. A term used to describe tactics for implementing the method of difference by substituting an intersubject or between-subjects comparison with a within-subject comparison. It is a single-subject design that is comparable to the experimental control group design. That is, each individual receives a different condition but simultaneously. Student 1 receives baseline conditions while Subject 2 receives the experimental condition. Often these conditions are rotated in a changeover phase. The design has obvious weaknesses (all designs have some), but when the tactic is used with others (e.g., combined designs, or a counter balancing logic), the logic of the design may increase the overall validity of the findings for the question under investigation.

single-case experimental design. The expression *single-case experimental design* is used to refer to methods of using the experimental method to analyze the behavior of individual organisms as opposed to group experimental design that are used to analyze behavior distributions of groups. *See also* **reversal design and withdrawal design; multiple baseline design; experimental method.**

slope. The angle of a trend in the analysis of visual displays of data in which the determination of the steepness of the slope indicates a dimension of the effects of an intervention or independent variable. Thus, ascending or descending trends in data are described as having steep, gradual, variable, or stable trends.

social conversation units. Verbal behavior episodes between two or more individuals in which both or all parties complete a three-term verbal contingency and the reinforcer is the verbal behavior of another. See chapter 9 for elaboration. *See also* **academic conversational units; learn units.**

social reinforcement (peer and teacher). *See* **social reinforcement operations.**

social reinforcement operations. *See also* **praise, ignoring, approval, and disapproval.** Social reinforcement operations refer to the use of praise, approval, or attention in order to achieve positive reinforcement effects. Teachers or peer students may perform these operations are intentionally or unintentionally (e.g., designed or accidental). Social reinforcement is a class of conditioned reinforcers, but it is a natural reinforcer; that is, the reinforcement effect is acquires early on by pairings with unconditioned or conditioned reinforcers. Some students, however, need to undergo a special instructional history before praise functions to reinforce. The use of frequent contingent praise along with the dispensation of conditioned or unconditioned reinforcers, the avoidance of reprimands, and the avoidance of praise errors is a repertoire that teachers are typically trained in the classroom such that it has the attributes of contingency-shaped teaching behavior. See chapters 4–6 and 9.

social skills instruction. Curricula and pedagogy designed to teach students social habilitative repertoires. Such repertoires include those that engender more positive reinforcement from others and fewer punitive contingencies from others. There are research-based curricula by Hill-Walker, Charles Greenwood, and others that are designed to teach social repertoires. They may be used in combination with procedures developed by Chu in 1998 to evoke appropriate social behavior and replace inappropriate social behavior.

speaker behavior. The control of consequences whereby the speaker substitutes verbal behavior for nonverbal behavior. The speaker affects his or her reinforcement and punishment via the use of verbal behavior. The role of verbal behavior as a conception of the communicative use of language revolves around the perspective of the speaker. The speaker mands, tacts, and uses intraverbal or autoclitics and in general affects the behavior of a listener. The listener's behavior, in turn, reinforces the verbal behavior of the speaker. *See also* **verbal behavior; academic conversational units; learn units; mand; tact; autoclitics.**

special education. A branch of the educational profession concerned with the instruction of children with disabilities or exceptional "gifts." Special education provides expertise regarding handicapping conditions and ethical and legal issues associated with handicapping conditions or exceptionalities and has contributed to the development of expertise in the science of pedagogy. See the discussion in chapter 1 regarding the wide-scale use of behavioral tactics of pedagogy in special education.

split-plot designs. A type of statistical designs that partition different independent variables according to subgroups. The term originates from Fischer's use of statistics in agricultural research where plots of land were associated with agricultural interventions.

staff turnover. The expression *staff turnover* is a term from management or administration that refers to the ratio of staff members who leave after a short period of time (usually 1 year in schools) to total staff population. A school with 10 teachers that has a staff turnover ratio of 2 to 10 is one in which two teachers left or were terminated and 8 teachers were retained.

standard or celeration chart. A semilogarithmic visual display that standardizes the display of graphically presented data. The learning curve appears as a straight rather than curved line. Responses of various types and data from any rate experiments can be compared directly on a coloration chart. There are good arguments for its use throughout the science. Standard charts, however, are not readily readable by the untrained observer, and some find the behavior of interest to be difficult to visually inspect because the display include a wide range of possibilities that are included in the standard chart. The scale is multiplicative rather than additive.

statistical significance level. The level of "confidence" that the differences between two means in subgroups that belong to the same population differ from chance. A .05 significance level means that the obtained difference between two or more means would occur by chance in only 5 of 100 comparisons done with the samples from the same population. One could be 95% confident that the differences were real. The logic of its use in experiments with groups is that if all of the necessary experimental controls are in effect, then the degree to which such differences deviate from chance suggests that something of "significance" has occurred such that the difference in means can be attributed to the experimental intervention. Of course, to attribute the difference to the intervention, all of the attributes associated with Mill's method of disagreement or difference must have occurred. The use of statistics is a valuable tool when the objective of the science is the prediction of populations without specific concern for individuals in the population. *See also* **hypothetico-deductive statistical procedures.**

steps of programming. Individual frames in programmed instruction or to steps in a task analysis.

stimulus delay. A teaching operation whereby the student's access or view to sensory stimuli, targeted to gain discriminative stimulus control, is delayed until the student is attending to the location where the stimulus will appear. It is used when the student appears to be responding precipitously without attention to the stimulus or stimulus congregate. Response delay operations involve delaying the response while the stimulus is in full view.

stimulus equivalence. A phenomenon found in a body of research in which untaught relationships emerge as a function of the teaching of certain match to sample instructional procedures. Attempts to tie the phenomenon to explanations of new forms of untaught verbal behavior have led to several theories. The most promising of these is Relational Frame Therapy.

stimulus generalization principle and operations. (1) *Stimulus generalization principle.* Antecedent stimuli that have characteristics of a previously conditioned stimulus–response–consequence relationship have a higher probability of evoking a response. The presence of an unknown male, for example, may evoke the response "father" from a young child. Different type settings evoke a common response or different styles of script writing evoke a common reading response. The degree to which salient or distinguishing characteristics are conditioned to evoke responding determines the generalizability of antecedent stimuli. The term generality refers to a separate phenomenon and it is inaccurate to use generality as a synonym for generalization. (2) *Stimulus generalization operations.*

stimulus prompts. Cues or hints that are provided by a teacher, tutor, or teaching device to increase the probability that a student is attending to the target antecedent. *See also* **target-controlling antecedent; intrastimulus prompts; extrastimulus prompts.**

strategies of behavior analysis and pedagogy. An expression that refers to the guiding scientific findings or practices (e.g., method of disagreement) or the epistemology of the science (e.g., behavior selection) that guide the efforts of scientists or teachers who function as strategic scientists of behavior. Strategies are verbally mediated practices that determine the possible source or locus of variability in behavior, learning problems, or instructional difficulties. The particular questions that are likely to result in answers are still other incidents of strategies. They include questions such as are the sources of the instructional problem in the history, setting events, antecedent, type of response, or consequence? The resulting ways in which the questions suggest possible answers (e.g., experimental analysis, reference to the existing literature) are still other incidents of strategic analyses. The strategic control of the quest is a form of verbally governed behavior.

strategic science of instruction. *See also* **strategies of behavior analysis and pedagogy.** The strategic science of instruction or pedagogy refers to the science of teaching based on the science of behavior (e.g., applied and experimental analyses of behavior). Like the science of behavior, the science of instruction uses strategies based on the epistemology of behavior selection and the practices of the science of behavior not only to uncover new principles, but also in the delivery of ongoing instruction on a moment-to-moment basis. Strategies refer to the kinds of questions to be asked about instruction and how to go about finding the answers. Specific tactics from prior research are introduced as a result of strategic analyses. A strategic analysis of a student's difficulties may locate the problem as, for example, one of motivation. An establishing operation or tactic from prior research may be suggested by the strategic analysis, for example. See chapter 2 for a thorough discussion. *See also* **tactics of behavior analysis pedagogy.**

strategic scientist of behavior. One who bases his or her experimental practices on behavior analytic operations and the epistemology of behavior selection. The choice of measures and sources for the potential controlling variables are guided by the practices of the science of behavior described in detail in chapter 4.

structural curricula. The term *structural curricula* is used throughout the book to indicate the traditional subject matter of schools—mathematics, history, geography, and biology. The contents of these subjects are based frequently on structural similarities and they do not describe functional repertoires. When one learns to use the content and practice of a discipline to affect the behavior of others or one's own behavior, one is learning function; however, the content itself is a structural or topographical classification. *See also* **functional curricula.**

structuralist analysis of language. A structural analysis of language is a view and analysis of language in terms of the structure, commonalities, and differences within and across languages. Dictionaries are repositories of some aspects of the structural or descriptive analysis of language. A structural analysis identifies how language is used or the community of usage. It is distinguished from verbal behavior, an analysis of communicative behavior developed by Skinner in 1957 in his book *Verbal Behavior*. Verbal behavior is the study of the communicative function of verbal behavior. It is the extension of the science of behavior from nonverbal to verbal domains. A structural analysis provides a description of language and its standardization in a community whereas a functional analysis of communicative behavior describes the function at the time of usage by a speaker and listener or listeners.

student products (measurement of products). Permanent or relatively permanent products of student behaviors that are measured to test the effectiveness of the student behaviors and the related stimulus control. Examples include written products or other construction outcomes produced by students.

successive approximation principles and operations. (1) *successive approximation principle*. Responses or response classes are acquired by reinforcement of responses that approximate or lead to the emission of that behavior–consequence relation sought. The process leads successively to reinforcement only of responses that progress toward the target behavior. (2) *successive approximation operation*. Teachers or programs of instruction deliver successive approximation operations to occasion new antecedent–behavior–consequence relations. These operations incorporate tactics that function to bring about a target behavior by reinforcing the progressive approximation of subobjectives or terminal objectives. The tactics associated with the principle are often referred to as shaping tactics. The accomplished use of successive approximations can lead to a student acquiring a new response/reinforcer relationship without error. The use of prompts or cues or modeling is commonly incorporated in the process. *See also* **prompts; schedules of reinforcement; task analysis; criterion-referenced objectives of instruction.**

supervision repertoires. Those repertoires of a principal or supervising teacher that directly involve pedagogy and curriculum related tasks as distinguished from administrative repertoires. Supervisor repertoires include tasks such as observing teachers; training teachers *in situ;* developing teacher modules (e.g., reading material, quizzes, performance criteria); tutoring teachers; reviewing pupil or teacher data for intervention purposes; summarizing and analyzing individual, classroom, and total school data; and providing verbally mediated strategies, to name only a few tasks. See chapter 9 for a detailed description. *See also* **management by behavior analysis; supervisor rate.**

supervisor modules. Developed for and by supervisors in CABAS schools to arrange their in-service instruction. As in the case of teacher modules, supervisors have modules that are arranged according to areas of pedagogical and scientific expertise. Supervisors who have already completed many teacher modules often target goals for the school or for individual or groups of students or teachers such as a research project or the development of a new curricular thrust. Chapter 9 includes an in-depth description of supervisors' modules. *See also* **teacher modules; Personalized System of Instruction (PSI); participatory management; Comprehensive Application of Behavior Analysis to Schooling (CABAS).**

supervisor rate. The number of supervisor tasks completed per hour for given period of time (i.e., daily, weekly, monthly, yearly). The measure is used in the CABAS model of schooling. The tasks are predefined and criterion-referenced with backup products. Such tasks must be associated generically with student learning or the provision of necessary services for students. Supervisors log these tasks and these data are posted along with teacher data. Like the data for teachers, the supervisor data function to guide the behavior of supervisors according to the effects of that behavior on teachers and students. See chapter 9 for a thorough treatment of the research associated with this measure and the operations involved.

supportive or nonadversarial supervision. Used herein to describe the nature of supervisor and teacher relationships in CABAS schools or other schools devoted to the comprehensive and cybernetic use of the science of pedagogy and the learning of students to guide the schooling process. Supervisors are to serve as mentors or teachers for novice teachers and colleagues for senior teachers. Supervisors are accountable for the teaching repertoire of teachers just as teachers are for their pupils. Teachers are taught, assisted, and monitored continuously using positive reinforcement operations. Supervisors adhere to the tenets of participatory management and the performance of supervisors is on public view (graphs) just as the performance of the teachers and students. Teaching takes precedence and teachers are esteemed, rewarded, and publicly recognized contingent on their pedagogical and scientific expertise. Supervisors avoid coercive traps and engender mutually and positively reinforcing relationships between teachers and themselves. Consultants assist in monitoring and maintaining mutually reinforcing relationships. See chapter 9 for a description of the key operations for supervision.

systematic replication experiments. *See also* **direct replication experiments.** A systematic replication is an experiment that is similar to an original experiment except that one or more of the original variables is/are changed. The type of student, a component of the intervention, or setting may be varied from the original experiment.

T

tact. One of several categories of verbal operants. It may take numerous forms (e.g., vocal or gestural, one or many words or gestures), but it functions to make contact with the nonverbal environment and is reinforced by a generalized reinforcer. "Pure" tacts are under the control of nonverbal antecedents. In the latter cases, an event or condition evokes a form of verbal behavior that has been correlated with the event and a generalized reinforcer in the past. A preschooler sees the picture of an elephant and says "pachyderm" and a listening adult approves (e.g., "Wow, you're a bright girl"). *See also* **mand; autoclitics; verbal behavior.**

tactics of behavior analysis pedagogy. The operations, or terms for those operations, that specify procedures to implement strategies or principles of the science. Numerous tactics may exist to implement the strategy or principle. For example, tactics such as social reinforcement, token dispensation, graphic displays, access to preferred activities, or the interspersal of a known discrimination may function as tactics for reinforcement strategies. A series of probe tactics may serve to isolate a controlling instructional variable, in which case the tactic was an operation to realize a strategic analysis of controlling variables. Tactics are often derived from the applied literature or from the particular *in situ* scientific practices used to trace a source of a learning problem (e.g., reversal or multiple baseline designs). New tactics also sometimes derived *in situ* to serve as probable operations for engendering a strategy or principle. See the lists of tactics in chapters 5 and 6.

target-controlling antecedent. The expression *target stimulus control* refers to the stimulus or attributes of a complex stimulus congregate for which the instruction is designed to evoke a target response by the student. It is the antecedent stimulus that the teacher wishes to have the student respond in the operant that is being taught.

target student. The expression *target student* is used to differentiate the student for which an instructional intervention is applied or the particular student of interest. It is synonymous with the expression the student of interest.

task analysis. The process of breaking down components of a task such that the individual parts are identified and taught individually. These individual parts are then taught such that the entire task is eventually mastered. Short-term and long-term instructional goals refer to the same process whereby the mastery of short-term objectives lead to the mastery of the long-term objective. The process is tied to the principle of successive approximation.

teacher antecedents. The presentation of target stimuli by teachers or the presentation of antecedent stimuli to gain attention to the target stimulus or to occasion the correct response to target stimuli.

teacher-contingent consequation. An expression describing a situation in which the teacher has provided, delivered, or arranged for the student to receive either reinforcement, response correction, or planned ignoring as a direct consequence for a student. The teacher response was the correct one for the response; thus it was contingent. Had the teacher provided the wrong consequence (e.g., reinforced when she should have corrected), the teacher behavior would be described as noncontingent. The word consequation as a noun and the verb to consequate originated in the field of behavior analysis to describe the effects or operations of consequences on behavior without regard to punishing or reinforcing effects or to encompass both. *See also* **teacher performance rate/accuracy (TPRA) observation.**

teacher modules. In the CABAS model of schooling, teacher modules refer to the arrangement or division of instruction provided to teachers in the PSI approach to in-service training of teachers, supervisors, teacher assistants, and other school personnel. Each module contains three components: (a) a component devoted to in-class skills or contingency-shaped teaching behavior, (b) a component devoted to mastery of the vocabulary of the science or verbal behavior about the science, and (c) a component devoted to the use of the science to solve instructional problems or verbally mediated behavior. The components of the modules include performance criteria with students, performance criteria concerned with describing the science and using the science to describe student–teacher behavior, and using the science to solve teaching and supervisory problems. See chapters 3, 5, 6, and 9 for a thorough discussion.

teacher operations. Ones that are used to teach antecedent–behavior–consequence relationships to teachers. They include all of the operations associated with the learn unit from the teacher's perspective. *See also* **learn units; contingency-shaped teaching behavior; verbally mediated repertoire.**

teacher performance rate/accuracy (TPRA) observation. A protocol for performing direct observation of teachers that simultaneously incorporates the correct and incorrect responses of students as well as teachers. The procedures include collection of data on teacher's presentation of learn units in terms of rate and accuracy (e.g., accurate or inaccurate teaching) of teaching. TPRA data are visually displayed as (1) number per minute of correct and incorrect teacher responding or (2) an algebraic TPRA score, determined by subtracting incorrect performance from correct and dividing the latter by elapsed time. Student responses collected during the observation are displayed as rate correct and incorrect. The TPRA may be done with (a) an entire class of students and a teacher, (b) the teacher and a subgroup of the class, and (c) the teacher and one student. The procedures can include vocal learn units, textural and written learn units, or combinations thereof. The procedure has proved to be useful both to train and to assist teachers. See chapter 9 for a detailed description of the procedures. *See also* **learn units; teaching as communication; contingency-shaped teaching behavior; Comprehensive Application of Behavior Analysis to Schooling (CABAS); incorrect teacher operation (teacher errors); correct teacher operations (accurate teacher operations).**

teaching as communication. In Direct Instruction, presentations of instructional steps or response opportunities (and presumably consequences) are referred to as teaching communications. The learn unit may also be described as an episode or conversational unit of verbal behavior whereby both the teacher and the student function both as speaker and listener.

teaching machine. Developed by B. F. Skinner in 1968, the teaching machine in its various permutations was designed to deliver automated programmed instruction. Its history parallels that of programmed instruction and is outlined in an article by Vargas and Vargas in 1991 and described by Skinner in 1968. The modern personal computer has the potential to function as an electronic teaching apparatus but only if the principles, strategies, and tactics of programming instruction are applied. Teaching machines were used to teach several subjects but was used prominently in introductory behavioral psychology at Harvard University and at Hamilton College. Various models may be viewed at the Cambridge Center for Behavioral Studies.

teaching operations undertaken to induce generalized stimulus control. For example, teaching an antecedent–behavior–consequence across different teachers or settings is one tactic. Another involves "loose training" whereby slight variations in the antecedent are used to induce generalized stimulus control. The process of general case instruction is a well-tested procedure for inducing generalized stimulus control. *See also* **general case instruction (multiple exemplar instruction).** The general case effect suggests that perhaps the generalization phenomenon is a function of the degree to which salient components of stimulus control behavior; thus, some have suggested that behavior may be attributable to essential stimulus control.

teaching repertoires. Teaching repertoires are described in this book as (a) contingency-shaped repertoires, (b) verbal behavior about the science repertoires, and (c) verbally mediated repertoires. Each repertoire and their relationships are described in detail in chapter 3. *See also* **contingency-shaped behavior; verbal behavior about the science; verbally mediated repertoire.**

technology of the science of teaching. The tactics of teaching and programming curricula described in this book. The expression probably had its origin with Skinner's 1968 book titled *The Technology of Teaching*. Like other technologies, those associated with teaching are derived from a basic science just as engineering technologies are derived from basic sciences. *See also* **strategic science of instruction; strategic scientist of behavior; tactics of behavior analysis pedagogy.**

terminal objectives (long-term objectives). Represents the culmination of a curricular program. Terminal objectives describe mastery, fluency, or both with regard to the terminal goal of a program of study.

textual learn units. Learn units in which the antecedent is text. The response may be textual (i.e., written or typed) or it may be vocal.

theory. Theory in the science of behavior refers to hypotheses about general principles associated with environment/behavior relationships. Theory in other areas of psychology is devoted to hypotheses about psychological constructs, or intervening variables, and refers to relationship between responses and inferred entities such as personality, attribution, or intelligence. A science of behavior eschews the latter type of theory but encourages the former type of theory.

thinking (behavior selection perspective). Thinking is a form of verbal behavior, for example, whereby one acts as one's own speaker and listener, typically without speaking aloud. One may also think musically or spatially (i.e., covertly think music, covertly see designs). Thinking is a form of behavior, not a cause of behavior. While it may function to affect nonthinking behavior, the external environment ultimately affects thinking. See chapter 7 for an elaboration.

three-term contingency. *See also* **contingencies of reinforcement and punishment (three-term contingences).** The three-term contingency refers to the antecedent–behavior–consequence unit. An operant is the incidence of a learned three-term contingency; that is, the three components are intertwined in a predictable manner. The three-term contingency is affected by instructional history, the setting event, and setting stimuli, particularly motivational conditions (e.g., deprivation or satiation). The three-term contingency trial is an instructional unit that has all three terms as part of each instructional presentation. The learn unit incorporates an interlocking and episodic relationship between the three-term contingency for the student and for the teacher. See chapter 2 for an extended discussion.

three-term contingency trial. A term used prior to the adoption of the term *learn unit*. *See also* **learn units; three-term contingency.**

time out from reinforcement. *See also* **negative punishment or type II punishment operation.** Time out from reinforcement is a tactic that draws on the principle of negative punishment also referred to as Type II punishment. It is an effective and acceptable form of dealing with noxious and interfering behavior provided that the time in setting contains reinforcement opportunities for the student. If high rates of reinforcement operations and a diversity of reinforcers are present in an ongoing fashion, the procedure will be effective, easy to use, and minimally disruptive. Briefly, removing the student from the instructional setting immediately after occurrences of a target behavior or response class and reinserting the student after a short period of prescribed appropriate behavior will function to reduce and eliminate the target behavior. Reinforcement for alternative and incompatible responses must be available once the student is reinserted. The occurrences of high rates of valid and accurately delivered learn units results in an adequately reinforcing environment. Time out settings where the student can see and/or view the delivery of learn units and reinforcement are more desirable provided no harm can result to peers and instructional staff. Time out periods of 5 min are usually sufficient, and with young children often only a few seconds arenecessary. If the instructional setting is adequate, time out from reinforcement will not be necessary after only a few instances or days of use. Time out from reinforcement or other negative punishment operations such as planned ignoring are preferable usually to the use of positive punishment or Type I operations such as reprimands or disapproval (*see* **reprimands**). The types and ethical issues associated with the use of time out are discussed in the basic sources cited throughout this text.

token economy. A tactic developed in the applied science for delaying the consumption of reinforcement while maintaining the effects of immediate consumption. This procedure allowed responding not to be interrupted by consumption for long pagoda of time. Tokens are items (points, tokens, check marks, counts) that function as generalized reinforcers. Their reinforcement value or effect is a function of the kind and diversity of backup reinforcers purchasable with the tokens and the instructional history and setting events. They make the individualized classroom possible and are used as means of shaping self-management, self-monitoring, and self-reinforcement. See chapters 6–8.

transducer of behavior. The expression *transducer of behavior* refers to a person whose data are direct expressions of responding. On the cumulative recorder, the response of a pigeon, for example, is transferred directly to a cumulative record by mechanical or electrical transduction. The result is a direct record of the response. Humans should observe behavior in a manner that approximates the direct behavior recording function of a cumulative recorder. The data are to be a direct transduction of the response with no inferences or ratings of the behavior. *See also* **reliability; interobserver agreement.**

transition time. A term used in educational research to refer to the time between engagement in instructional presentations or responding. It is also referred to as "down time." The less the transition time, the more instruction occurs. When the teacher measures learn units, the transition time can be more precisely calculated as latency in time between learn units.

tutorial sessions. An instructional session involves a tutor and a tutee. The tutor presents learn units and the tutee responds. Tutees also ask questions and receive other assistance from tutors. Tutorial sessions that involve trained student tutors are often more effective than ones involving a teacher and a student. Tutorial sessions can be scripted and applied with numerous students on a class-wide basis.

tutoring. The process whereby an individual teaches another individual one or more three-term contingency relationships. When the tutor is taught accurate learn unit operations, tutoring by peers or older students is a very effective procedure to change the number of learn units received and the achievement of objectives. It is a frequent form of instruction in classrooms that individualize instruction comprehensively. See the discussions of tutoring in chapter 10. The tutor frequently benefits as much or more from the tutoring process as does the tutee.

U

unambiguous exemplar. Faultless target antecedent presentations by a teacher or teaching device and faultless forms of target antecedent stimulus. Thus, antecedents are presented consistently alike and in a clear, succinct, and unconfusing manner when they are unambiguous. Unambiguous antecedents evoke behavior in a manner that does not allow confusion with other previously trained or untrained antecedents. Unambiguous antecedents may refer to stimulus discriminatives that are already in the student's repertoire or that refer to stimuli targeted to gain the statue of antecedent stimulus control. See discussions of antecedent presentations in chapters 5 and 16. *See also* **faultless presentations.**

utilitarianism. A philosophy associated with the writings of William James and other pragmatists. Its relevance to considerations of curriculum include the proviso that objectives of curricula should consider not only the benefits for the individual, but the benefits or utility of what is learned for the culture at large. *See also* **habilitation.**

V

vaganotic measurement. Johnston and Pennypacker in 1981 coined the term *vaganotic* to contrast absolute unit or idemnotic measures (e.g., real behaviors in real time) with measures that were ratings or scaled. Thus, rather than counting the use of positive reinforcement operations by a teacher (a form of idemnotic measurement), a researcher who used vaganotic measurement would rate the teacher's use of reinforcement on a scale of 1 to 5, for example; vaganotic measures also include projective test items wherein a response on the test, or a series of responses, is said to be a manifestation of an unobservable intervening variable (e.g., personality type). *See also* **rating scales; idemnotic measurement; projective measurement; hypothetical constructs (intervening variables, mentalistic constructs).**

validity of research findings. The degree to which the findings are replicated with other subjects, settings, or behaviors. A valid research finding is one that holds up in both systematic and direct replication experiments.

variability of behavior. In the science of behavior, much research is devoted to determining the source of variability in behavior. The more stable the behavior, the greater the knowledge about the controlling variable. Variability refers to the degree of deviance that a visual display shows from a trend or mean. The functional relationship of independent variables and dependent variables is best detected when baseline performance shows little variability. The occurrence of variability in the baseline or an intervention phase suggests that other variables need to be isolated and tested. *See also* discussions of variability in chapter 2. *See also* **rate of responding; reliability; ascending trend; descending trend.**

variable. The term used to specify stimuli, events, and conditions associated with an experiment. Some variables are controlled for as in the selection of subjects, conditions, and settings. This is done via the logic of the experimental method. The dependent variable is the behavior measured and the independent variable is the treatment or teaching intervention variable. *See also* **independent variable; dependent variables; method of disagreement (method of difference); within-subject designs (intra-subject designs); between-subject research designs (intersubject designs).**

variable trend. A trend of a linear visual display of data is variable when the data fluctuate widely from the trend line or when no ascending or descending trend is discernible. See **descending trend; ascending trend; variability of behavior.**

verbal behavior. Skinner's 1957 conceptual extension of the science of behavior to communicative behavior is termed verbal behavior. Verbal behavior or communicative behavior is social in that it always involves a speaker and listener or writer and reader, and the effects of the behavior determine its function. The functions of communicative behavior include mand functions, tact functions, autoclitic functions, and audience functions. These functions are under both verbal and nonverbal antecedent controls that in turn affect their categorization. A word or group of words or a linguistic form may have several communicative functions depending on the context and consequence of its usage. The word water may function as a mand (a speaker commands a listener to produce water for the person commanding) or as a tact (the speaker makes contact with water verbally and is reinforced by a generalized reinforcer). An autoclitic quantifies, qualifies, specifies, or in came way affects the effect of another form of verbal behavior for the speaker or writer. Intraverbal behavior is verbal behavior controlled by other verbal behavior (e.g., reciting a poem). Verbal behavior involves speaker and listener functions sometimes residing in the same person as in the process of self-editing. Behavior that is governed by verbal repertoires or verbal behavior functions was termed rule-governed behavior whereas behavior shaped directly by its nonverbal contingencies wan termed contingency-shaped behavior. In addition to Skinner's 1957 book, the reader is directed to Catania's 1984 treatment of verbal behavior in *Learning*. *See also* chapter 7; **mand; tact; academic conversational units; autoclitics; speaker behavior; listener behavior.**

verbal behavior about the science. An expression used to summarize the use of the vocabulary of the science of verbal behavior. It refers to a repertoire that incorporates the description of student and teacher behavior, the environment, and curricula consistent with the science and epistemology of behaviorism. It refers to learning and using the verbal community of the science to describe behavior and pedagogical operations and to analyze curricular goals consistent with the precision of the science.

verbal community of the science. *See also* **verbal behavior about the science.** The vocabulary of the science of behavior is beat characterized as a verbal community. The operant nucleus, contexts, and known environment/behavior relationships are described by a vocabulary. The precision of the vocabulary and its consistent use result in a community of practitioners who can communicate strategies and tactics such that they can be applied by anyone who has the verbal repertoire of the science. The use of verbal mediation repertoires by a teacher, who is a strategic scientist, is an example of the ways in which the verbal community of the science solves instructional problems. Teaching interactions and student responses are described with a vocabulary that instantly communicates the situation to fellow practitioners.

verbally mediated repertoire. The term *verbally mediated repertoire* builds on a term originated by Ernest Vargas in 1991 (i.e., verbally governed behavior) to replace Skinner's 1957 term *rule governed*. It describes behavior that is mediated or directed by verbal behavior rather than the direct contingencies of behavior (e.g., contingency-shaped behavior). The performance of a task that is controlled by verbal instructions (written, oral, or gestural) is under the control of verbal contingencies. Ethical behavior may also be under verbal control in certain situations. We expanded the term in this book to encompass certain strategies for solving instructional problems by teachers or supervisors. It is also used to describe repertoires associated with teaching students to use methods of science, tenacity, authority, and self-management to solve problems under the control of verbal mediation. See the extensive treatment of verbal mediation in chapters 7, 9, and 10. *See also* **verbally mediated repertoire; modes of curriculum; functional curricula.**

verbal mediation. A term that expands on Skinner's usage of rule-governed behavior. Verbal behavior that governs or guides nonverbal behavior in lieu of the natural contingencies is an incidence of behavior under the control of verbal mediation. Erecting an edifice or performing a set of operations guided by what has been read or heard is an example. It is distinguished from contingency-shaped behavior in that the verbal antecedents replace the nonverbal antecedents. Individuals performing a series of operations under verbal mediation can perform those operations without having ever experienced the direct contingencies. It plays a critical role in stages of education, problem solving, and the transmission of scientific, literary, ethical, or any of a number of kinds or specific forms of expertise. *See also* **verbally mediated repertoire** for an elaboration on the concept to operationalize instructional expertise made feasible by certain repertoires of verbal behavior.

visual display. *See* **also graphic display.** The expression visual display refers to graphic displays of the behavior of individuals. Visual displays and an instructional history of reading such displays are critical to the strategic science of behavior. *See also* **rate as primary datum; rate of responding; ascending trend; descending trend; variability of behavior; variable trend; slope.**

vocal learn units. Learn units with both a vocal antecedent and vocal responses. A variation is one in which a vocal antecedent is to result in a textural response as in a spelling test where the teacher dictates a word and the student writes the response. More specifically, the latter is a vocal textural learn unit.

W

within-subject designs (intrasubject designs). An intrasubject design is an experimental design in which the control and experimental conditions are tested with the same individual or group. In single-subject designs, each individual receives repeated sessions under baseline and experimental conditions. In group designs, each group receives the control or baseline condition and the experimental condition (independent variable) or individuals in the group receive each condition in a counterbalanced arrangement. *See also* **method of disagreement (method of difference); between-subject research designs (intersubject designs).**

writer as own reader. Writers functions as their own readers when they read and self-edit their own writing as if they were the intended eventual reader or audience. One reads one's own writing with the level of reader expertise that engenders reinforcement and correction of one's own writing consistent with the expertise of an eventual audience. It is an extension of the notion of one speaking as one's own listener done without overt vocalization (i.e., thinking). This repertoire, and the range of expertise with the repertoire, can serve as a useful categorization of student repertoires and curricular objectives. See chapters 5–7 for a discussion of pedagogical operations and curricular component a of the writer as reader. The writer as reader is a functional curricular goal that can be taught across numerous structural curricular goals.

Z

zeitgeist. A German word that translates as "spirit of the times." The use of the term period of enlightenment to describe the prevailing intellectual climate of a period of time in history is an example of a particular zeitgeist.

Index